Physics-Compatible Finite Element Methods for Scalar and Tensorial Advection Problems

Christoph Lohmann

Physics-Compatible Finite Element Methods for Scalar and Tensorial Advection Problems

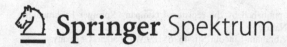
Springer Spektrum

Christoph Lohmann
Dortmund, Germany

Dissertation Technische Universität Dortmund, Fakultät für Mathematik, 2019

Erstgutachter: Prof. Dr. Dmitri Kuzmin
Zweitgutachter: Prof. Dr. Matthias Möller
Tag der Disputation: 08. Mai 2019

ISBN 978-3-658-27736-9   ISBN 978-3-658-27737-6   (eBook)
https://doi.org/10.1007/978-3-658-27737-6

Springer Spektrum

This Springer Spektrum imprint is published by the registered company Springer Fachmedien Wiesbaden GmbH part of Springer Nature.
The registered company address is: Abraham-Lincoln-Str. 46, 65189 Wiesbaden, Germany

Acknowledgments

First and foremost, I would like to express my sincere gratitude to my advisor Prof. Dr. Dmitri Kuzmin for his guidance and continuous support during my PhD studies. He gave me the opportunity to work on this exciting topic and to attend many scientific conferences. His time invested in various fruitful discussions is highly appreciated. I also would like to thank him for proofreading this thesis and for providing me with valuable feedback.

Furthermore, I would like to take this opportunity to thank Prof. Dr. Stefan Turek for significantly contributing to my academic education and for already hiring me as an undergraduate student to work at the "Institute of Applied Mathematics (LS III)". In this context, I like to express my deep gratitude to all former and current members of LS III. It was a great pleasure to be part of your group and to make new friendships! Thank you very much for all your social and moral support during the past years! Special thanks also go to the whole FEAT team for always helping me out in case of software problems.

In addition, I acknowledge the financial support provided by the German Research Association (DFG) without which this work would have been impossible. At this point, a special thanks goes to Omid Ahmadi who worked with me on the DFG project that this thesis is mainly concerned with.

I am very grateful to Dr. John N. Shadid for making my exciting stay at Sandia National Laboratories possible. It was a fruitful collaboration and a warm atmosphere with you and Dr. Sibusiso Mabuza!

I also thank Prof. Dr. Matthias Möller and Prof. Dr. Matthias Röger for kindly agreeing to act as reviewers and examiners in the defense of this thesis.

Finally, I am profoundly grateful to my family and friends for their invaluable support during my PhD studies.

Thank you so much!

Contents

Abstract

This thesis presents the author's contributions to the development and analysis of bound-preserving numerical methods for hyperbolic equations. A special focus is placed on physics-compatible approaches to simulation of fiber suspension flows. In this application, a second order tensor field is frequently introduced to approximate the orientation distribution of rigid slender particles in a Newtonian carrier fluid. To simulate its evolution in a physically admissible and stable manner, the unit trace property must be preserved and eigenvalues must stay nonnegative. The methodology developed in this work leads to numerical approximations which are guaranteed to satisfy these requirements under suitable time step restrictions.

Before considering the evolution of tensor quantities, the state of the art in the field of algebraic flux correction (AFC) for continuous finite element discretizations of scalar hyperbolic equations is reviewed. Existing algorithms and the underlying theory are supplemented with new results. Stabilization techniques based on the AFC approach modify the entries of finite element matrices so as to make the scheme bound-preserving. Low order approximations with desired properties can easily be constructed by adding discrete diffusion operators. In high order nonlinear extensions, the accuracy of the solution is improved by removing redundant diffusion in smooth regions. Two ways to perform such corrections are described in this thesis: (i) fractional-step algorithms which add bound-preserving antidiffusive fluxes to a low order predictor and (ii) monolithic approaches which incorporate nonlinear corrections into the residual of the low order method. Both kinds of AFC schemes are backed by proofs of generalized discrete maximum principles (DMPs), which imply preservation of global bounds in particular. Transient transport problems are discretized in time by the θ-scheme. Upper bounds for admissible time steps are derived from sufficient conditions for the validity of DMPs. For the monolithic AFC discretization of the steady state advection equation, results on the existence of a (unique) solution and corresponding a priori error estimates are presented.

A major highlight of this work is the extension of AFC tools to the numerical treatment of symmetric tensor quantities. The proposed algorithms constrain the eigenvalue range of evolving tensor fields by imposing local discrete maximum principles on the maximal and minimal eigenvalues. Using this design principle and corresponding generalizations of the theoretical framework, robust property-preserving tensor limiters are introduced following the analysis and design of their scalar counterparts. The proofs of generalized DMP properties employ spectral decompositions and positive semidefinite programming tools.

In the last part of this work, a frequently used model of fluid-fiber flows is considered. The involved nonhomogeneous transport equation for the second order orientation tensor is discretized using an operator splitting approach. Under suitable assumptions and time step restrictions, preservation of physical properties is shown for the forward and backward Euler time discretizations of the tensorial ODE that governs the local orientation dynamics. Finally, a numerical method for solving the fully coupled problem is developed on the basis of customized solvers for individual subproblems. Numerical results for the fiber suspension flow through an axisymmetric contraction illustrate the potential of the proposed solution strategy.

1 Introduction

1.1 Motivation

In the field of computational fluid dynamics, many applications of practical interest require the use of robust discretization techniques equipped with adaptive control mechanisms for pointwise solution values, cell averages, slopes, fluxes, and/or other quantities of interest. The design of bound-preserving high-resolution schemes for hyperbolic conservation laws is a particularly challenging task because exact solutions may exhibit steep gradients (or discontinuities) in narrow *layers* which cannot be resolved properly on the given mesh. The lack of control over the range of possible solution values may give rise to spurious oscillations and unrealistic results. In the worst case scenario, an unphysical solution behavior may cause the simulation to blow up. In contrast to this, property-preserving algorithms guarantee the physical admissibility of approximate solutions but the corresponding constraints may dictate the use of low order approximations at least in close proximity to troublesome layers.

The foundations of numerical methods satisfying certain discrete maximum principles (DMPs) were laid by Lax [Lax54]. His finite difference algorithm for one dimensional compressible fluid flows paved the way for the development of the *Lax-Friedrichs scheme*. It guarantees the boundedness of the solution in terms of initial values and has remained the main workhorse of computational gas dynamics for several decades. According to the Godunov theorem [God59], linear bound-preserving methods of this kind can be at most first order accurate. Consequently, more accurate constrained solutions can only be produced by nonlinear algorithms. To circumvent Godunov's order barrier with the aim of achieving sharp and non-oscillatory resolution of shock waves, Boris and Book [BB73] applied nonlinear conservative antidiffusive corrections to a low order predictor. This far-reaching idea is the key ingredient of their *flux-corrected transport* (FCT) algorithm and many other nonlinear high-resolution schemes. A fully multidimensional version of FCT was introduced by Zalesak [Zal79] and combined with finite element (FE) discretizations by Parrott and Christie [PC86]. The work of Löhner et al.

© Springer Fachmedien Wiesbaden GmbH, part of Springer Nature 2019
C. Lohmann, *Physics-Compatible Finite Element Methods for Scalar and Tensorial Advection Problems*, https://doi.org/10.1007/978-3-658-27737-6_1

[Löh+87] extended explicit FE-FCT schemes to unstructured grids and systems of conservation laws. The development of edge-based finite element methods based on various extensions of *total variation diminishing* (TVD) finite difference schemes [Har83; Har84] was also actively pursued in the late 1980s and 1990s [AD89; Sel87a; Sel87b; Cou+98; Lyr95; Lyr+94]. A general framework for the design of bound-preserving finite element approximations was introduced in [Kuz07] and named *algebraic flux correction* (AFC). Many numerical algorithms considered in this work belong to the family of AFC schemes.

The main goal of this work is the construction of physics-compatible numerical methods for simulating fiber suspension flows. In this application, the hydrodynamic behavior of a fluid-fiber mixture is governed by the Navier-Stokes equations with a generalized stress tensor which depends on the orientation distribution of solid slender particles. To keep computational costs acceptable, this distribution function is frequently approximated by a symmetric second order tensor quantity whose eigenvalues are nonnegative and sum up to unity. If the numerical method does not preserve these properties, the problem is not well-defined and the simulation breaks down. The evolution of the positive semidefinite *orientation tensor* is governed by an advection equation with a nonhomogeneous right hand side. The following four approaches to approximating the solution of this *Folgar-Tucker equation* can be found in the literature:

- The easiest way to model the rheological behavior of a fiber suspension is to assume that all fibers are fully aligned with the velocity field and to calculate fiber-induced stresses using the corresponding orientation tensor [Lip+88]. Obviously, the computational cost associated with the numerical treatment of fibers is minimized by this approach and the properties of an orientation tensor cannot be violated. The obtained simulation results frequently exhibit a good qualitative agreement with experimental data because fibers with large aspect ratios are, indeed, likely to align themselves with the flow direction. However, the assumption of fully aligned fibers is justified only for simple flow fields. In general, it may result in a poor approximation of the evolving orientation tensor.

- The hyperbolic nature of the Folgar-Tucker equation has led Papanastasiou and Alexandrou [PA87] and Rosenberg, Denn, and Keunings [Ros+90] to solve it using the *method of characteristics* which transforms an advection equation with a source term into an ordinary differential equation (ODE) to be integrated in time along the trajectories of Lagrangian fibers. The absence of space derivatives in the initial value problem eliminates the need for using bound-preserving space discretizations. However, recirculating flows of fiber suspensions cannot be simulated in this way because all characteristics have to start on the inflow boundary. Moreover, the use of streamlined finite element meshes, as proposed by Papanastasiou and Alexandrou [PA87], is feasible only for very simple stationary flow fields. On general meshes, backtracking along the characteristics may also produce unphysical solutions if high order interpolation techniques are employed. Linear interpolation is bound-preserving but introduces numerical diffusion which is as strong as that of first order Eulerian advection schemes.

- Standard Galerkin discretizations of advection problems on general fixed meshes do not impose any unrealistic restrictions on the structure of the flow field. However, the use of continuous finite elements without proper stabilization may lead to ill-posed discrete problems, numerical instabilities, spurious oscillations, and violations of global bounds that define the range of physically admissible solution values. As a consequence, positive semidefiniteness of the orientation tensor is generally not preserved by such approximations. VerWeyst [Ver98] and VerWeyst and Tucker [VT02] discretized the Folgar-Tucker equation using the streamline upwind/Petrov-Galerkin (SUPG) method [BH82] which localizes ripples to a small neighborhood of steep gradients but may fail to prevent the formation of small undershoots/overshoots. Therefore, suitable eigenvalue corrections are generally required to avoid unphysical orientation states. In addition, the amount of SUPG stabilization depends on a heuristic parameter which has a strong impact on the accuracy of the numerical results and requires careful tuning.

- Reddy and Mitchell [RM01] simulated fiber suspension flows using a discontinuous Galerkin (DG) discretization of the evolution equation for the orientation tensor. Following the work of Lasaint and Raviart [LR74], the amount of upwinding in the numerical flux was adjusted using a free parameter. In contrast to the continuous FE approach with SUPG stabilization, the DG treatment of the advective term preserves the physical properties of an orientation tensor at least in the case of the piecewise constant approximation with full upwinding. However, the order of this property-preserving scheme is as low as $\frac{1}{2}$ even for smooth solutions on general meshes, whereas the use of higher order finite elements or a poor choice of the upwinding parameter may result in a loss of positive semidefiniteness.

A well-designed numerical method for solving the Folgar-Tucker equation should be generally applicable and produce sufficiently accurate approximations which are guaranteed to preserve physical properties of orientation tensors without any post-processing. None of the above mentioned approaches meets all of these design criteria at once. To fill this gap, we constrain the continuous and piecewise linear Galerkin discretization of the evolution equation for the orientation tensor using algebraic flux correction. Making use of operator splitting techniques, we decompose the problem at hand into a homogeneous advection equation and an ODE model of local orientation dynamics. Using spectral decompositions and positive semidefinite programming tools, we construct nonlinear high resolution schemes which preserve the trace and positive semidefiniteness of the orientation tensor in the process of advection. The numerical solution of the ODE describing rotation of fibers under the influence of velocity gradients is also shown to be physics-compatible under suitable assumptions regarding the involved closures. The two-step algorithm for solving the Folgar-Tucker equation is used to evolve the orientation tensor when it comes to simulating of three dimensional fiber suspension flows. The simulation tool developed for this purpose is applied to the axisymmetric contraction problem considered in [Lip+88; Ver98; VT02].

1.2 Outline

In our presentation, we begin with the standard Galerkin discretization of scalar advection-reaction equations. Then we introduce the AFC methodology for scalar fields and extend it to symmetric tensors. The level of complexity is gradually increased until all building blocks of a physics-compatible numerical algorithm for the non-Newtonian mixture model of fiber suspension flows become available. Each subproblem is analyzed individually and customized bound-preserving methods are constructed for the corresponding (initial-)boundary value problems. The main focus of this work is on the design of finite element methods satisfying discrete maximum principles. In this context, we split our presentation of AFC schemes into sections devoted to the numerical treatment of stationary and transient problems. As a numerical time integrator, we use the two-level θ-scheme. The presented analysis can be readily extended to strong stability preserving (SSP) Runge-Kutta methods [Got+01; Got+11] which deliver DMP satisfying solutions under the same time step restrictions as the forward Euler method.

More specifically, the objectives and contents of individual chapters are as follows:

Chapter 2 is concerned with the derivation and analysis of partial differential equations (PDEs) that are commonly used in mathematical models of incompressible fluid dynamics. They are inferred from physical conservation laws under suitable assumptions and simplifications. In particular, we introduce the stationary and transient versions of linear advection-reaction equations. These model problems are representatives of hyperbolic PDEs which call for the use of bound-preserving discretizations to be presented in subsequent chapters. For this reason, qualitative properties of analytical solutions are analyzed in some detail. In the stationary case, uniqueness can be shown under the assumption that the bilinear form of the variational formulation is coercive. This property is, therefore, required in various parts of this work. Using the method of characteristics, we prove the fact that the solution to an advection-reaction equation is nonnegative whenever the source term and the initial/boundary data are nonnegative. The boundedness of the solution by the extrema of given data is shown for the homogeneous advection problem.

A brief introduction to the Galerkin finite element method is given in Chapter 3. The hyperbolic model problems are discretized in space using continuous piecewise (multi-)linear finite elements. In the variational forms of the PDEs under consideration, we impose inflow boundary conditions in a weak sense by using the given boundary data to calculate the advective fluxes. The use of such weakly imposed Dirichlet boundary conditions (also called *flux boundary conditions*) is common practice in numerical methods for hyperbolic PDEs and is preferred for that reason. Most results of this work can be easily adapted to the case of strongly enforced Dirichlet boundary conditions. The semi-discrete counterpart of the time dependent advection-reaction equation is discretized in time using the θ-scheme. In Section 3.4, the need for bound-preserving corrections is illustrated by numerical examples in which the Galerkin finite element approximations to steady state solutions of advection problems are polluted by spurious oscillations.

Readers who are familiar with the basic equations of fluid dynamics and the Galerkin finite element method for numerical solution of PDEs may want to skip these introductory parts. However, they are advised to peruse Section 1.3 which summarizes symbols and notational conventions to be used throughout this work.

Chapter 4 introduces the reader to the algebraic flux correction (AFC) methodology. Adopting the design principles of FCT approaches proposed by Boris and Book [BB73] and Zalesak [Zal79], AFC schemes for finite elements constrain the standard or stabilized Galerkin approximation in a way which leaves it unchanged in smooth regions but replaces it with a non-oscillatory low order approximation in a neighborhood of steep fronts. The aim is to minimize the amount of numerical dissipation, while preserving physical properties of the exact solution like nonnegativity or boundedness. In the case of a piecewise (multi-)linear approximation, the finite element solution is globally bounded by its values at the vertices of the mesh. Therefore, it suffices to ensure the validity of suitably defined discrete maximum principles for the nodal values. A typical DMP proof for the solution of a fully discrete problem is based on a set of sufficient conditions concerning the off-diagonal matrix entries, its row sums, and the right hand side of the algebraic system for the degrees of freedom (Section 4.2). These conditions can be readily enforced by manipulating the matrices of the Galerkin discretization. As explained in Section 4.2.3.1, Godunov's order barrier implies that the resulting linear methods can be at most first order accurate. To improve the accuracy of the bound-preserving finite element solution, two kinds of nonlinear and conservative correction procedures are currently employed in AFC schemes. The first one is based on a predictor-corrector strategy and traces its origins to the work of Zalesak [Zal79]. In this version, low order solution values are corrected by adding antidiffusive fluxes that are limited to prevent violations of local bounds (Section 4.4). To avoid solving a high dimensional optimization problem, the involved inequality constraints are replaced with localized sufficient conditions which guarantee boundedness in terms of the low order solution values under worst case assumptions. The resulting simplifications lead to explicit formulas for calculating the correction factors for flux limiting. The theoretical analysis presented in this work furnishes the first rigorous proof of consistency for antidiffusive corrections of this kind. The second approach to bound-preserving limiting in AFC schemes was originally proposed in [Kuz07]. In contrast to predictor-corrector approaches, it incorporates limited antidiffusive terms into the residual of the low order method. The definition of correction factors should guarantee that the resulting scheme reduces to the low order method at every local extremum and to the unmodified Galerkin solution in smooth regions. In this monolithic approach, the problem to be solved becomes nonlinear and the existence of a unique solution is generally not guaranteed. However, recent advances in the theoretical analysis of AFC schemes [Bar+16] make it possible to prove the existence of a (unique) solution to the steady state problem and obtain an a priori error estimate for arbitrarily chosen correction factors. Furthermore, all methods to be presented in this chapter are backed by proofs of discrete maximum principles, possibly under CFL-like time step restrictions.

In Chapter 5, the theoretical foundations of AFC schemes are extended to symmetric tensor fields and new approaches to bound-preserving limiting of tensor-valued fluxes are explored. In contrast to the scalar case, it is not obvious which characteristics of the advected tensor should be controlled when it comes to the design and analysis of physics-compatible numerical schemes. Since this work is aimed at numerical simulation of fiber suspension flows, we impose an upper bound on the maximal eigenvalue and a lower bound on the minimal eigenvalue. In particular, the latter constraint implies preservation of positive semidefiniteness. Further motivation for imposing local bounds on the range of eigenvalues (rather than principal invariants or components of the tensor field) is provided in Section 5.1.2. Adopting the principle of eigenvalue range preservation as a design criterion of primary importance for our tensorial extensions, we prove its validity for a linear method which evolves each component of the tensor field as a scalar quantity using the same bound-preserving low order scheme (Section 5.3). After proving the physical admissibility of the low order approximation, we proceed to the presentation of generalized FCT algorithms in Section 5.4. To reduce the amount of numerical diffusion while preserving the eigenvalue range, we exploit the concept of Löwner ordering and define scalar-valued correction factors for antidiffusive fluxes using exact or approximate solutions of small positive semidefinite programming problems for auxiliary tensors. Numerical studies are performed in Section 5.6 to illustrate the pros and cons of different ways to calculate the FCT correction factors. In Section 5.5, we introduce a monolithic limiting approach for tensor quantities. Rotating the tensor-valued correction terms into the principal axis system of the solution, different eigenvalues can be constrained in a synchronized or segregated manner. The synchronized limiting strategy corresponds to using the same scalar correction factor for all components of the tensor. It preserves a constant trace and is amenable to theoretical analysis. The eigenvalue range preservation properties for stationary and time dependent transport problems can be shown following the proofs of discrete maximum principles for the scalar case. Segregated limiters use tensorial correction factors that are likely to produce more accurate results but, regrettably, may fail to preserve the unit trace of an orientation tensor in applications to fiber suspension flows. Moreover, proofs of eigenvalue range preservation become more involved and are currently available only for steady state problems.

Chapter 6 deals with the simulation of fiber suspension flows using a non-Newtonian mixture model in which the effective stress tensor depends on the orientation distribution of the evolving particles. The incompressible Navier-Stokes equations for the velocity and pressure of the mixture are solved using the discrete projection scheme presented in Section 6.2. The second order orientation tensor approximating the orientation distribution is advected using the methodology developed in Chapter 5. Orientation changes caused by rotation of fibers are taken into account by solving the space independent Folgar-Tucker equation at each node of the spatial mesh. The ability of this model to deliver physics-compatible solutions depends on the choice of the *closure* for the fourth order orientation tensor which appears in the ODE system for the independent components of its second order counterpart. In Section 6.3, we derive necessary conditions for such closures to be physically admissible. For closures satisfying these requirements, we prove that the backward and forward Euler discretizations of the problem at hand preserve the properties of orientation tensors under suitable time step restrictions. Combining the ODE solver with an AFC discretization of the advection problem, we construct a physics-compatible scheme for the numerical treatment of the Folgar-Tucker equation. Numerical studies of the fiber suspension flow through an axisymmetric contraction are performed in Section 6.4 to illustrate the practical use of proposed methods.

1.3 Notation

A number of notational conventions are used in this work for conciseness and convenience. For example, bold uppercase letters like $\mathbf{V}$ and $\bar{\mathbf{V}}$ denote tensor quantities in $\mathbb{R}^{d \times d}$, $d \in \{2, 3\}$. Bold lowercase letters like $\mathbf{v} \in \mathbb{R}^d$ are reserved for vectors in the d dimensional space.

The indices k, ℓ, m, n, o, and p refer to the physical space dimension and attain values in the range $\{1, \ldots, d\}$. The degrees of freedom of the finite element space approximation are labeled using the indices i, j, and l. For the sake of brevity, the upper and lower limits of summation are omitted in many cases. For instance, we write $\sum_{j \neq i}$ instead of $\sum_{j=1, j \neq i}^{N}$, where N denotes the total number of degrees of freedom.

The systems of algebraic equations to be considered in this work stem from continuous and piecewise (multi-)linear finite element discretizations. A fully discrete problem for a generic scalar quantity has the form $\mathcal{A}u = g$ where $\mathcal{A} \in \mathbb{R}^{N \times N}$ and $u, g \in \mathbb{R}^N$. The degrees of freedom $u_1, \ldots, u_N$ are the values of the numerical solution at the vertices of the mesh. Due to the compact support property of Lagrange basis functions, the system matrix $\mathcal{A}$ is sparse, i.e., $a_{ij} = 0$ for $j \notin \mathcal{N}_i$, where $\mathcal{N}_i$ denotes the nodal stencil of index i (as defined in Section 3.1.1). It follows that sums involving entries of $\mathcal{A}$ reduce to sums of nonvanishing terms, e.g.,

$$\sum_{j \neq i} a_{ij} = \sum_{j \in \mathcal{N}_i \setminus \{i\}} a_{ij} \quad \forall i \in \{1, \ldots, N\}.$$

In a similar vein, the i-th equation of the linear system can be written as $\sum_j a_{ij} u_j = g_i$ or $\sum_{j \in \mathcal{N}_i} a_{ij} u_j = g_i$. For the sake of brevity, we frequently use the global sum notation but bear in mind that the matrices are sparse and summation reduces to that over indices in $\mathcal{N}_i$.

A summary of frequently used symbols can be found in Tables 1.1–1.4.

Table 1.1: Symbols referring to physical domain.

Symbol	Description
d	Number of space dimensions/size of a tensor
k, ℓ, m, n, o, p	Indices of a space direction $1, \ldots, d$ (if used as subscript)
Ω	Physical domain, $\Omega \subseteq \mathbb{R}^d$
$\mathbf{v}$	Velocity field, $\mathbf{v} : \Omega \to \mathbb{R}^d$
a	Diffusion coefficient, $a : \Omega \to \mathbb{R}_0^+$
c	Reactivity parameter, $c : \Omega \to \mathbb{R}_0^+$
f	Scalar-valued source term, $f : \Omega \to \mathbb{R}$
$\mathbf{F}$	Tensor-valued source term, $\mathbf{F} : \Omega \to \mathbb{S}_d$
$\partial\Omega$	Boundary of physical domain Ω
$\mathbf{n}$	Unit outward normal vector of $\partial\Omega$
Γ_{in}	Inflow boundary of $\partial\Omega$, i.e., $\Gamma_{\mathrm{in}} := \{\mathbf{s} \in \partial\Omega \mid \mathbf{v}(\mathbf{s}) \cdot \mathbf{n}(\mathbf{s}) < 0\}$
Γ_{out}	Outflow boundary of $\partial\Omega$, i.e., $\Gamma_{\mathrm{out}} := \{\mathbf{s} \in \partial\Omega \mid \mathbf{v}(\mathbf{s}) \cdot \mathbf{n}(\mathbf{s}) > 0\}$
$\mathbf{V}, \bar{\mathbf{V}}, \mathbf{U}, \bar{\mathbf{U}}, \ldots$	Tensors in $\mathbb{R}^{d \times d}$ (possibly symmetric)
v_k	k-th smallest eigenvalue of $\mathbf{V}$, i.e., $v_1 \leqslant \ldots \leqslant v_d$
$v_{k\ell}$	entry of $\mathbf{V} = (v_{k\ell})_{k,\ell=1}^d$ with row index k and column index ℓ
$\mathbf{v}_k$	k-th column vector of $\mathbf{V}$
$\tilde{\mathbf{V}}$	Diagonal form $\tilde{\mathbf{V}} = \mathrm{diag}(v_1, \ldots, v_d)$ of $\mathbf{V}$ with sorted eigenvalues
$\mathbf{V} = \mathbf{Q}\tilde{\mathbf{V}}\mathbf{Q}^\top$	Spectral decomposition of $\mathbf{V}$, i.e., $\mathbf{V}\mathbf{q}_k = v_k\mathbf{q}_k$ for all $k \in \{1, \ldots, d\}$

Table 1.2: Symbols referring to finite element discretization.

Symbol	Description
$\mathcal{T}_h$	Triangulation (mesh, grid)
E	Number of elements $K \in \mathcal{T}_h$
N	Dimension of finite element space/total number of nodes
M	Number of degrees of freedom excluding nodes on the Dirichlet boundary
i, j, l	Indices of degrees of freedom $1, \ldots, N$
e	Index of an element
n	Index of a time step if used as superscript
K^e	Element of $\mathcal{T}_h$ with index $e \in \{1, \ldots, E\}$
v_{ij}	Edge connecting two nodes $\mathbf{x}_i, \mathbf{x}_j \in K$ of $K \in \mathcal{T}_h$
Υ	Set of edges v_{ij}
$\mathcal{N}_i$	Nodal stencil
$\mathcal{N}^e$	Element stencil
$\mathbf{x}_i$	Nodal point with index $i \in \{1, \ldots, N\}$
$\mathcal{E}_i$	Element patch containing $\mathbf{x}_i$
u_h	Scalar-valued finite element function
u_i	Nodal value of u_h at $\mathbf{x}_i$
$\mathbf{U}_h$	Tensor-valued finite element function
$\mathbf{U}_i$	Nodal value of $\mathbf{U}_h$ at $\mathbf{x}_i$
$\mathcal{A}$	Finite element system matrix
$\mathcal{M}$	Finite element mass matrix
$\mathcal{K}$	Finite element convection matrix
$\mathcal{D}$	Artificial diffusion matrix
g	Generic right hand side vector
$\underline{g}$	Right hand side vector of a time dependent problem
$\mathcal{V}'$	Matrix $\mathcal{V} \in \mathbb{R}^{N \times N}$ restricted to first M rows
v'	Vector $v \in \mathbb{R}^N$ restricted to first M components

Table 1.3: Sets and norms.

Symbol	Description
$\mathbb{N}$	Natural numbers (without zero)
$\mathbb{R}$	Real numbers
$\mathbb{R}_0^+$	Nonnegative real numbers
$\mathbb{R}^{d \times d}$	$d \times d$ dimensional tensors with real tensor entries
$\mathbb{S}_d$	Subspace of symmetric $d \times d$ dimensional tensors
$\mathbb{S}_{d,+}$	Subspace of symmetric positive semidefinite $d \times d$ dimensional tensors
$\mathbb{A}_d$	Subspace of symmetric positive semidefinite $d \times d$ dim. tensors with unit trace, i.e., $\mathbb{A}_d := \left\{ \mathbf{A} \in \mathbb{S}_{d,+} \,\middle|\, \mathrm{tr}(\mathbf{A}) = 1 \right\}$
$C^k(\Omega)$	Space of functions that are k times continuously differentiable on Ω
$L^k(\Omega)$	Lebesgue space, i.e., $L^k(\Omega) := \left\{ u : \Omega \to \mathbb{R} \,\middle|\, u \text{ is Lebesgue measurable and } \int_\Omega |u|^p < \infty \right\}$
$W^{k,p}(\Omega)$	Sobolev space, i.e., space of functions whose weak derivatives of order up to k are in $L^p(\Omega)$
$H^k(\Omega)$	Sobolev space for the special case $p = 2$, i.e., $H^k(\Omega) := W^{k,2}(\Omega)$
$\mathrm{Lip}(\Omega)$	Space of scalar Lipschitz continuous functions
$\|\cdot\|_{L^p(\Omega)}$	Norm on the space $L^p(\Omega)$
$\|\cdot\|_{W^{k,p}(\Omega)}$	Norm on the space $W^{k,p}(\Omega)$
$\|\cdot\|_{H^k(\Omega)}$	Norm on the space $H^k(\Omega)$
$\|\cdot\|_p$	ℓ_p-vector norm
$\|\cdot\|_{\mathrm{F}}$	Frobenius norm
B^d	Open unit ball in $\mathbb{R}^d$, i.e., $B^d := \left\{ \mathbf{x} \in \mathbb{R}^d \,\middle|\, \|\mathbf{x}\|_2 < 1 \right\}$
∂B^d	Unit sphere in $\mathbb{R}^d$, i.e., $\partial B^d := \left\{ \mathbf{x} \in \mathbb{R}^d \,\middle|\, \|\mathbf{x}\|_2 = 1 \right\}$

Table 1.4: Symbols referring to AFC methodology.

Symbol	Description
$u_i^{\max}$	Local upper bound for u_i
$u_i^{\min}$	Local lower bound for u_i
α_{ij}	Scalar correction factor corresponding to edge v_{ij}
$\mathbf{S}_{ij}$	Tensorial correction factor corresponding to edge v_{ij}
$\mathscr{A}_{ij}$	Limiting operator corresponding to edge v_{ij}
f_{ij}	Scalar antidiffusive flux corresponding to edge v_{ij}
$\mathbf{F}_{ij}$	Tensorial antidiffusive flux corresponding to edge v_{ij}
L	Superscript referring to a low order approximation
H	Superscript referring to a high order approximation

2 Equations of fluid dynamics

In this chapter, we derive the basic equations of fluid dynamics following
Hirsch [Hir07]. The equations that describe the macroscopic behavior of the
fluid are based on fundamental assumptions like the conservation of mass
and momentum. Their derivation is based on the Eulerian point of view,
where the behavior of a (scalar) quantity in a spatially fixed control volume
is considered. In contrast to this approach, e.g., Donea and Huerta [DH05,
Chapter 1] focus on the Lagrangian framework and exploit the Reynolds
transport theorem to derive the same conservation laws.

After illustrating the derivation of a general conservation law using fundamental assumptions and the divergence theorem of Gauss, we consider the
continuity equation governing the conservation of mass. Next, we introduce
the general advection-diffusion-reaction equation and discuss properties of
the solution for hyperbolic models corresponding to the limit of vanishing
diffusion. In particular, we formulate conditions for solvability and uniqueness, as well as maximum principles that guarantee boundedness of exact
solutions by the initial and/or boundary conditions. These theoretical results
provide valuable criteria for the development of physics-compatible numerical
algorithms in subsequent chapters.

In the last section, we briefly introduce the incompressible Navier-Stokes
equations which are commonly used for the simulation of fluids and represent
a system of conservation laws for mass and momentum. In Chapter 6, we
come back to these equations and present generalizations to non-Newtonian
models of fiber suspension flows. Following a bottom-up approach, we
develop numerical methods for solution of increasingly complex problems by
piecing together subproblem solvers that are backed by theoretical analysis
and preserve important qualitative properties of exact solutions.

2.1 Scalar conservation laws

2.1.1 Physics

We consider a generic scalar quantity $u = u(\mathbf{x}, t)$, where $\mathbf{x}$ is a spatial location and $t \geqslant 0$ denotes the time variable. For instance, u may represent the density or energy of the observed fluid. To describe and simulate the evolution of u in a domain $\Omega \subset \mathbb{R}^d$ with a Lipschitz boundary $\partial\Omega$, we derive a mathematical form of physical balance laws that may influence the value of u at a given location in space. For this purpose, let $\Omega' \subset \Omega$ be an arbitrary fixed control volume bounded by the control surface $\partial\Omega' \subset \bar{\Omega}$. Then the total amount of u contained in Ω' at time t is given by

$$\int_{\Omega'} u(\mathbf{x}, t) \, \mathrm{d}\mathbf{x}, \tag{2.1}$$

while the temporal variation of u is represented by

$$\frac{\mathrm{d}}{\mathrm{d}t} \int_{\Omega'} u(\mathbf{x}, t) \, \mathrm{d}\mathbf{x}.$$

The value of (2.1) changes when fluxes $\mathbf{f} = \mathbf{f}(u(\mathbf{x}, t), \mathbf{x}, t)$ transport the conserved quantity u across the control surface $\partial\Omega'$, i.e., if the contribution of the flux perpendicular to the surface is nonvanishing. If the flux is parallel to the surface, it does not change the value of the integral. More precisely, the temporal change of the integral due to incoming fluxes is given by [Hir07]

$$-\oint_{\partial\Omega'} \mathbf{f}(u(\mathbf{s}, t), \mathbf{s}, t) \cdot \mathbf{n}(\mathbf{s}) \, \mathrm{d}\mathbf{s},$$

where $\mathbf{n} = \mathbf{n}(\mathbf{s}) \in \mathbb{R}^d$, $\mathbf{s} \in \partial\Omega'$, is the outward-pointing unit normal vector of the surface $\partial\Omega'$. The minus sign is used because the rate at which the total amount of u in Ω' *increases* due to transport across the boundary is given by the scalar product of $\mathbf{f}$ and the *inward-pointing* normal $-\mathbf{n}$. In addition, the value of (2.1) may change due to sinks and sources in the interior, denoted by q_V. The corresponding contribution can be expressed by

$$\int_{\Omega'} q_V(\mathbf{x}, t) \, \mathrm{d}\mathbf{x}.$$

In summary, the variation of the total mass of u is governed by the integral conservation law [Hir07, Eq. (1.1.1)]

$$\frac{\mathrm{d}}{\mathrm{d}t} \int_{\Omega'} u(\mathbf{x}, t) \, \mathrm{d}\mathbf{x} + \oint_{\partial\Omega'} \mathbf{f}(u(\mathbf{s}, t), \mathbf{s}, t) \cdot \mathbf{n}(\mathbf{s}, t) \, \mathrm{d}\mathbf{s} = \int_{\Omega'} q_V(\mathbf{x}, t) \, \mathrm{d}\mathbf{x} \quad \forall \Omega' \subset \Omega. \tag{2.2}$$

According to (2.2), the change of the total amount of u in the control volume Ω' depends on surface integrals of the effective flux. These integrals can be transformed into volume integrals using the following theorem.

Theorem 2.1 (Divergence theorem of Gauss). *Let $\mathbf{f}(\mathbf{x})$ be a continuously differentiable vector field defined on a neighborhood of a compact domain Ω' with a piecewise smooth boundary $\partial\Omega'$. Then we have*

$$\int_{\Omega'} \mathrm{div}\big(\mathbf{f}(\mathbf{x})\big)\,\mathrm{d}\mathbf{x} = \oint_{\partial\Omega'} \mathbf{f}(\mathbf{s}) \cdot \mathbf{n}(\mathbf{s})\,\mathrm{d}\mathbf{s}. \tag{2.3}$$

If this theorem is applicable, (2.2) can be transformed into the equivalent integral form

$$\frac{\mathrm{d}}{\mathrm{d}t} \int_{\Omega'} u(\mathbf{x},t)\,\mathrm{d}\mathbf{x} + \int_{\Omega'} \mathrm{div}\big(\mathbf{f}(u(\mathbf{x},t),\mathbf{x},t)\big)\,\mathrm{d}\mathbf{x} = \int_{\Omega'} q_V(\mathbf{x},t)\,\mathrm{d}\mathbf{x} \quad \forall\Omega' \subset \Omega. \tag{2.4}$$

Equation (2.4) is valid for all control volumes $\Omega' \subset \Omega$. Hence, the residual of the integrand must vanish almost everywhere, which leads to the *differential form of the conservation law* [Hir07, Eq. (1.1.4)]

$$\frac{\partial u(\mathbf{x},t)}{\partial t} + \mathrm{div}\big(\mathbf{f}(u(\mathbf{x},t),\mathbf{x},t)\big) = q_V(\mathbf{x},t) \quad \forall\mathbf{x} \in \Omega. \tag{2.5}$$

Due to the presence of derivatives in (2.5), the flux $\mathbf{f}$ must be differentiable in the strong sense for a classical solution to exist, i.e., we must have $\mathbf{f} \in C^1(\Omega, \mathbb{R}^d)$. In contrast to this, formulation (2.4) is well-posed under the weaker regularity assumption $\mathrm{div}(\mathbf{f}) \in L^1(\Omega)$, whereas the integral conservation law (2.2) does not contain any derivatives of $\mathbf{f}$.

In what follows, we define the fluxes and sources for some conserved quantities that commonly occur in mathematical models of continuum mechanics. The modeling of these terms is based on the laws of physics and experimental observations. The resulting (systems of) partial differential equations (PDEs) provide a mathematical description of underlying conservation laws. While this description is not exact, it can be used for simulation purposes. Numerical approximation techniques introduce additional errors but the resulting computational models produce reasonably accurate predictions for many applications of practical interest.

2.1.1.1 Continuity equation

In continuum mechanics, the differential form of the mass conservation law is known as the *continuity equation*. This first order hyperbolic PDE describes the *advection* of a scalar conserved quantity u by a given velocity field $\mathbf{v} = \mathbf{v}(\mathbf{x}, t)$. The *advective flux* $\mathbf{f} = \mathbf{v}u$ determines the rate at which u enters or leaves an arbitrary control volume per unit area and unit time. In the absence of other effects that may change the mass balance, the source term q_V vanishes. Under these assumptions, the differential form of the mass conservation law reads [Hir07, Eq. (1.2.2); DH05, Eq. (1.11)]

$$\frac{\partial u(\mathbf{x}, t)}{\partial t} + \mathrm{div}\big(\mathbf{v}(\mathbf{x}, t)u(\mathbf{x}, t)\big) = 0 \qquad \forall \mathbf{x} \in \Omega. \tag{2.6}$$

This conservation law is also called *convection* or *advection equation*. In mathematical models of continuum mechanics, the physical meaning of the density function u is mass per unit volume and its evolution is governed by the continuity equation (2.6). It reduces to the incompressibility constraint $\mathrm{div}\big(\mathbf{v}(\mathbf{x}, t)\big) = 0$ if the density is constant.

2.1.1.2 Heat equation

Another important special case of (2.5) is the *heat equation*, a parabolic PDE which represents a simplified form of a diffusion equation in the context of mass transfer modeling. If the transport of heat by external velocity fields is neglected, the sole mechanism by which u is transmitted across the boundary of a control volume is the microscopic Brownian motion of molecules. According to the Fourier law of heat conduction and Fick's law of mass diffusion, molecular mixing transports the quantities of interest from regions of high concentration into regions of low concentration. Therefore, we define $\mathbf{f} = -a\,\mathrm{grad}(u)$, where $a \geqslant 0$ is the diffusion coefficient. The so defined diffusive flux vanishes for constant states; otherwise it is oriented in the direction of lower concentrations. Substituting it into (2.5), we obtain the *heat equation*

$$\frac{\partial u(\mathbf{x}, t)}{\partial t} - \mathrm{div}\big(a(\mathbf{x}, t)\,\mathrm{grad}(u(\mathbf{x}, t))\big) = 0 \qquad \forall \mathbf{x} \in \Omega. \tag{2.7}$$

In models of heat and mass transfer, u represents the temperature or concentration (mass density) of chemical species.

2.1.1.3 Advection-diffusion-reaction equation

To formulate a generic conservation law that includes (2.6) and (2.7) as special cases, we substitute the combined flux $\mathbf{f} = \mathbf{v}u - a\,\mathrm{grad}(u)$ and the 'reactive' term $q_V = f - cu$, $c > 0$ into (2.5). The result is the *conservative form* of the *advection-diffusion-reaction equation*

$$\frac{\partial u}{\partial t} + \text{div}(\mathbf{v}u) - \text{div}\big(a\,\text{grad}(u)\big) + cu = f \qquad \text{in } \Omega. \qquad (2.8)$$

Here and below, we no longer indicate the dependence of the involved functions and parameters on $\mathbf{x} \in \Omega$ and $t \geqslant 0$ for the sake of simplicity. Applying the product rule to the convective term, the generic transport equation can be written in the *non-conservative form*

$$\frac{\partial u}{\partial t} + \mathbf{v} \cdot \text{grad}(u) - \text{div}\big(a\,\text{grad}(u)\big) + \big(\text{div}(\mathbf{v}) + c\big)u = f \qquad \text{in } \Omega. \qquad (2.9)$$

In what follows, we are only interested in the hyperbolic limit and, therefore, set $a = 0$. Furthermore, we substitute c for $\text{div}(\mathbf{v}) + c$ to simplify the notation. The properties of solutions to the resulting model problem and its steady state counterpart play an important role in the analysis and development of numerical methods. Therefore, we begin with a summary of known theoretical results that will be used in subsequent chapters. In particular, we formulate regularity assumptions under which the (initial-)boundary value problems are well-posed in a weak sense and prove the validity of maximum principles which guarantee boundedness of analytical solutions by the data.

2.1.2 Well-posedness

This section is devoted to the analysis of existence and uniqueness of the solution to (2.9) with negligible diffusivity ($a = 0$). For this purpose, we first consider the stationary version of the hyperbolic PDE model and formulate requirements which are sufficient for the existence of a unique solution. Next, we consider the unsteady case in which at most one solution exists under less restrictive assumptions.

2.1.2.1 Steady problem

This section follows the presentation of [DE11, Section 2.1], which is in turn based on the work of Ern and Guermond [EG06]. For simplicity, we restrict our analysis to advection-reaction equations with homogeneous inflow boundary conditions and refer to [DE11, Section 2.1.6] for the general case.

Imposing homogeneous Dirichlet boundary conditions at the inlet and assuming a steady state equilibrium, we consider the hyperbolic boundary value problem

$$\mathbf{v} \cdot \mathrm{grad}(u) + cu = f \qquad \text{in } \Omega, \tag{2.10a}$$

$$u = 0 \qquad \text{on } \Gamma_{\text{in}}, \tag{2.10b}$$

where $c : \Omega \to \mathbb{R}$ is the local reaction rate, $f : \Omega \to \mathbb{R}$ is the local intensity of source/sink terms, and $\Gamma_{\text{in}} := \{\mathbf{s} \in \partial\Omega \mid \mathbf{v}(\mathbf{s}) \cdot \mathbf{n}(\mathbf{s}) < 0\}$ denotes the inflow part of $\partial\Omega$ depending on the given velocity field $\mathbf{v} : \bar{\Omega} \to \mathbb{R}^d$. For later use, we also define the outflow boundary $\Gamma_{\text{out}} := \{\mathbf{s} \in \partial\Omega \mid \mathbf{v}(\mathbf{s}) \cdot \mathbf{n}(\mathbf{s}) > 0\}$.

Herein, we analyze the continuous problem (2.10) by introducing appropriate function spaces for the solution u and the parameters $\mathbf{v}$, c, and f. Furthermore, the meaning of the inflow boundary condition (2.10b) for the solution u is discussed. While we focus on the non-conservative form of the PDE, the conservative form leads to very similar results.

To begin with, let us assume that the given parameters of problem (2.10) satisfy

$$\mathbf{v} \in \big[\mathrm{Lip}(\Omega)\big]^d, \qquad c \in L^\infty(\Omega), \qquad f \in L^2(\Omega), \tag{2.11}$$

where $\mathrm{Lip}(\Omega)$ denotes the space of Lipschitz continuous functions: A function $\varphi : \Omega \to \mathbb{R}$ is called *Lipschitz continuous* if there exists a *Lipschitz constant* $L > 0$ such that

$$\big|\varphi(\mathbf{x}) - \varphi(\bar{\mathbf{x}})\big| \leqslant L\|\mathbf{x} - \bar{\mathbf{x}}\|_2 \qquad \forall \mathbf{x}, \bar{\mathbf{x}} \in \Omega. \tag{2.12}$$

Then a vector-valued function $\mathbf{v} \in \big[\mathrm{Lip}(\Omega)\big]^d$ satisfies

$$\big\|\mathbf{v}(\mathbf{x}) - \mathbf{v}(\bar{\mathbf{x}})\big\|_2^2 = \sum_k \big|v_k(\mathbf{x}) - v_k(\bar{\mathbf{x}})\big|^2 \leqslant \sum_k L_{v_k}^2 \|\mathbf{x} - \bar{\mathbf{x}}\|_2^2 \leqslant d L_{\mathbf{v}}^2 \|\mathbf{x} - \bar{\mathbf{x}}\|_2^2$$

$$\forall \mathbf{x}, \bar{\mathbf{x}} \in \Omega,$$

where $L_{\mathbf{v}} := \max_k L_{v_k}$ and $L_{v_k} > 0$ is the Lipschitz constant of the k-th component of $\mathbf{v}$, $k \in \{1, \ldots, d\}$. According to [BS07, Chapter 1], the assumption $\mathbf{v} \in \big[\mathrm{Lip}(\Omega)\big]^d$ implies

$$\mathbf{v} \in \big[W^{1,\infty}(\Omega)\big]^d, \tag{2.13a}$$

$$\big\|\mathrm{grad}(v_k)\big\|_{[L^\infty(\Omega)]^d} \leqslant L_{v_k} \leqslant L_{\mathbf{v}} \qquad \forall k \in \{1, \ldots, d\}. \tag{2.13b}$$

Another basic requirement that we need for the well-posedness of (2.10) is the coercivity condition

$$\exists c_0 > 0 : \qquad L^\infty(\Omega) \ni \Lambda := c - \tfrac{1}{2}\operatorname{div}(\mathbf{v}) \geqslant c_0 \quad \text{a.e. in } \Omega. \qquad (2.14)$$

For a function $\varphi \in L^2(\Omega)$, we say that $\mathbf{v} \cdot \operatorname{grad}(\varphi) \in L^2(\Omega)$ if and only if [DE11]

$$\int_\Omega \varphi \operatorname{div}(\mathbf{v}\psi) \leqslant C_\varphi \|\psi\|_{L^2(\Omega)} \qquad \forall \psi \in C_0^\infty(\Omega).$$

This motivates the following definition of the function space in which a solution u may exist.

Definition 2.2 ([DE11, Definition 2.1]). The *graph space* is defined by

$$V := \{\varphi \in L^2(\Omega) \mid \mathbf{v} \cdot \operatorname{grad}(\varphi) \in L^2(\Omega)\}. \qquad (2.15)$$

It is equipped with the following scalar product and associated *graph norm*:

$$(\varphi, \psi)_V := (\varphi, \psi)_{L^2(\Omega)} + \big(\mathbf{v} \cdot \operatorname{grad}(\varphi), \mathbf{v} \cdot \operatorname{grad}(\psi)\big)_{L^2(\Omega)} \qquad \forall \varphi, \psi \in V, \qquad (2.16)$$

$$\|\varphi\|_V^2 = (\varphi, \varphi)_V \qquad \forall \varphi \in V. \qquad (2.17)$$

We will see that V is defined as the least restrictive Hilbert space such that a (specific) weak formulation of (2.10) is well-defined if (2.14) holds. Under assumption (2.11), there exists a unique weak solution $u \in V$ to (2.10). Consequently, the existence of a solution belonging to more restrictive Hilbert spaces like $H^1(\Omega) \subseteq V$ requires further assumptions.

Di Pietro and Ern [DE11, Proposition 2.2] have shown that V is a Hilbert space. To show that a weak solution u of (2.10) can be sought in V, we define a trace operator, which maps functions of V to functions that are measurable on the boundary. That is, the value range of the trace operator is given by

$$L^2(|\mathbf{v} \cdot \mathbf{n}|; \partial\Omega) := \Big\{\varphi \text{ measurable on } \partial\Omega \;\Big|\; \int_{\partial\Omega} |\mathbf{v} \cdot \mathbf{n}|\varphi^2 < \infty\Big\}$$

with the associated (semi-)norm

$$\|\varphi\|_{L^2(|\mathbf{v}\cdot\mathbf{n}|;\partial\Omega)}^2 := \int_{\partial\Omega} |\mathbf{v} \cdot \mathbf{n}|\varphi^2 \qquad \forall \varphi \in L^2(|\mathbf{v} \cdot \mathbf{n}|; \partial\Omega).$$

Lemma 2.3 ([DE11, Lemma 2.5]). *If* $\mathrm{dist}(\Gamma_{\mathrm{in}}, \Gamma_{\mathrm{out}}) > 0$, *the trace operator*

$$\gamma : C^0(\bar{\Omega}) \to L^2(|\mathbf{v} \cdot \mathbf{n}|; \partial\Omega), \qquad \varphi \mapsto \varphi|_{\partial\Omega} \qquad (2.18)$$

extends continuously to V. *That is, there exists* $C_\gamma > 0$ *such that*

$$\|\gamma(\varphi)\|_{L^2(|\mathbf{v}\cdot\mathbf{n}|;\partial\Omega)} \leqslant C_\gamma \|\varphi\|_V \qquad \forall \varphi \in V \qquad (2.19)$$

and integration by parts can be performed using the formula

$$\int_\Omega \mathbf{v} \cdot \mathrm{grad}(\varphi)\psi + \mathbf{v} \cdot \mathrm{grad}(\psi)\varphi + \mathrm{div}(\mathbf{v})\varphi\psi = \int_{\partial\Omega} \mathbf{v} \cdot \mathbf{n}\gamma(\varphi)\gamma(\psi) \quad \forall \varphi, \psi \in V. \tag{2.20}$$

Proof. The proof can be found in [DE11, Section 2.1.5]. $\square$

To prove the existence of a weak solution to (2.10), we consider the *weak formulation*

$$\text{Find } u \in V \text{ s.t.} \qquad a(u, \varphi) = \int_\Omega f\varphi \quad \forall \varphi \in V, \qquad (2.21\mathrm{a})$$

$$a(\psi, \varphi) := \int_\Omega c\psi\varphi + \int_\Omega \mathbf{v} \cdot \mathrm{grad}(\psi)\varphi + \int_{\partial\Omega} (\mathbf{v} \cdot \mathbf{n})_-\psi\varphi \quad \forall \varphi, \psi \in V. \tag{2.21b}$$

In the last term, we omitted the trace operator γ for the sake of simplicity and restricted the integration to the inflow boundary by using the negative part $(\mathbf{v} \cdot \mathbf{n})_-$ of the normal velocity. The decomposition of a generic scalar into its positive and negative components is defined by

$$v = v_+ - v_- := \max(v, 0) - \max(-v, 0) \qquad \forall v \in \mathbb{R}. \qquad (2.22)$$

The integrals occurring in (2.21) are well-defined due to the definition of V, the fact that $f \in L^2(\Omega)$, and Lemma 2.3. Furthermore, the following properties hold.

Lemma 2.4 ([DE11]). *The bilinear form* $a : V \times V \to \mathbb{R}$ *defined by (2.21b) is bounded in* $V \times V$ *as follows:*

$$|a(\psi, \varphi)| \leqslant \sqrt{1 + \|c\|^2_{L^\infty(\Omega)}} \|\psi\|_V \|\varphi\|_{L^2(\Omega)} + C_\gamma^2 \|\psi\|_V \|\varphi\|_V \quad \forall \varphi, \psi \in V. \tag{2.23}$$

Furthermore, it is coercive in the following sense:

$$a(\varphi, \varphi) \geqslant c_0 \|\varphi\|^2_{L^2(\Omega)} + \tfrac{1}{2}\|\varphi\|^2_{L^2(|\mathbf{v}\cdot\mathbf{n}|;\partial\Omega)} \qquad \forall \varphi \in V. \tag{2.24}$$

Proof. The first estimate is a direct consequence of Lemma 2.3 and the Cauchy-Schwarz inequality because

$$\left| \int_\Omega \left(c\psi + \mathbf{v} \cdot \mathrm{grad}(\psi) \right) \varphi \right|$$

$$\leqslant \left\| c\psi + \mathbf{v} \cdot \mathrm{grad}(\psi) \right\|_{L^2(\Omega)} \|\varphi\|_{L^2(\Omega)}$$

$$\leqslant \left(\|c\|_{L^\infty(\Omega)} \|\psi\|_{L^2(\Omega)} + \left\| \mathbf{v} \cdot \mathrm{grad}(\psi) \right\|_{L^2(\Omega)} \right) \|\varphi\|_{L^2(\Omega)}$$

$$\leqslant \max \left(\|c\|_{L^\infty(\Omega)}, 1 \right) \left(\|\psi\|_{L^2(\Omega)} + \left\| \mathbf{v} \cdot \mathrm{grad}(\psi) \right\|_{L^2(\Omega)} \right) \|\varphi\|_{L^2(\Omega)}$$

$$\leqslant \sqrt{1 + \|c\|_{L^\infty(\Omega)}^2} \, \|\psi\|_V \|\varphi\|_{L^2(\Omega)},$$

$$\left| \int_{\partial\Omega} (\mathbf{v} \cdot \mathbf{n})_- \psi\varphi \right| \leqslant \|\psi\|_{L^2(|\mathbf{v}\cdot\mathbf{n}|;\partial\Omega)} \|\varphi\|_{L^2(|\mathbf{v}\cdot\mathbf{n}|;\partial\Omega)} \leqslant C_\gamma^2 \|\psi\|_V \|\varphi\|_V.$$

The second property follows from

$$a(\varphi,\varphi) = \int_\Omega c\varphi^2 + \int_\Omega \mathbf{v} \cdot \mathrm{grad}(\varphi)\varphi + \int_{\partial\Omega} (\mathbf{v} \cdot \mathbf{n})_- \varphi^2$$

$$= \int_\Omega c\varphi^2 - \tfrac{1}{2} \int_\Omega \mathrm{div}(\mathbf{v})\varphi^2 + \tfrac{1}{2} \int_{\partial\Omega} (\mathbf{v} \cdot \mathbf{n})\varphi^2 + \int_{\partial\Omega} (\mathbf{v} \cdot \mathbf{n})_- \varphi^2 \quad (2.25)$$

$$= \int_\Omega \Lambda\varphi^2 + \tfrac{1}{2} \int_{\partial\Omega} |\mathbf{v} \cdot \mathbf{n}|\varphi^2 \geqslant c_0 \|\varphi\|_{L^2(\Omega)}^2 + \tfrac{1}{2} \int_{\partial\Omega} |\mathbf{v} \cdot \mathbf{n}|\varphi^2,$$

where we have used (2.20) and the coercivity condition (2.14). □

Due to the 'coercivity' of $a(\cdot, \cdot)$, there exists at most one solution u of the weak problem (2.21). However, property (2.24) does *not* imply the coercivity of $a(\cdot, \cdot)$ in V. Therefore, the Lax-Milgram Lemma cannot be exploited to show the existence of a solution.

Each solution to (2.21) satisfies (2.10) in a weak sense due to the following proposition.

Proposition 2.5 ([DE11, Proposition 2.7]). *If $u \in V$ solves (2.21), it is a weak solution of (2.10), that is,*

$$\mathbf{v} \cdot \mathrm{grad}(u) + cu = f \qquad \textit{a.e. in } \Omega, \qquad\qquad (2.26\mathrm{a})$$

$$u = 0 \qquad \textit{a.e. on } \Gamma_{\mathrm{in}}. \qquad\qquad (2.26\mathrm{b})$$

Proof. Due to (2.21), we have

$$\int_\Omega \left(cu + \mathbf{v} \cdot \mathrm{grad}(u) - f \right) \varphi = 0 \qquad \forall \varphi \in C_0^\infty(\Omega)$$

and (2.26a) holds because $C_0^\infty(\Omega)$ is dense in $L^2(\Omega)$. Furthermore, (2.21) with $\varphi = u$ leads to

$$0 = \int_{\partial\Omega} (\mathbf{v} \cdot \mathbf{n})_- u^2 = \int_{\Gamma_{\text{in}}} |\mathbf{v} \cdot \mathbf{n}| u^2,$$

due to (2.26a) and $\mathbf{v} \cdot \mathbf{n} < 0$ on Γ_{in}. This implies the validity of (2.26b)   □

The following theorem ensures the existence and uniqueness of a solution to (2.21), which is a weak solution of the steady advection-reaction equation by Proposition 2.5.

Theorem 2.6 ([DE11, Theorem 2.9]). *Problem* (2.21) *is well-posed.*

Proof. The proof of this theorem can be found in [DE11, Section 2.1.5]. It exploits the well-posedness of the problem with strongly enforced boundary conditions and then shows that this solution must be the unique solution of (2.21).   □

In the absence of the coercivity condition (2.14), the solution of (2.10) is possibly not unique as the following example illustrates.

Example 2.7. Let us consider the homogeneous steady advection equation ($c = f = 0$) on the unit disc $\Omega = B^2 := \left\{ \mathbf{x} \in \mathbb{R}^2 \mid \|\mathbf{x}\|_2 < 1 \right\}$ with the velocity field $\mathbf{v} = (-x_2, x_1)^\top \in \mathbb{R}^2$, i.e.,

$$\mathbf{v} \cdot \text{grad}(u) = \text{div}(\mathbf{v}u) = 0 \qquad \text{in } \Omega. \tag{2.27}$$

No boundary conditions are to be prescribed on $\partial\Omega = \left\{ \mathbf{x} \in \mathbb{R}^2 \mid \|\mathbf{x}\|_2 = 1 \right\}$ because

$$\mathbf{v}(\mathbf{s}) \cdot \mathbf{n}(\mathbf{s}) = \mathbf{v}(\mathbf{s}) \cdot \mathbf{s} = 0 \qquad \forall \mathbf{s} \in \partial\Omega.$$

Since each radial symmetric function is a solution of (2.27), the solution is obviously not unique.

In [DE11, Section 2.1.6], the general steady advection-reaction equation (2.10) with inhomogeneous inflow boundary data $u_{\text{in}} : \Gamma_{\text{in}} \to \mathbb{R}$ is considered. The boundary value problem is given by

$$\mathbf{v} \cdot \text{grad}(u) + cu = f \qquad \text{in } \Omega, \tag{2.28a}$$

$$u = u_{\text{in}} \qquad \text{on } \Gamma_{\text{in}}. \tag{2.28b}$$

Extending u_{in} trivially to $\partial\Omega \setminus \Gamma_{\text{in}}$, we assume $u_{\text{in}} \in L^2(|\mathbf{v} \cdot \mathbf{n}|; \partial\Omega)$. Then (2.28) holds in a weak sense if $u \in V$ is the solution of

$$\text{Find } u \in V \text{ s.t.} \qquad a(u, \varphi) = \int_\Omega f\varphi + \int_{\partial\Omega} (\mathbf{v} \cdot \mathbf{n})_- u_{\text{in}}\varphi \quad \forall \varphi \in V. \quad (2.29)$$

Di Pietro and Ern [DE11, Theorem 2.12] have proved the well-posedness of (2.29) under the above assumptions.

If the coercivity condition (2.14) does not hold, the above theory does not apply. As we will see in Section 2.1.3.1, existence of a solution may still be guaranteed under certain assumptions regarding the velocity field.

2.1.2.2 Unsteady problem

We now derive energy estimates for the weak solution of the time dependent counterpart of the steady advection-reaction equation following [DE11, Section 3.1.1]. We restrict our attention to the problem

$$\frac{\partial u}{\partial t} + \mathbf{v} \cdot \text{grad}(u) + cu = f \qquad \text{in } \Omega \times (0, T), \quad (2.30a)$$

$$u = u_{\text{in}} \qquad \text{on } \Gamma_{\text{in}} \times (0, T), \quad (2.30b)$$

$$u(\cdot, 0) = u_0 \qquad \text{in } \Omega \quad (2.30c)$$

with homogeneous boundary conditions $u_{\text{in}} = 0$, where $T > 0$ is the final time and $u_0 \in L^2(\Omega)$ denotes the initial data for the time dependent solution u. Furthermore, we impose the minimal regularity requirements

$$\mathbf{v} \in \text{Lip}(\Omega)^d, \qquad c \in L^\infty(\Omega), \qquad f \in C^0([0, T]; L^2(\Omega)),$$

where the assumption on f means that f is continuous in time with $f(t) \in L^2(\Omega)$ for all $t \in [0, T]$. The function $u \in C^0([0, T]; V) \cap C^1([0, T]; L^2(\Omega))$ is a weak solution to (2.30) if

$$\int_\Omega \frac{\partial u}{\partial t}\varphi + a(u, \varphi) = \int_\Omega f\varphi + \int_{\partial\Omega} (\mathbf{v} \cdot \mathbf{n})_- u_{\text{in}}\varphi \quad \forall \varphi \in V, t \in (0, T],$$
$$(2.31a)$$

$$u(\cdot, 0) = u_0 \qquad \qquad \text{in } \Omega \quad (2.31b)$$

holds, where the notation $u \in C^1([0, T]; L^2(\Omega))$ means that $u : [0, T] \to L^2(\Omega)$ is continuously differentiable in time. In contrast to the analysis of the steady advection-reaction equation, the following energy estimate holds without the need for the coercivity condition (2.14).

Lemma 2.8. *If* $u \in C^0([0,T];V) \cap C^1([0,T];L^2(\Omega))$ *solves* (2.31) *with* $u_{\mathrm{in}} = 0$, *then*

$$\|u(t)\|^2_{L^2(\Omega)} \leqslant e^{\frac{t}{\varsigma}}\left(\|u_0\|^2_{L^2(\Omega)} + \varsigma T \max_{s\in[0,T]}\|f(s)\|^2_{L^2(\Omega)}\right) \qquad \forall t \in [0,T] \quad (2.32)$$

holds, where $\varsigma := \left(T^{-1} + 2\|\Lambda\|_{L^\infty(\Omega)}\right)^{-1}$.

Proof. The proof follows [DE11, Lemma 3.2], where the test function $\varphi = u(t)$ is substituted into the weak form (2.31a) and the integral over time is considered. To prove this result, we need the estimates

$$
\begin{aligned}
a\big(u(t),u(t)\big) &= \int_\Omega c u(t)^2 + \int_\Omega \mathbf{v} \cdot \mathrm{grad}\big(u(t)\big)u(t) + \int_{\partial\Omega}(\mathbf{v}\cdot\mathbf{n})_- u(t)^2 \\
&= \int_\Omega \big(c - \tfrac{1}{2}\,\mathrm{div}(\mathbf{v})\big)u(t)^2 + \tfrac{1}{2}\int_{\partial\Omega}|\mathbf{v}\cdot\mathbf{n}|u(t)^2 \\
&= \int_\Omega \Lambda u(t)^2 + \tfrac{1}{2}\int_{\partial\Omega}|\mathbf{v}\cdot\mathbf{n}|u(t)^2 \\
&\geqslant \int_\Omega \Lambda u(t)^2 \geqslant -\|\Lambda\|_{L^\infty(\Omega)}\|u(t)\|^2_{L^2(\Omega)},
\end{aligned}
$$

$$(2.33)$$

$$
\begin{aligned}
\big(f(t),u(t)\big)_{L^2(\Omega)} &\leqslant \|f(t)\|_{L^2(\Omega)}\|u(t)\|_{L^2(\Omega)} \\
&\leqslant \frac{T}{2}\|f(t)\|^2_{L^2(\Omega)} + \frac{1}{2T}\|u(t)\|^2_{L^2(\Omega)},
\end{aligned}
$$

where we have used the fact that $\mathbf{v}\cdot\mathbf{n} \geqslant 0$ on Γ_{out} and $T > 0$. This yields

$$
\begin{aligned}
\frac{\mathrm{d}}{\mathrm{d}t}\|u(t)\|^2_{L^2(\Omega)} &= 2\big(f(t),u(t)\big)_{L^2(\Omega)} - 2a\big(u(t),u(t)\big) \\
&\leqslant T\|f(t)\|^2_{L^2(\Omega)} + T^{-1}\|u(t)\|^2_{L^2(\Omega)} + 2\|\Lambda\|_{L^\infty(\Omega)}\|u(t)\|^2_{L^2(\Omega)} \quad (2.34) \\
&\leqslant T \max_{s\in[0,T]}\|f(s)\|^2_{L^2(\Omega)} + \varsigma^{-1}\|u(t)\|^2_{L^2(\Omega)}
\end{aligned}
$$

by virtue of $u_{\mathrm{in}} = 0$. Thus, taking advantage of

$$
\begin{aligned}
\frac{\mathrm{d}}{\mathrm{d}t}\left(e^{-\frac{t}{\varsigma}}\|u(t)\|^2_{L^2(\Omega)}\right) &= e^{-\frac{t}{\varsigma}}\left(\frac{\mathrm{d}}{\mathrm{d}t}\|u(t)\|^2_{L^2(\Omega)} - \varsigma^{-1}\|u(t)\|^2_{L^2(\Omega)}\right) \\
&\leqslant T e^{-\frac{t}{\varsigma}} \max_{s\in[0,T]}\|f(s)\|^2_{L^2(\Omega)}
\end{aligned}
$$

and integrating in time results in

$$\|u(t)\|^2_{L^2(\Omega)} \leqslant e^{\frac{t}{\varsigma}}\left(\|u(0)\|^2_{L^2(\Omega)} + T \max_{s\in[0,T]}\|f(s)\|^2_{L^2(\Omega)} \int_0^t e^{-\frac{s}{\varsigma}}\,\mathrm{d}s\right)$$
$$\leqslant e^{\frac{t}{\varsigma}}\left(\|u_0\|^2_{L^2(\Omega)} + \varsigma T \max_{s\in[0,T]}\|f(s)\|^2_{L^2(\Omega)}\right)$$

because

$$\int_0^t e^{-\frac{s}{\varsigma}}\,\mathrm{d}s = \varsigma - \varsigma e^{-\frac{t}{\varsigma}} \leqslant \varsigma \qquad \forall t,\varsigma > 0. \qquad \square$$

For long time intervals such that $t \approx T \gg \|\Lambda\|^{-1}_{L^\infty(\Omega)}$, the exponential term $e^{\frac{t}{\varsigma}}$ dominates the energy estimate (2.32) due to

$$T\varsigma^{-1} = T\left(T^{-1} + 2\|\Lambda\|_{L^\infty(\Omega)}\right) = 1 + 2T\|\Lambda\|_{L^\infty(\Omega)} \gg 1$$

and Lemma 2.8 loses its significance. Improved estimates can be obtained if the coercivity condition (2.14) is applicable.

Remark 2.9. If condition (2.14) is valid as well, estimate (2.33) can be replaced by

$$a\big(u(t), u(t)\big) \geqslant \int_\Omega \Lambda u(t)^2 \geqslant c_0\|u(t)\|^2_{L^2(\Omega)}.$$

Instead of (2.34), we obtain the estimate

$$\frac{\mathrm{d}}{\mathrm{d}t}\|u(t)\|^2_{L^2(\Omega)} + 2c_0\|u(t)\|^2_{L^2(\Omega)}$$
$$\leqslant 2\big(f(t), u(t)\big)_{L^2(\Omega)} \leqslant 2\|f(t)\|_{L^2(\Omega)}\|u(t)\|_{L^2(\Omega)}$$
$$\leqslant 2c_0\|u(t)\|^2_{L^2(\Omega)} + (2c_0)^{-1}\|f(t)\|^2_{L^2(\Omega)} \qquad \forall t \in [0,T],$$

which holds by virtue of the Cauchy-Schwarz and Young's inequalities. In other words, we have [DE11]

$$\frac{\mathrm{d}}{\mathrm{d}t}\|u(t)\|^2_{L^2(\Omega)} \leqslant (2c_0)^{-1}\|f(t)\|^2_{L^2(\Omega)} \qquad \forall t \in [0,T].$$

Integration in time yields the energy estimate

$$\|u(t)\|^2_{L^2(\Omega)} \leqslant \|u_0\|^2_{L^2(\Omega)} + (2c_0)^{-1}t \max_{s\in[0,T]}\|f(s)\|^2_{L^2(\Omega)} \qquad \forall t \in [0,T]. \quad (2.35)$$

Lemma 2.8 and the stability estimate (2.35) guarantee that at most one solution to (2.30) exists. Under the assumptions of sufficiently smooth data, existence of u can be established using the method of characteristics presented in the next section.

Stability analysis leading to similar results for inhomogeneous boundary conditions can be found, for example, in [QV94, Section 14.3.1]. The stability estimates presented therein were derived for the Galerkin finite element approximation but also hold for the weak solution to the continuous problem (2.31).

2.1.3 Maximum principles

The purpose of this section is the derivation of maximum principles for the steady and unsteady advection-reaction equation. The theory developed in frequently cited studies like [GT15; PW12; Spe81] is restricted to elliptic, parabolic, and hyperbolic problems of second order and does not apply to first order hyperbolic equations. To the author's best knowledge, the unpublished guidebook by Kuzmin [Kuz10, Section 3.1.3] is the only work that partially fills this gap. The main theoretical results regarding maximum principles for hyperbolic problems are summarized in this section.

2.1.3.1 Steady problem

To keep things simple, it is assumed that the data $\mathbf{v}$, c, and g as well as the solution u are sufficiently smooth so that all derivatives and integrals that appear in the below expressions exist.

Theoretical studies of linear first order hyperbolic PDEs are typically performed using the *method of characteristics*. Before giving a formal definition of characteristics, we define the *trajectory* of a material particle as a curve in Ω such that the velocity vector is tangent to this curve at any point along the curve.

Definition 2.10. A curve $\gamma : [0, S] \to \bar{\Omega}$, $S > 0$ with $\gamma'(s) = \mathbf{v}(\gamma(s))$ for all $s \in [0, S]$ is called a *trajectory*.

In the stationary case, the curve γ follows the *streamlines* of the velocity field and can be interpreted as the trajectory of a Lagrangian particle whose velocity at pseudo-time s is given by $\mathbf{v}(\gamma(s))$. Due to this property, the solution of an advection-reaction equation can be determined by solving an ordinary differential equation (ODE) [Kuz10, Section 3.1.3.1] along the streamline that connects a given point to the inflow boundary. If the solution is known at any point on the streamline, its changes along the streamline are uniquely determined by the solution of an initial value problem backwards or forwards in time. The method of characteristics yields the following analytical formula for the solution of (2.28a) along the streamlines.

Lemma 2.11. *For each trajectory* $\gamma : [0, S] \to \bar{\Omega}$, *the solution of* (2.28a) *along* γ *reads [DH05, Eq. (3.13)]*

$$u\big(\gamma(s)\big) = \varrho(s)^{-1}\Big(u\big(\gamma(0)\big) + \int_0^s \varrho(s')f\big(\gamma(s')\big)\,\mathrm{d}s'\Big) \qquad \forall s \in [0, S], \quad (2.36)$$

where $\varrho(s) := \exp\big(\int_0^s c(\gamma(s'))\,\mathrm{d}s'\big)$.

Proof. Differentiation of (2.36) with respect to s yields

$$\frac{\mathrm{d}u\big(\gamma(s)\big)}{\mathrm{d}s} = \varrho(s)^{-1}\varrho(s)f\big(\gamma(s)\big)$$

$$- \varrho'(s)\varrho(s)^{-2}\Big(u\big(\gamma(0)\big) + \int_0^s \varrho(s')f\big(\gamma(s')\big)\,\mathrm{d}s'\Big)$$

$$= f\big(\gamma(s)\big) - c\big(\gamma(s)\big)\varrho(s)^{-1}\Big(u\big(\gamma(0)\big) + \int_0^s \varrho(s')f\big(\gamma(s')\big)\,\mathrm{d}s'\Big)$$

$$= f\big(\gamma(s)\big) - c\big(\gamma(s)\big)u\big(\gamma(s)\big).$$

On the other hand, we have

$$\frac{\mathrm{d}u\big(\gamma(s)\big)}{\mathrm{d}s} = \frac{\mathrm{d}\gamma(s)}{\mathrm{d}s} \cdot \big(\mathrm{grad}\,u(\mathbf{x})\big)\big|_{\mathbf{x}=\gamma(s)} = \mathbf{v}\big(\gamma(s)\big) \cdot \big(\mathrm{grad}\,u(\mathbf{x})\big)\big|_{\mathbf{x}=\gamma(s)}$$

and (2.28a) holds for the solution defined by (2.36).   □

The solution of (2.10) can be determined using (2.36) on each trajectory in an independent manner. Thus, u is possibly discontinuous in the crosswind direction if, for instance, the inflow boundary condition is discontinuous.

To exploit Lemma 2.11 for the derivation of maximum principles, it is required that each location in the domain can be reached by a trajectory starting on the inflow boundary.

Definition 2.12. A velocity field $\mathbf{v}$ is called *perfusing* (with respect to Ω) if for every $\mathbf{x} \in \Omega$ there exists a trajectory $\gamma_{\mathbf{x}} : [0, S_{\mathbf{x}}] \to \bar{\Omega}$ such that $\gamma_{\mathbf{x}}(0) \in \Gamma_{\mathrm{in}}$ and $\gamma_{\mathbf{x}}(S_{\mathbf{x}}) = \mathbf{x}$.

Furthermore, if every location in the domain can be reached by just one trajectory starting on the inflow boundary, the function defined by (2.36) is the unique solution of the steady advection-reaction equation even if the coercivity condition (2.14) does not hold.

Moreover, the assumption of a perfusing velocity field makes it possible to prove the following maximum principles.

Theorem 2.13. *If* v *is perfusing, the solution* u *of* (2.28) *is bounded by the extrema of the prescribed boundary data* u_{in} *in the following manner*

$$\max_{\mathbf{x} \in \Omega} u(\mathbf{x}) \leqslant \max_{\mathbf{s} \in \Gamma_{\text{in}}} u_{\text{in}}(\mathbf{s}) \qquad \text{if } c = 0, \ f \leqslant 0 \text{ in } \Omega, \qquad (2.37\text{a})$$

$$\min_{\mathbf{x} \in \Omega} u(\mathbf{x}) \geqslant \min_{\mathbf{s} \in \Gamma_{\text{in}}} u_{\text{in}}(\mathbf{s}) \qquad \text{if } c = 0, \ f \geqslant 0 \text{ in } \Omega, \qquad (2.37\text{b})$$

$$\max_{\mathbf{x} \in \Omega} u(\mathbf{x}) \leqslant \max\big(0, \max_{\mathbf{s} \in \Gamma_{\text{in}}} u_{\text{in}}(\mathbf{s})\big) \qquad \text{if } c \geqslant 0, \ f \leqslant 0 \text{ in } \Omega, \qquad (2.37\text{c})$$

$$\min_{\mathbf{x} \in \Omega} u(\mathbf{x}) \geqslant \min\big(0, \min_{\mathbf{s} \in \Gamma_{\text{in}}} u_{\text{in}}(\mathbf{s})\big) \qquad \text{if } c \geqslant 0, \ f \geqslant 0 \text{ in } \Omega. \qquad (2.37\text{d})$$

Result (2.37a) was originally shown in [Kuz10, Theorem 3.7].

Proof. Theorem 2.13 is a direct consequence of Lemma 2.11. To show (2.37a), let $\gamma_{\mathbf{x}}$ be the trajectory through an arbitrary $\mathbf{x} \in \Omega$. Its existence is guaranteed by Definition 2.12 and the assumption that v is perfusing. In the case $c = 0$, we have $\varrho(s) = 1$ for all $s \in [0, S_{\mathbf{x}}]$ and

$$u(\mathbf{x}) \doteq u\big(\gamma_{\mathbf{x}}(S_{\mathbf{x}})\big) = u\big(\gamma_{\mathbf{x}}(0)\big) + \int_0^{S_{\mathbf{x}}} f\big(\gamma_{\mathbf{x}}(s')\big) \, \mathrm{d}s' \leqslant u\big(\gamma_{\mathbf{x}}(0)\big) = u_{\text{in}}\big(\gamma_{\mathbf{x}}(0)\big)$$

by virtue of $f \leqslant 0$. Similarly, statement (2.37c) holds since $\varrho(s) \geqslant 1$ for all $s \in [0, S_{\mathbf{x}}]$ and

$$u(\mathbf{x}) = u\big(\gamma_{\mathbf{x}}(S_{\mathbf{x}})\big) \leqslant \max\Big(0, u\big(\gamma_{\mathbf{x}}(0)\big) + \int_0^{S_{\mathbf{x}}} \varrho(s') f\big(\gamma_{\mathbf{x}}(s')\big) \, \mathrm{d}s'\Big)$$

$$\leqslant \max\Big(0, u_{\text{in}}\big(\gamma_{\mathbf{x}}(0)\big)\Big).$$

The remainder of Theorem 2.13 can be shown using the same arguments. $\square$

Unfortunately, a general velocity field may fail to satisfy the requirement of Theorem 2.13. Possible circumstances causing v not to be perfusing are
1. partially vanishing velocity fields;
2. sources in the velocity field (leading to trajectories starting in the interior of the domain; e.g., $\Omega = B^2 := \big\{\mathbf{x} \in \mathbb{R}^2 \mid \|\mathbf{x}\|_2 < 1\big\}$ and $\mathbf{v} = (x_1^2, x_2^2)^\top$);

3. vortices in the velocity field (leading to closed trajectories; cf. Example 2.7).

Inequalities (2.37c) and (2.37d) still hold in the case of $\mathbf{v}(\mathbf{x}) = 0$ for some $\mathbf{x} \in \Omega$ if the coercivity condition (2.14) is valid: Then $c(\mathbf{x}) \geqslant c_0 > 0$ and, for instance, (2.37c) holds due to

$$\mathbf{v}(\mathbf{x}) \cdot \text{grad}\big(u(\mathbf{x})\big) + c(\mathbf{x})u(\mathbf{x}) = c(\mathbf{x})u(\mathbf{x}) = f(\mathbf{x}) \quad \Longrightarrow \quad u(\mathbf{x}) = \frac{f(\mathbf{x})}{c(\mathbf{x})} \leqslant 0.$$

In the case of (2.37a) and (2.37b), vanishing velocity fields may render problem (2.28) ill-posed.

The requirements of Theorem 2.13 can be weakened if only the nonnegativity of the solution is of interest [Kuz10, Theorem 3.8].

Theorem 2.14. *If* $\mathbf{v}$ *is perfusing, the solution* u *of* (2.28) *satisfies*

$$\max_{\mathbf{x} \in \Omega} u(\mathbf{x}) \leqslant 0 \qquad \textit{if } f \leqslant 0 \textit{ in } \Omega, \, u_{\text{in}} \leqslant 0 \textit{ on } \Gamma_{\text{in}}, \qquad (2.38\text{a})$$

$$\min_{\mathbf{x} \in \Omega} u(\mathbf{x}) \geqslant 0 \qquad \textit{if } f \geqslant 0 \textit{ in } \Omega, \, u_{\text{in}} \geqslant 0 \textit{ on } \Gamma_{\text{in}}. \qquad (2.38\text{b})$$

Proof. For the sake of simplicity, we focus on (2.38b). Let $\gamma_{\mathbf{x}}$ be the trajectory through an arbitrary $\mathbf{x} \in \Omega$, whose existence for perfusing velocity fields is guaranteed by Definition 2.12. We have

$$\varrho(s), \; \varrho(s)f\big(\gamma_{\mathbf{x}}(s)\big) \geqslant 0 \quad \forall s \in [0, S_{\mathbf{x}}], \qquad \Gamma_{\text{in}} \ni u\big(\gamma_{\mathbf{x}}(0)\big) \geqslant 0.$$

Then the result follows by exploiting (2.36). $\qquad \square$

This theorem implies uniqueness of a solution to problem (2.28) [Kuz10, Corollary 3.6].

Remark 2.15. The results of Theorems 2.13 and 2.14 hold for arbitrary domains and boundary conditions. They can be localized to a subdomain $\Omega' \subset \Omega$ by considering the steady advection-reaction equation (2.28a) in Ω' with inflow boundary conditions inferred from the 'global' solution u.

For instance, if $c = 0$ and $f \leqslant 0$ hold only in the subdomain Ω' (cf. (2.37a)), then no maximum can be generated in the interior. Hence, the maximal solution value must be attained on the inflow boundary of Ω'.

2.1.3.2 Unsteady problem

The method of characteristics, as presented in the previous section, can also be used to derive maximum principles in the case of the time dependent advection-reaction equation (2.30). For the stationary problem, a pseudo time was introduced for tracers moving along the streamlines and the solution of the corresponding initial value problems was determined via backtracking to the inflow boundary. For time dependent problems, no artificial parameterization is required and time integration along the characteristics may be more intuitive.

To avoid repetitions, we do not derive the applicable maximum principles from scratch as before. Instead, the results of Section 2.1.3.1 are exploited by treating the time dependent advection-reaction equation (2.30) as a steady advection-reaction equation on the space-time cylinder $\Omega \times (0, t)$ with the velocity field $(\mathbf{v}^\top, 1)^\top$. By definition of the space-time velocity field, issues like vanishing velocity fields and closed trajectories do not occur in the unsteady case, and the problem is well-posed for any $\mathbf{v} \in \mathrm{Lip}(\Omega)^d$. Moreover, the solution values at a time instant $t > 0$ are bounded by the data on the inflow boundary of $\Omega \times (0, t)$ which is given by $\Omega \times \{0\} \cup \tilde{\Gamma}_{\mathrm{in}}(t)$, where

$$\tilde{\Gamma}_{\mathrm{in}}(t) := \left\{ (\mathbf{s}, \tau) \in \partial\Omega \times (0, t] \mid \mathbf{v}(\mathbf{s}, \tau) \cdot \mathbf{n}(\mathbf{s}) < 0 \right\} \tag{2.39}$$

and $\mathbf{n} : \partial\Omega \to \mathbb{R}^d$ is the unit outward normal vector of Ω as before.

Theorem 2.16. *The solution u of (2.30) at $t \in (0, T]$ is bounded by the extrema of the initial condition u_0 and the prescribed boundary data u_{in} in the following manner*

$$\max_{\mathbf{x} \in \Omega} u(\mathbf{x}, t) \leqslant \max\left(\max_{\mathbf{x} \in \Omega} u_0(\mathbf{x}), \max_{(\mathbf{s}, \tau) \in \tilde{\Gamma}_{\mathrm{in}}(t)} u_{\mathrm{in}}(\mathbf{s}, \tau)\right) \qquad \text{if } c = 0,\ f \leqslant 0 \text{ in } \Omega,$$
$$\tag{2.40a}$$

$$\min_{\mathbf{x} \in \Omega} u(\mathbf{x}, t) \geqslant \min\left(\min_{\mathbf{x} \in \Omega} u_0(\mathbf{x}), \min_{(\mathbf{s}, \tau) \in \tilde{\Gamma}_{\mathrm{in}}(t)} u_{\mathrm{in}}(\mathbf{s}, \tau)\right) \qquad \text{if } c = 0,\ f \geqslant 0 \text{ in } \Omega,$$
$$\tag{2.40b}$$

$$\max_{\mathbf{x} \in \Omega} u(\mathbf{x}, t) \leqslant \max\left(0, \max_{\mathbf{x} \in \Omega} u_0(\mathbf{x}), \max_{(\mathbf{s}, \tau) \in \tilde{\Gamma}_{\mathrm{in}}(t)} u_{\mathrm{in}}(\mathbf{s}, \tau)\right) \qquad \text{if } c \geqslant 0,\ f \leqslant 0 \text{ in } \Omega,$$
$$\tag{2.40c}$$

$$\min_{\mathbf{x} \in \Omega} u(\mathbf{x}, t) \geqslant \min\left(0, \min_{\mathbf{x} \in \Omega} u_0(\mathbf{x}), \min_{(\mathbf{s}, \tau) \in \tilde{\Gamma}_{\mathrm{in}}(t)} u_{\mathrm{in}}(\mathbf{s}, \tau)\right) \qquad \text{if } c \geqslant 0,\ f \geqslant 0 \text{ in } \Omega.$$
$$\tag{2.40d}$$

Proof. This is a consequence of Theorem 2.13 applied to the above space-time setting. □

Furthermore, positivity preservation is guaranteed if u_0 and u_{in} are non-negative functions.

Theorem 2.17. *The solution u of* (2.30) *satisfies*

$$\max_{\mathbf{x} \in \Omega} u(\mathbf{x}) \leqslant 0 \qquad \textit{if } u_0, f \leqslant 0 \textit{ in } \Omega, u_{\text{in}} \leqslant 0 \textit{ on } \Gamma_{\text{in}}, \qquad (2.41\text{a})$$

$$\min_{\mathbf{x} \in \Omega} u(\mathbf{x}) \geqslant 0 \qquad \textit{if } u_0, f \geqslant 0 \textit{ in } \Omega, u_{\text{in}} \geqslant 0 \textit{ on } \Gamma_{\text{in}}. \qquad (2.41\text{b})$$

Proof. This is a consequence of Theorem 2.14 applied to the above space-time setting. □

2.2 Incompressible Navier-Stokes equations

2.2.1 Physics

In Section 2.1.1, we derived the continuity equation by balancing the inflow and outflow of a conserved quantity in a control volume Ω'. In mathematical models of continuum mechanics, this equation embodies the law of mass conservation. Another fundamental law of Newtonian mechanics is the conservation of momentum, which will be used in this section to derive the incompressible Navier-Stokes equations. This nonlinear system of PDEs describes a great variety of fluid flows by coupling the continuity equation with a momentum conservation law. Below we derive the vector-valued momentum equation following [Hir07] again.

The momentum density $\mathbf{u} = \rho \mathbf{v}$ of a physical system is a vector-valued quantity defined as the product of the mass density ρ and the velocity $\mathbf{v}$. According to Newton's laws of motion, the variation of the total momentum is given by the sum of all forces acting on it. Therefore, scalar conservation laws of the form (2.5) can be formulated for each component of the conserved quantity. In the absence of external forces, the evolution of momentum is governed by the convective flux $\mathbf{F} = \rho \mathbf{v} \otimes \mathbf{v}$ of the *inviscid Burgers equation*

$$\frac{\partial(\rho \mathbf{v})}{\partial t} + \nabla \cdot (\rho \mathbf{v} \otimes \mathbf{v}) = \mathbf{0}, \qquad (2.42)$$

where $(\nabla \cdot \mathbf{V})_k := \sum_{\ell=1}^{d} \partial_{x_\ell} v_{k\ell}$, $k \in \{1, \ldots, d\}$. In general, this force-free conservation law does not hold because momentum can be destroyed and/or created by forces acting inside a control volume or on its surface. The influence of body and surface forces is taken into account by introducing the vector-valued source terms $\mathbf{q}_V(\mathbf{x}, t) \in \mathbb{R}^d$ and tensor-valued fluxes $\mathbf{Q}_S(\mathbf{x}, t) \in \mathbb{R}^{d \times d}$. The contribution of body forces is taken into account using $\mathbf{q}_V = \rho \mathbf{g}$, where $\mathbf{g}$ typically represents the gravitational force density. The main surface forces are due to viscous friction on the control surface and pressure exerted by the surrounding fluid. The net extra flux is given by the Cauchy stress tensor $\mathbf{Q}_S = \boldsymbol{\sigma} = -p\mathbf{I} + \boldsymbol{\tau} \in \mathbb{R}^{d \times d}$, where $p \geqslant 0$ is the pressure of the fluid and $\boldsymbol{\tau} \in \mathbb{R}^{d \times d}$ is the viscous shear stress tensor. Adding $\mathbf{q}_V$ and $\nabla \cdot \mathbf{Q}_S$ to the inviscid Burgers equation, we obtain the general _momentum equation_ [Hir07, Eq. (1.3.11)]

$$\frac{\partial(\rho\mathbf{v})}{\partial t} + \nabla \cdot (\rho\mathbf{v} \otimes \mathbf{v}) = \rho\mathbf{g} - \operatorname{grad}(p) + \nabla \cdot \boldsymbol{\tau}. \tag{2.43}$$

For Newtonian fluids, $\boldsymbol{\tau}$ is given by [Hir07, Eq. (1.3.7)]

$$\boldsymbol{\tau} = \lambda \operatorname{div}(\mathbf{v})\mathbf{I} + 2\mu\mathbf{D},$$

where $\mathbf{D} = \frac{1}{2}(\nabla\mathbf{v} + \nabla\mathbf{v}^\top)$ is the strain rate tensor with $(\nabla\mathbf{v})_{k\ell} := \partial_{x_\ell} v_k$, μ is the dynamic viscosity of the fluid, and $\lambda = -\frac{2}{3}\mu$ is determined by Stokes relation [Hir07, Eq. (1.3.8)]

$$2\mu + 3\lambda = 0.$$

If the considered fluid is incompressible, we can assume that $\rho \equiv \operatorname{const} > 0$. Then the continuity equation (2.6) with $u = \rho$ leads to

$$0 = \frac{\partial\rho}{\partial t} + \operatorname{div}(\mathbf{v}\rho) = \frac{\partial\rho}{\partial t} + \operatorname{div}(\mathbf{v})\rho + \mathbf{v} \cdot \operatorname{grad}(\rho) = \operatorname{div}(\mathbf{v})\rho \qquad \text{in } \Omega$$

and the velocity field is solenoidal. Hence, the viscous shear stress is given by $\boldsymbol{\tau} = 2\mu\mathbf{D}$ leading to the well-known system of _incompressible Navier-Stokes equations_

$$\rho\frac{\partial\mathbf{v}}{\partial t} + \rho(\mathbf{v} \cdot \nabla)\mathbf{v} = \rho\mathbf{g} - \operatorname{grad}(p) + \nabla \cdot \boldsymbol{\tau}, \qquad \operatorname{div}(\mathbf{v}) = 0, \tag{2.44}$$

where $\mathbf{v} \cdot \nabla := \sum_{\ell=1}^{d} v_\ell \partial_{x_\ell}$. Theoretical properties of this model are discussed, for instance, in [Tem77; Gal11].

3 Discretization

The main objective of this chapter is to introduce the discretization of the advection-reaction equation and a numerical method for calculating approximate solutions. In this work, we only consider the Galerkin discretization using (multi-)linear and continuous finite elements. Other discretizations like the discontinuous Galerkin method, finite volume methods, or finite difference methods are beyond the scope of this work.

After introducing the key ingredients of finite element approximations, we focus on the Galerkin discretization of the steady advection-reaction equation (2.28). Then we briefly discuss the numerical treatment of its time dependent counterpart and describe the discretization of the time derivative using the θ-scheme. Numerical experiments at the end of this chapter illustrate the need for using limiting techniques like the ones presented in Chapter 4.

3.1 Finite elements

In what follows, we focus on the spatial discretization of a generic PDE. Following [QV94, Chapter 3], we introduce the basic concept of a finite element discretization whose key ingredients are given by
1. a triangulation of the domain,
2. a finite dimensional subspace of the test and trial space, and
3. basis functions with compact support.

In this context, we only consider the continuous Galerkin method with (multi-)linear finite elements. The underlying triangulation is given by a triangular or quadrilateral mesh (tetrahedra or hexahedra in three space dimensions).

3.1.1 Triangulation

Let Ω be a open, bounded, and connected subset of $\mathbb{R}^d$ such that $\bar{\Omega}$ can be exactly represented by a union of a finite number of polyhedra. Then errors due to the approximation of the domain can be neglected.

© Springer Fachmedien Wiesbaden GmbH, part of Springer Nature 2019
C. Lohmann, *Physics-Compatible Finite Element Methods for Scalar and Tensorial Advection Problems*, https://doi.org/10.1007/978-3-658-27737-6_3

Definition 3.1 ([QV94, Eq. (3.1.2, 3.1.3, 3.1.5)]). A finite set $\mathcal{T}_h$ is called *triangulation of* $\bar{\Omega}$ if $\bar{\Omega} = \bigcup_{K \in \mathcal{T}_h} K$ and

1. each *element* or *cell* $K \in \mathcal{T}_h$ is a closed polygon/polyhedron with $\mathring{K} \neq \emptyset$;
2. $\mathring{K}^1 \cap \mathring{K}^2 = \emptyset$ for each pair of distinct elements $K^1, K^2 \in \mathcal{T}_h$;
3. the *mesh size* $h > 0$ is defined by $h := \max_{K \in \mathcal{T}_h} h_K$, where $h_k = \operatorname{diam}(K)$ for any $K \in \mathcal{T}_h$.

The following terminology is introduced to describe triangulations in a concise and unambiguous manner:

- The vertices $\mathbf{x}_i \in K$ of $K \in \mathcal{T}_h$ are called *nodes*;

- The *edge* connecting two nodes $\mathbf{x}_i, \mathbf{x}_j \in K$ is denoted by $v_{ij} \subset K$, while the set of all edges is given by $\Upsilon := \{v_{ij} \mid \mathbf{x}_i, \mathbf{x}_j \in K, K \in \mathcal{T}_h\}$;

- The triangulation $\mathcal{T}_h$ contains a total of $E = |\mathcal{T}_h| \in \mathbb{N}$ elements and a total of $N \in \mathbb{N}$ nodes;

- The number of *non-Dirichlet nodes*, i.e., of vertices which are not located on the inflow boundary Γ_{in}, is denoted by $M < N$ (cf. Section 3.2.3);

- The *element stencil* $\mathcal{N}^e \subseteq \{1, \dots, N\}$ contains the indices of all nodes belonging to element K^e, $e \in \{1, \dots, E\}$. This means that $\mathbf{x}_i \in K^e$ for all $i \in \mathcal{N}^e$;

- The *nodal stencil* $\mathcal{N}_i \subseteq \{1, \dots, N\}$ contains the indices of nodes belonging to the same elements as $\mathbf{x}_i$. For each $j \in \mathcal{N}_i$, there exists an element $K^e \in \mathcal{T}_h$ such that $\mathbf{x}_i, \mathbf{x}_j \in K^e$.

- The numbers of elements containing node $\mathbf{x}_i$ are stored in the set $\mathcal{E}_i \subseteq \{1, \dots, E\}$;

- The patches $\varpi_i, \varpi^e \subseteq \Omega$ of elements containing node i and sharing a node with element K^e, respectively, are denoted by

$$\varpi_i := \bigcup_{e \in \mathcal{E}_i} K^e \quad \forall i \in \{1, \dots, N\}, \qquad \varpi^e := \bigcup_{i \in \mathcal{N}^e} \varpi_i \quad \forall e \in \{1, \dots, E\};$$

- The diameter of the inscribed circle of an element $K \in \mathcal{T}_h$ is denoted by $\rho_K > 0$, i.e., $\rho_K := \sup\{\operatorname{diam}(S) \mid S \subseteq K \text{ is a ball}\}$ [QV94].

Additionally, we assume that the triangulation $\mathcal{T}_h$ is admissible in the following sense.

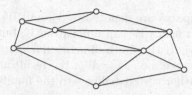
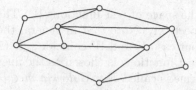

(a) Admissible triangulation.   **(b)** Non-admissible triangulation.

Figure 3.1: Two dimensional admissible and non-admissible triangulations with triangular elements. The right triangulation is non-admissible because 'hanging nodes' exist.

Definition 3.2. The triangulation $\mathcal{T}_h$ is called *admissible* if the intersection $F = K^1 \cap K^2$ of two distinct elements $K^1, K^2 \in \mathcal{T}_h$ is either empty, i.e., $F = \emptyset$, or a common face, side, or vertex of K^1 and K^2 (cf. Fig. 3.1).

For error estimation purposes, we also assume that the following properties are preserved in the process of mesh refinement.

Definition 3.3 ([QV94, Definition 3.4.1]). A family of triangulations $(\mathcal{T}_h)_{h>0}$ is called *(shape) regular* if there exists a constant $\sigma_1 \geqslant 1$ such that

$$\max_{K \in \mathcal{T}_h} \frac{h_K}{\rho_K} \leqslant \sigma_1 \qquad \forall h > 0. \tag{3.1}$$

Definition 3.4 ([QV94, Definition 6.3.1]). A family of triangulations $(\mathcal{T}_h)_{h>0}$ is called *quasi-uniform* if it is regular and moreover there exists a constant $\sigma_2 \in (0, 1]$ such that

$$\min_{K \in \mathcal{T}_h} h_K \geqslant \sigma_2 h \qquad \forall h > 0. \tag{3.2}$$

This definition implies that there exists a constant $\sigma_3 \geqslant 1$ such that [QV94, Remark 6.2.3]

$$\max_{K \in \mathcal{T}_h} \frac{h^d}{|K|} \leqslant \sigma_3 \qquad \forall h > 0. \tag{3.3}$$

As already mentioned, we restrict our analysis to triangular or quadrilateral meshes in two space dimensions ($d = 2$) and meshes containing only tetrahedra and convex hexahedra in the three dimensional case ($d = 3$).

3.1.2 Piecewise-polynomial subspace

The next step toward the construction of a finite element approximation is to define suitable, finite dimensional subspaces of the graph space V and of the space of test functions W. These spaces are denoted by V_h and W_h, respectively. Note that in general the test space W can differ from V, but in our case V and W coincide (cf. (2.29)). If the finite dimensional space of test functions is chosen as the ansatz space, i.e., $V_h = W_h$, the resulting scheme is called *Ritz-Galerkin method*. Otherwise, it belongs to the family of *Petrov-Galerkin methods*.

In general, we distinguish between finite dimensional spaces V_h which are subspaces of V and those which are not:

Definition 3.5. A finite element space V_h is called *conforming* if $V_h \subseteq V$ and *non-conforming* otherwise.

One of the most common ways to construct a conforming finite element approximation on $\mathcal{T}_h$ is to define V_h as the space of globally continuous functions which reduce to polynomials on each element. For example, the space of linear finite elements is given by [QV94, Eq. (3.2.4)]

$$V_h := \left\{ v_h \in C^0(\bar{\Omega}) \mid v_h|_K \in \mathbb{P}_1(K) \ \forall K \in \mathcal{T}_h \right\}, \tag{3.4a}$$

where $\mathbb{P}_1$ is the space of linear polynomials in the variables $x_1, \ldots, x_d$. For a reasonable definition of this space, it is necessary that all elements of the triangulation be triangular (or tetrahedral in three dimensions). If the triangulation consists of convex quadrilaterals (or hexahedra if $d = 3$), bilinear/trilinear finite elements are commonly employed. The corresponding finite dimensional subspace is defined by [QV94, Eq. (3.2.5)]

$$V_h := \left\{ v_h \in C^0(\bar{\Omega}) \mid v_h|_K \circ T_K \in \mathbb{Q}_1([0,1]^d) \ \forall K \in \mathcal{T}_h \right\}, \tag{3.4b}$$

where $\mathbb{Q}_1$ is the space of polynomials that are of degree less than or equal to 1 with respect to each variable. A function $\hat{v} \in \mathbb{Q}_1(\hat{K})$ defined on the reference element $\hat{K} = [0,1]^d$ is associated with $\hat{v} \circ T_K^{-1}$ via an invertible transformation $T_K : \hat{K} \to K$, $T_K \in \left[\mathbb{Q}_1(\hat{K}) \right]^d$. By definition, the space $\mathbb{Q}_1$ differs from $\mathbb{P}_1$ in that polynomials like $x_1 x_2$ belong to $\mathbb{Q}_1$ but not to $\mathbb{P}_1$.

Due to the following proposition, both approximations produce $V_h \subseteq H^1(\Omega)$ and the finite element spaces are called H^1-*conforming*.

Proposition 3.6 ([QV94, Proposition 3.2.1]). *A function $v : \Omega \to \mathbb{R}$ belongs to $H^1(\Omega)$ if and only if*

1. *$v|_K \in H^1_\bullet(K)$ for all $K \in \mathcal{T}_h$;*

2. *for each $F = K_1 \cap K_2 \neq \emptyset$, $K_1, K_2 \in \mathcal{T}_h$, the traces of $v|_{K_1}$ and $v|_{K_2}$ coincide on F.*

Proof. The proof of this proposition can be found in [QV94, Proposition 3.2.1]. □

For a more detailed introduction to general finite element approximations, the interested reader is referred, e.g., to [Cia02].

3.1.3 Degrees of freedom and basis functions

After selecting the finite dimensional spaces $V_h \subseteq V$ and $W_h \subseteq W$, the numerical approximation $u_h \in V_h$ of the solution u is substituted into the weak formulation of the PDE. A system of (differential)-algebraic equations for the nodal values of u_h is obtained by using test functions $\varphi_h \in W_h$. For (multi-)linear finite elements, the dimension of the trial and test space coincides with the number of vertices N. Representing the solution $u_h \in V_h$ as a linear combination

$$u_h(\mathbf{x}) = \sum_{i=1}^{N} u_i \varphi_i(\mathbf{x}) \qquad \forall \mathbf{x} \in \Omega \tag{3.5}$$

of *basis functions* $(\varphi_i)_{i=1}^{N}$ and using the basis functions of the space W_h as test functions in the weak formulation, the finite element discretization produces a (possibly nonlinear) system of N equations for N unknowns. The coefficients of the finite element approximation u_j are called *degrees of freedom* and stored in the vector $u = (u_i)_{i=1}^{N}$. Whereas the exact solution of the continuous problem is also denoted by u in this work, the intended meaning should be clear from the context.

A poor choice of the basis functions of the ansatz and test spaces V_h and W_h can make the resulting problem ill-conditioned or incur inordinately high computational costs. To reduce the numerical effort, the third and last ingredient of the finite element approach is the set of properly chosen basis functions with compact support such that the matrices of linear systems for the corresponding degrees of freedom are sparse and can be 'inverted' in an efficient manner.

For (multi-)linear finite elements, the global *Lagrange basis functions* are uniquely defined by (cf. Fig. 3.2)

$$\varphi_i(\mathbf{x}_j) = \delta_{ij} = \begin{cases} 1 & : j = i, \\ 0 & : j \neq i \end{cases} \qquad \forall i, j \in \{1, \ldots, N\}, \tag{3.6a}$$

$$\varphi_i \in V_h \qquad\qquad \forall i \in \{1, \ldots, N\}, \tag{3.6b}$$

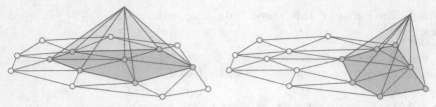

(a) Basis function in interior of domain.   **(b)** Basis function near boundary.

Figure 3.2: Support of two dimensional linear basis functions. The index set $\mathcal{E}_i$ contains all indices of elements shown in gray, while all indices of gray nodes form the nodal stencil $\mathcal{N}_i$.

where δ_{ij} is the Kronecker symbol and the nodes of the triangulation are given by $\mathbf{x}_1, \ldots, \mathbf{x}_N$ [QV94]. The so defined basis functions satisfy

$$\text{supp}(\varphi_i) = \varpi_i = \bigcup_{e \in \mathcal{E}_i} K^e \qquad \forall i \in \{1, \ldots, N\}, \tag{3.7}$$

$$u_h(\mathbf{x}_i) = \sum_{j=1}^{N} u_j \varphi_j(\mathbf{x}_i) = \sum_{j=1}^{N} u_j \delta_{ij} = u_i \qquad \forall i \in \{1, \ldots, N\}, \tag{3.8}$$

$$\min_{j \in \mathcal{N}^e} u_j \leqslant u_h(\mathbf{x}) \leqslant \max_{j \in \mathcal{N}^e} u_j \qquad \forall \mathbf{x} \in K^e, e \in \{1, \ldots, E\}. \tag{3.9}$$

These properties are exploited by the methods presented in Chapters 4 and 5.

3.2 Steady problem

3.2.1 Galerkin method

The use of (3.4a) or (3.4b) as finite dimensional trial and test spaces $V_h = W_h$ in the weak formulation (2.29) leads to the Ritz-Galerkin method

$$\text{Find } u_h \in V_h \text{ s.t.} \qquad a(u_h, \varphi_h) - \int_\Omega f\varphi + \int_{\partial\Omega} (\mathbf{v} \cdot \mathbf{n})_- u_{\text{in}}\varphi_h \quad \forall \varphi_h \in W_h. \tag{3.10}$$

The approximations $V_h \subseteq V$ and $W_h \subseteq W$ are conforming because $V_h = W_h \subseteq H^1(\Omega) \subseteq V$ thanks to (2.13a), the definition of the graph space V, and Proposition 3.6.

Due to the linearity of (3.10), $u_h \in V_h$ is a solution of the discrete problem if and only if (3.10) holds for each basis function $\varphi_1, \ldots, \varphi_N$ of V_h. By (3.5), the resulting linear system for the degrees of freedom reads

$$\mathcal{A}u = g \qquad \Longleftrightarrow \qquad \sum_{j=1}^{N} a_{ij}u_j = g_i \qquad \forall i \in \{1, \ldots, N\}, \qquad (3.11\text{a})$$

where the system matrix $\mathcal{A} = (a_{ij})_{i,j=1}^{N}$ and the right hand side vector $g = (g_i)_{i=1}^{N}$ are given by

$$a_{ij} := a(\varphi_j, \varphi_i) \hspace{3cm} \forall i, j \in \{1, \ldots, N\}, \qquad (3.11\text{b})$$

$$g_i := \int_{\Omega} f\varphi_i + \int_{\partial\Omega} (\mathbf{v} \cdot \mathbf{n})_{-} u_{\text{in}}\varphi_i \qquad \forall i \in \{1, \ldots, N\}. \qquad (3.11\text{c})$$

Remark 3.7. If the coercivity condition (2.14) holds, then the system matrix $\mathcal{A}$ is positive definite. That is, there exists a constant $C_M > 0$ such that

$$\sum_{i,j=1}^{N} v_i a_{ij} v_j \geqslant C_M \sum_{i=1}^{N} v_i^2 > 0 \qquad \forall v \in \mathbb{R}^N \setminus \{0\}. \qquad (3.12)$$

This property follows from (2.14), (2.25), and

$$\sum_{i,j=1}^{N} v_i a_{ij} v_j = a(v_h, v_h) \geqslant c_0 \|v_h\|_{L^2(\Omega)}^2 + \tfrac{1}{2}\int_{\partial\Omega} |\mathbf{v}\cdot\mathbf{n}|v_h^2 > 0 \quad \forall v \in \mathbb{R}^N \setminus \{0\},$$

where $v_h \in V_h$ is the piecewise linear finite element function with degrees of freedom $v = (v_i)_{i=1}^{N}$.

The positive definiteness of $\mathcal{A}$ automatically implies its invertibility: Otherwise there would exist $v \in \mathbb{R}^N \setminus \{0\}$ such that $\mathcal{A}v = 0$ in contradiction to (3.12). Since $\mathcal{A}$ is invertible, the discrete problem (3.11a) has a unique solution and is well-posed.

3.2.2 Error analysis

So far, we have constructed a method to calculate an approximation u_h to the exact solution u of the steady advection-reaction equation (2.28). For this discretization to produce reasonable results in practice, u_h should converge to u in an appropriate manner when the mesh is refined, that is, for $h \searrow 0$. A useful convergence criterion is presented in the following theorem.

Theorem 3.8. *Let $u_h \in V_h$ be the numerical solution to (3.10) and $u \in H^2(\Omega)$ be the exact solution of (2.28). Under the coercivity condition (2.14), the following error estimate holds*

$$\|u_h - u\|_{L^2(\Omega)} \leqslant Ch\|u\|_{H^2(\Omega)}, \tag{3.13}$$

where $C > 0$ is independent of h.

Proof. In Section 4.5.1.3, we analyze a nonlinear method in which the Galerkin approximation of (2.28) serves as a high order target. Therefore, the validity of Theorem 3.8 can be shown as in the proof of Theorem 4.72 by neglecting the term $\sqrt{d_h(u_h; I_h u, I_h u)}$. $\square$

3.2.3 Boundary conditions

In our definition of the bilinear form $a(\cdot, \cdot)$ and in the right hand side of the weak formulation, Dirichlet boundary conditions are imposed weakly on the inflow boundary. This weak treatment is often used for hyperbolic systems and is compatible with the coercivity of the bilinear form. For pure diffusion problems, this approach to weak imposition of Dirichlet boundary data is not feasible anymore and a penalty term has to be introduced to ensure the validity of the coercivity condition (3.12). The penalized version is called *Nitsche's method* and goes back to [Nit71].

Another commonly used technique enforces Dirichlet boundary conditions in a strong and interpolatory sense by manipulating (3.11): To satisfy the Dirichlet constraint $u_i = u_h(\mathbf{x}_i) = u_{\text{in}}(\mathbf{x}_i)$ for an inflow boundary node $\mathbf{x}_i$, the i-th row of the system matrix is overwritten by that of the identity matrix and g_i is replaced by $u_{\text{in}}(\mathbf{x}_i)$. Without loss of generality, we assume that indices smaller than or equal to M correspond to degrees of freedom that *are not* determined by the boundary conditions. In the case of the advection-reaction equation, M coincides with the number of nodes located in the interior of Ω or on $\partial\Omega \setminus \Gamma_{\text{in}}$, while nodes with indices greater than M lie on the inflow boundary Γ_{in}. Adopting this node numbering convention, we incorporate the strongly imposed interpolatory boundary conditions into (3.11) by replacing $\mathcal{A}$ and g with $\tilde{\mathcal{A}}$ and $\tilde{g}$ that are defined as follows:

$$\tilde{a}_{ij} = \begin{cases} a_{ij} & : i \leqslant M, \\ \delta_{ij} & : i > M \end{cases} \qquad \forall i, j \in \{1, \ldots, N\}, \tag{3.14a}$$

$$\tilde{g}_i = \begin{cases} g_i & : i \leqslant M, \\ \tilde{a}_{ii} u_{\text{in}}(\mathbf{x}_i) & : i > M \end{cases} \qquad \forall i \in \{1, \ldots, N\}. \tag{3.14b}$$

The positive definiteness of the system matrix $\mathcal{A}$ can be preserved if the modified rows are rescaled in an appropriate manner.

Theorem 3.9. *The system matrix $\tilde{\mathcal{A}}$ stays positive definite if strongly enforced Dirichlet boundary conditions are implemented by setting*

$$\tilde{a}_{ij} = \begin{cases} a_{ij} & : i \leqslant M, \\ \delta_{ij} \frac{N-M}{C_M} \sum_{l=1}^{M} a_{lj}^2 & : i > M \end{cases} \qquad \forall i,j \in \{1,\ldots,N\}. \qquad (3.15)$$

Proof. The matrix $(\tilde{a}_{ij})_{i,j=1}^{M}$ of coefficients associated with non-Dirichlet nodes satisfies

$$\sum_{i,j=1}^{M} v_i \tilde{a}_{ij} v_j = \sum_{i,j=1}^{M} v_i a_{ij} v_j \geqslant C_M \sum_{i=1}^{M} v_i^2 \qquad \forall v \in \mathbb{R}^M \setminus \{0\}$$

by (3.12). For any $v \in \mathbb{R}^N \setminus \{0\}$, Young's inequality yields

$$\sum_{i,j=1}^{N} v_i \tilde{a}_{ij} v_j = \sum_{i,j=1}^{M} v_i a_{ij} v_j + \sum_{j=M+1}^{N} \tilde{a}_{jj} v_j^2$$

$$+ \sum_{i=1}^{M} \sum_{j=M+1}^{N} \sqrt{\frac{C_M}{N-M}} v_i a_{ij} \sqrt{\frac{N-M}{C_M}} v_j$$

$$\geqslant C_M \sum_{i=1}^{M} v_i^2 + \sum_{j=M+1}^{N} \tilde{a}_{jj} v_j^2$$

$$- \sum_{i=1}^{M} \sum_{j=M+1}^{N} \left(\frac{C_M}{2(N-M)} v_i^2 + a_{ij}^2 \frac{N-M}{2C_M} v_j^2 \right)$$

$$= \frac{C_M}{2} \sum_{i=1}^{M} v_i^2 + \frac{N-M}{2C_M} \sum_{j=M+1}^{N} \left(\sum_{i=1}^{M} a_{ij}^2 \right) v_j^2 > 0.$$

Hence, $\tilde{\mathcal{A}}$ is positive definite.   □

Strongly imposed boundary conditions can also be implemented using L^2-projections instead of interpolation conditions for the Dirichlet boundary data.

3.3 Unsteady problem

The purpose of this section is to discuss the numerical treatment of the
unsteady advection-reaction equation (2.30) using finite elements for the
space discretization and a finite difference approximation for time derivatives.
In this work, we discretize unsteady PDEs in space and time using the *method
of lines*. That is, we begin with the space discretization and apply a numerical
time integrator to the resulting semi-discrete problem, which leads us to its
fully discrete counterpart.

3.3.1 Galerkin method

This section is based on [QV94, Section 14.3], where the space discretization
of the time dependent advection-reaction equation (2.30) is considered. The
semi-discrete problem with weakly enforced Dirichlet boundary conditions is
formulated by approximating V and W as before and considering the weak
formulation (2.31) [QV94, Eq. (14.3.11)]

$$\int_\Omega \frac{\partial u_h}{\partial t}\varphi_h + a(u_h, \varphi_h) = \int_\Omega f\varphi_h + \int_{\partial\Omega} (\mathbf{v}\cdot\mathbf{n})_- u_{\text{in}}\varphi_h$$
$$\forall\varphi_h \in W_h,\, t \in (0,T], \qquad (3.16\text{a})$$
$$u_h(\cdot, 0) = u_{0,h} \qquad \text{on } \Omega, \qquad (3.16\text{b})$$

where $u_{0,h} \in V_h$ is an appropriate approximation of the initial condition u_0.
Similarly to (3.5), the time dependent solution u_h is given by

$$u_h(\mathbf{x}, t) = \sum_{i=1}^N u_i(t)\varphi_i(\mathbf{x}) \qquad \forall\mathbf{x} \in \Omega,\, t \in [0,T] \qquad (3.17)$$

with time dependent degrees of freedom $u = u(t) = \big(u_i(t)\big)_{i=1}^N$. In contrast
to the assumptions considered in Section 2.1.2.2, we allow c to be time
dependent but still assume that $\mathbf{v}$ is stationary for simplicity.

The semi-discrete formulation (3.16) leads to the system of differential-
algebraic equations

$$\mathcal{M}\frac{du}{dt} - \mathcal{K}u = g \qquad (3.18\text{a})$$

$$\iff \quad \sum_{j=1}^N m_{ij}\frac{du_j}{dt} - \sum_{j=1}^N k_{ij}u_j = g_i \qquad \forall i \in \{1,\dots,N\}, \qquad (3.18\text{b})$$

where $\mathcal{M} = (m_{ij})_{i,j=1}^{N}$ denotes the *(consistent) mass matrix* defined by

$$m_{ij} := \int_\Omega \varphi_i \varphi_j \qquad \forall i,j \in \{1,\dots,N\}. \tag{3.19}$$

The right hand side g is defined by (3.11c) while the coefficients of the *convection matrix* $\mathcal{K}$ are given by $k_{ij} = -a(\varphi_j, \varphi_i)$, that is, $\mathcal{K} = -\mathcal{A}$.

Quarteroni and Valli [QV94, Eq. (14.3.16)] have shown that the finite element approximation u_h obtained by solving (3.16) satisfies the a priori error estimate

$$\max_{t \in [0,T]} \left\| u(t) - u_h(t) \right\|_{L^2(\Omega)} + \left(\int_0^T \left\| u(t) - u_h(t) \right\|_{L^2(|\mathbf{v}\cdot\mathbf{n}|;\partial\Omega)}^2 \, dt \right)^{\frac{1}{2}}$$
$$\leqslant C \left(\| u_0 - u_{0,h} \|_{L^2(\Omega)} + h \right) \quad (3.20)$$

if $u_0 \in H^1(\Omega)$, $u \in L^2([0,T]; H^2(\Omega))$, and $\partial_t u \in L^2([0,T]; H^1(\Omega))$. Furthermore, they presented energy estimates, which are similar to the semi-discrete counterparts of (2.32) and (2.35) [QV94, Eqs. (14.3.12) and (14.3.13)].

3.3.2 Time integrator

The second step of the method of lines applies a numerical time integrator to (3.18) leading to a fully discretized system of equations.

For the sake of simplicity, we only focus on the two-level θ-scheme for the temporal discretization. It is at most second order accurate and represents a handy generalization of three popular time-stepping schemes (*forward Euler*: $\theta = 0$, *Crank-Nicolson*: $\theta = \frac{1}{2}$, *backward Euler*: $\theta = 1$). Since other time integrators, including strong stability preserving (SSP) Runge-Kutta methods [Got+01; Got+11], can be analyzed similarly, we use the simple θ-scheme for the time discretization of our model problem and restrict our analysis to fully discrete problems produced by this particular time integrator.

The θ-scheme replaces time derivatives by the two-level finite difference approximation. The remaining terms are discretized using a linear combination of evaluations at the time levels t^n and $t^{n+1} = t^n + \Delta t$, where $\Delta t > 0$ denotes the time increment. Then the fully discretized form of system (3.18) reads

$$\mathcal{A}u^{n+1} + \mathcal{B}u^n = g, \tag{3.21a}$$

$$\begin{aligned}\mathcal{A} &:= \mathcal{M} - \Delta t\theta \mathcal{K}^{n+1}, & g &:= \Delta t\big(\theta g^{n+1} + (1-\theta)g^n\big), \\ \mathcal{B} &:= -\mathcal{M} - \Delta t(1-\theta)\mathcal{K}^n.\end{aligned} \tag{3.21b}$$

The superscripts $n+1$ and n refer to the time levels at which the corresponding matrices and vectors are defined. At each time step, a linear system with the matrix $\mathcal{A}$ has to be solved to calculate the solution u^{n+1} to the fully discrete problem. For the sake of simplicity, we only consider constant time increments $\Delta t > 0$ so that $t^n = n\Delta t$ for all $n \in \mathbb{N}$.

Remark 3.10. The system matrix $\mathcal{A}$ of (3.21) is positive definite if there exists a constant $c_0 > 0$ such that

$$1 + \Delta t\theta\left(c - \tfrac{1}{2}\operatorname{div}(\mathbf{v})\right) \geqslant c_0 \quad \text{a.e. in } \Omega$$

holds for the given values of $\theta \in [0, 1]$ and $\Delta t > 0$. This can be shown as in Remark 3.7.

3.4 Numerical examples

In this section, we consider Galerkin approximations to different test configurations for the steady advection-reaction equation. A matter of particular interest is the occurrence of artificial oscillations violating bounds that determine the range of exact solution values. We will see that the Galerkin method does not satisfy discrete maximum principles and the numerical solution may exhibit overshoots and undershoots. This behavior motivates the need for using the bound-preserving algorithms to be presented and analyzed in the next chapter.

The stationary two dimensional problem under investigation is the *circular convection* benchmark introduced by Hubbard [Hub07]. In the absence of source terms, the convective transport of the inflow boundary profile u_{in} by the perfusing velocity field $\mathbf{v} = (-x_2, x_1)^{\top}$ produces a solution which is radially symmetric with respect to the bottom left corner of the domain $\Omega = (0, 1)^2$. In accordance with the theoretical analysis based on the method of characteristics, the solution of the pure advection equation remains constant along the streamlines. We extend this boundary value problem by introducing a source term f and a reactive term cu with constant reactivity parameter $c \geqslant 0$. This yields

$$\mathbf{v} \cdot \operatorname{grad}(u) + cu = f \qquad \text{in } \Omega := (0, 1)^2, \tag{3.22a}$$

$$u = u_{\text{in}} \qquad \text{on } \Gamma_{\text{in}} := [0, 1] \times \{0\} \cup \{1\} \times [0, 1]. \tag{3.22b}$$

Substituting the manufactured solution

$$u(\mathbf{x}) = \begin{cases} u_{\text{in}}(\|\mathbf{x}\|_2, 0) & : \|\mathbf{x}\|_2 \leqslant 1, \\ u_{\text{in}}\left(1, \sqrt{\|\mathbf{x}\|_2^2 - 1}\right) & : \|\mathbf{x}\|_2 > 1 \end{cases} \qquad \forall \mathbf{x} \in \Omega \tag{3.23}$$

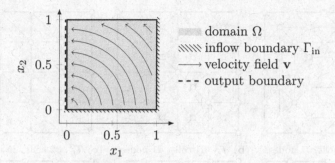

Figure 3.3: Circular convection: Geometry of domain Ω, velocity field $\mathbf{v} = (-x_2, x_1)^\top$, and inflow boundary Γ_{in}.

into (3.22a), we define the source term $f := cu$. The unique exact solution (3.23) of the so defined test problem depends only on the distance $r = \|\mathbf{x}\|_2$ from the origin and on the shape of the inflow boundary condition u_{in} (see Fig. 3.3).

By definition, the solution u is globally bounded by the extrema of u_{in} and satisfies maximum principles. Note that, strictly speaking, Theorem 2.13 provides bounds for the solution only in the case $c = 0$ and $f = 0$. However, due to the special choice of $f = cu$, the exact solution given by (3.23) is bounded above and below by the extrema of the inflow boundary function u_{in} regardless of the choice of c. In view of the fact that the velocity field is solenoidal, problem (3.22) is well-posed and the associated bilinear form $a(\cdot, \cdot)$ is coercive if $\Lambda = c - \frac{1}{2}\operatorname{div}(\mathbf{v}) = c > 0$. For $u \in H^2(\Omega)$, the Galerkin approximation converges to u in $\|\cdot\|_{L^2(\Omega)}$ with at least first order by Theorem 3.8. However, the reactive term may vanish and numerical solutions to pure advection equations may be of interest in practical applications. In this case, the above analysis breaks down and the convergence of the Galerkin approximation is not guaranteed.

We discretize problem (3.22) using linear Lagrange finite elements on different uniform and distorted triangular meshes. For example, the uniform triangulations $\mathcal{T}_h^1$, $\mathcal{T}_h^2$, and $\mathcal{T}_h^3$ on level 2 are presented in Figs. 3.4a–3.4c. Their distorted counterparts $\tilde{\mathcal{T}}_h^1$, $\tilde{\mathcal{T}}_h^2$, and $\tilde{\mathcal{T}}_h^3$ are generated by randomly moving the nodes of the uniform meshes after each refinement step (cf. Figs. 3.4d–3.4f).

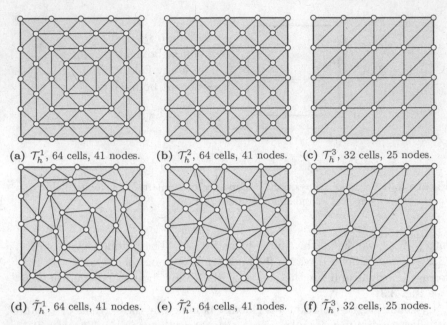

(a) $\mathcal{T}_h^1$, 64 cells, 41 nodes.   (b) $\mathcal{T}_h^2$, 64 cells, 41 nodes.   (c) $\mathcal{T}_h^3$, 32 cells, 25 nodes.

(d) $\tilde{\mathcal{T}}_h^1$, 64 cells, 41 nodes.   (e) $\tilde{\mathcal{T}}_h^2$, 64 cells, 41 nodes.   (f) $\tilde{\mathcal{T}}_h^3$, 32 cells, 25 nodes.

Figure 3.4: Different uniform and distorted triangulations of unit square on level 2.

The boundary data function u_{in} is defined so that the exact solution is given by

$$u(\mathbf{x}) = \exp\left(-100\big(\|\mathbf{x}\|_2 - 0.7\big)^2\right) \qquad \forall \mathbf{x} \in \Omega. \tag{3.24}$$

Due to the smoothness of u, numerical approximations are expected to be non-oscillatory. However, the Galerkin discretization of problem (3.22) can become ill-posed in the case $c = 0$. In our numerical study, it produces ripples violating the global bounds given by the extrema of the inflow boundary function (cf. Fig. 3.5). Interestingly enough, the results are highly sensitive to the employed triangulation. While the solution exhibits only small overshoots and undershoots on $\tilde{\mathcal{T}}_h^1$, $\mathcal{T}_h^2$, $\mathcal{T}_h^3$, and $\tilde{\mathcal{T}}_h^3$, global bounds are violated significantly on $\mathcal{T}_h^1$ and $\tilde{\mathcal{T}}_h^2$.

Table 3.1 summarizes the $\|\cdot\|_{L^2(\Omega)}$-errors and the *experimental orders of convergence* (EOC) determined by the formula

$$\text{EOC} = \log\big(\tfrac{e_h}{e_{2h}}\big) \log\big(\tfrac{h}{2h}\big)^{-1}, \qquad e_h = \|u - u_h\|_{L^2(\Omega)}$$

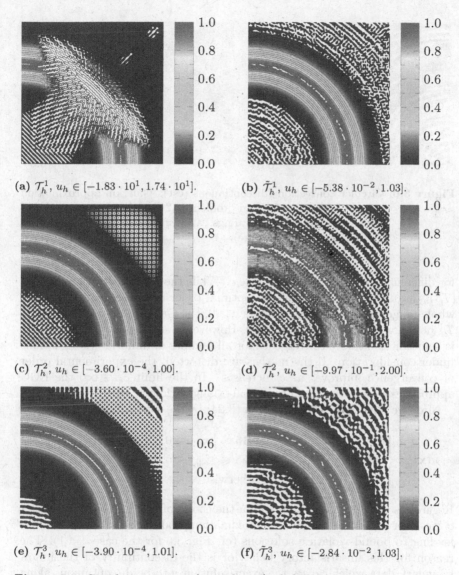

(a) $\mathcal{T}_h^1$, $u_h \in [-1.83 \cdot 10^1, 1.74 \cdot 10^1]$.

(b) $\tilde{\mathcal{T}}_h^1$, $u_h \in [-5.38 \cdot 10^{-2}, 1.03]$.

(c) $\mathcal{T}_h^2$, $u_h \in [-3.60 \cdot 10^{-4}, 1.00]$.

(d) $\tilde{\mathcal{T}}_h^2$, $u_h \in [-9.97 \cdot 10^{-1}, 2.00]$.

(e) $\mathcal{T}_h^3$, $u_h \in [-3.90 \cdot 10^{-4}, 1.01]$.

(f) $\tilde{\mathcal{T}}_h^3$, $u_h \in [-2.84 \cdot 10^{-2}, 1.03]$.

Figure 3.5: Circular convection (smooth test): Galerkin approximation to problem (3.22) with $c = 0$ on level 6 of different triangulations. Overshoots and undershoots are plotted in white.

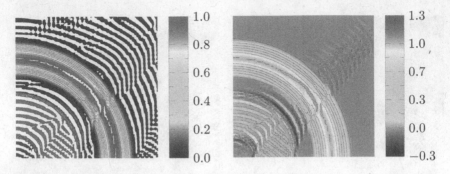

Figure 3.6: Circular convection (discontinuous test): Galerkin approximation $u_h \in [-3.05 \cdot 10^{-1}, 1.33]$ to problem (3.22) with $c = 1$ on level 6 of $\mathcal{T}_h^3$ using two different scalings. Overshoots and undershoots are plotted in white (left figure).

for different values of the reaction rate c. While the families of triangulations $(\mathcal{T}_h^2)_{h>0}$ and $(\mathcal{T}_h^3)_{h>0}$ provide the optimal order of convergence 2 for the whole range of reactivity parameters $c \geqslant 0$, the Galerkin approximation on $\mathcal{T}_h^1$ does not converge at all for $c = 0$. However, a small amount of reactivity is sufficient to achieve convergence for all families of uniform triangulations under consideration. If the meshes are distorted, the experimental order of convergence improves as c increases. In the limit $c \to \infty$, the EOC approaches 2 as in the case of uniform triangulations.

Next, we replace the exact solution (3.24) by the composition (cf. [Hub07])

$$u(\mathbf{x}) = \begin{cases} 1 & : 0.15 \leqslant \|\mathbf{x}\|_2 \leqslant 0.45, \\ \cos^2\left(10\pi\frac{\|\mathbf{x}\|_2 - 0.7}{3}\right) & : 0.55 \leqslant \|\mathbf{x}\|_2 \leqslant 0.85, \qquad \forall \mathbf{x} \in \Omega. \quad (3.25) \\ 0 & : \text{otherwise} \end{cases}$$

Regardless of the triangulation and of the choice of the reactivity parameter c, we observe that the Galerkin approximation produces spurious oscillations leading to bound-violating solutions (cf. Fig. 3.6 for the case $c = 1$). The reason for this unsatisfactory behavior is the discontinuity of the inflow boundary data which causes the exact solution u to be discontinuous along the corresponding streamlines. The largest undershoots and overshoots are generated around the discontinuities. In the limit $c \to \infty$, the Galerkin approximation of (3.22) converges to the oscillatory L^2-projection of the exact solution.

Table 3.1: Circular convection (smooth test): Convergence of $\| \cdot \|_{L^2(\Omega)}$-errors for Galerkin solution on uniform and distorted meshes.

(a)

	mesh level	$c = 0$		$c = 0.01$	
		error	EOC	error	EOC
$\mathcal{T}_h^1$	5	$1.96 \cdot 10^{-3}$		$1.95 \cdot 10^{-3}$	
	6	$1.05 \cdot 10^{0}$	-9.06	$4.77 \cdot 10^{-4}$	2.03
	7	$2.30 \cdot 10^{-1}$	2.19	$1.19 \cdot 10^{-4}$	2.01
	8	$1.87 \cdot 10^{0}$	-3.02	$2.96 \cdot 10^{-5}$	2.00
$\tilde{\mathcal{T}}_h^1$	5	$1.56 \cdot 10^{-2}$		$1.56 \cdot 10^{-2}$	
	6	$5.21 \cdot 10^{-3}$	1.58	$5.20 \cdot 10^{-3}$	1.58
	7	$2.07 \cdot 10^{-3}$	1.33	$2.06 \cdot 10^{-3}$	1.33
	8	$7.48 \cdot 10^{-4}$	1.47	$7.47 \cdot 10^{-4}$	1.47
$\mathcal{T}_h^2$	5	$5.22 \cdot 10^{-3}$		$5.22 \cdot 10^{-3}$	
	6	$1.24 \cdot 10^{-3}$	2.07	$1.24 \cdot 10^{-3}$	2.07
	7	$3.07 \cdot 10^{-4}$	2.02	$3.06 \cdot 10^{-4}$	2.02
	8	$7.65 \cdot 10^{-5}$	2.00	$7.63 \cdot 10^{-5}$	2.00
$\tilde{\mathcal{T}}_h^2$	5	$1.42 \cdot 10^{-1}$		$1.36 \cdot 10^{-1}$	
	6	$8.66 \cdot 10^{-2}$	0.71	$8.24 \cdot 10^{-2}$	0.73
	7	$6.57 \cdot 10^{-2}$	0.40	$5.58 \cdot 10^{-2}$	0.56
	8	$3.83 \cdot 10^{-2}$	0.78	$2.99 \cdot 10^{-2}$	0.00
$\mathcal{T}_h^3$	5	$3.40 \cdot 10^{-3}$		$3.40 \cdot 10^{-3}$	
	6	$7.93 \cdot 10^{-4}$	2.10	$7.93 \cdot 10^{-4}$	2.10
	7	$1.96 \cdot 10^{-4}$	2.02	$1.96 \cdot 10^{-4}$	2.02
	8	$4.87 \cdot 10^{-5}$	2.00	$4.87 \cdot 10^{-5}$	2.00
$\tilde{\mathcal{T}}_h^3$	5	$1.14 \cdot 10^{-2}$		$1.14 \cdot 10^{-2}$	
	6	$5.27 \cdot 10^{-3}$	1.11	$5.27 \cdot 10^{-3}$	1.11
	7	$1.97 \cdot 10^{-3}$	1.42	$1.97 \cdot 10^{-3}$	1.42
	8	$8.21 \cdot 10^{-4}$	1.27	$8.20 \cdot 10^{-4}$	1.27

Table 3.1: Circular convection (smooth test): Convergence of $\|\cdot\|_{L^2(\Omega)}$-errors for Galerkin solution on uniform and distorted meshes.

(b)

	mesh level	$c = 1$		$c = 100$	
		error	EOC	error	EOC
$\mathcal{T}_h^1$	5	$1.77 \cdot 10^{-3}$		$1.54 \cdot 10^{-3}$	
	6	$4.32 \cdot 10^{-4}$	2.03	$3.75 \cdot 10^{-4}$	2.04
	7	$1.07 \cdot 10^{-4}$	2.01	$9.32 \cdot 10^{-5}$	2.01
	8	$2.68 \cdot 10^{-5}$	2.00	$2.33 \cdot 10^{-5}$	2.00
$\tilde{\mathcal{T}}_h^1$	5	$7.11 \cdot 10^{-3}$		$2.10 \cdot 10^{-3}$	
	6	$2.51 \cdot 10^{-3}$	1.50	$5.27 \cdot 10^{-4}$	1.99
	7	$9.45 \cdot 10^{-4}$	1.41	$1.46 \cdot 10^{-4}$	1.85
	8	$3.33 \cdot 10^{-4}$	1.51	$4.48 \cdot 10^{-5}$	1.71
$\mathcal{T}_h^2$	5	$4.20 \cdot 10^{-3}$		$1.56 \cdot 10^{-3}$	
	6	$1.02 \cdot 10^{-3}$	2.04	$3.84 \cdot 10^{-4}$	2.02
	7	$2.53 \cdot 10^{-4}$	2.01	$9.56 \cdot 10^{-5}$	2.01
	8	$6.31 \cdot 10^{-5}$	2.00	$2.39 \cdot 10^{-5}$	2.00
$\tilde{\mathcal{T}}_h^2$	5	$1.18 \cdot 10^{-2}$		$1.96 \cdot 10^{-3}$	
	6	$5.08 \cdot 10^{-3}$	1.22	$5.18 \cdot 10^{-4}$	1.92
	7	$2.06 \cdot 10^{-3}$	1.30	$1.49 \cdot 10^{-4}$	1.79
	8	$8.26 \cdot 10^{-4}$	1.32	$4.64 \cdot 10^{-5}$	1.69
$\mathcal{T}_h^3$	5	$3.36 \cdot 10^{-3}$		$3.32 \cdot 10^{-3}$	
	6	$7.91 \cdot 10^{-4}$	2.09	$7.90 \cdot 10^{-4}$	2.07
	7	$1.95 \cdot 10^{-4}$	2.02	$1.95 \cdot 10^{-4}$	2.02
	8	$4.87 \cdot 10^{-5}$	2.00	$4.86 \cdot 10^{-5}$	2.00
$\tilde{\mathcal{T}}_h^3$	5	$9.02 \cdot 10^{-3}$		$4.07 \cdot 10^{-3}$	
	6	$3.67 \cdot 10^{-3}$	1.30	$1.02 \cdot 10^{-3}$	2.00
	7	$1.17 \cdot 10^{-3}$	1.64	$2.68 \cdot 10^{-4}$	1.92
	8	$4.37 \cdot 10^{-4}$	1.42	$7.64 \cdot 10^{-5}$	1.81

4 Limiting for scalars

This chapter presents the fundamentals of the algebraic flux correction (AFC) methodology for scalar quantities. The first section gives a short and informal introduction to the field of bound-preserving finite element methods. Then the key ingredients of such methods are summarized (Section 4.2) and low order approximations with desired properties are derived for the hyperbolic model equations (Section 4.3). The following Sections 4.4 and 4.5 describe advanced techniques for obtaining more accurate solutions. The presented approaches are finally validated numerically in Section 4.6 by applying them to commonly used benchmarks.

4.1 Introduction

The numerical studies in Section 3.4 have shown that the Galerkin approximation of hyperbolic PDEs may lead to oscillatory solutions. Especially in the case of small reactive terms and/or discontinuous solutions, the Galerkin approximation becomes polluted by spurious ripples. Overshoots and undershoots may give rise to unphysical results and violations of maximum principles depending on the given data (source terms, boundary and initial conditions, etc.).

Due to this deficiency, several approaches were proposed in the literature to stabilize the Galerkin discretization. For instance, the *streamline upwind/Petrov-Galerkin (SUPG)* method penalizes steep gradients in streamline direction by adding an appropriate stabilization term. This modification damps oscillations and localizes them to a close proximity of discontinuities. However, unphysical solutions may still occur.

The main idea of the AFC methodology is different. Instead of adding residual-based terms in the variational formulation, algebraic corrections of finite element matrices are performed to make the discretization bound-preserving. A physically admissible manipulation is the addition of artificial diffusion which smears unresolvable steep gradients by moving mass from one degree of freedom to another. The discrete diffusion operator can be designed to ensure that the nodal values of the finite element approximation

© Springer Fachmedien Wiesbaden GmbH, part of Springer Nature 2019
C. Lohmann, *Physics-Compatible Finite Element Methods for Scalar and Tensorial Advection Problems*, https://doi.org/10.1007/978-3-658-27737-6_4

are bounded by the maxima and minima of function values at neighboring nodes. Global extrema must be attained on the inflow boundary and high-frequency oscillations are ruled out.

The downside of using global algebraic corrections of this kind are strong smearing effects. Especially in the case of smooth solutions, significant amounts of artificial diffusion can be removed without generating any un-physical artifacts. In fact, local modifications of finite element matrices are sufficient to suppress spurious extrema, while preserving the high accuracy of the underlying target scheme in smooth regions.

In Sections 4.4 and 4.5, two different algebraic approaches for improving the accuracy of the solution and enforcing discrete maximum principles are explored. The first one is based on a predictor-corrector algorithm which applies artificial diffusion globally and performs local antidiffusive corrections of the resulting low order predictor. A local extremum diminishing limiting procedure guarantees that the corrected nodal values remain bounded by the maxima and minima of the bound-preserving low order approximation. Therefore, the availability of a sufficiently accurate predictor is essential to prevent the local bounds from becoming too restrictive. The second technique is a correction procedure that limits the artificial diffusion operators before adding them to the matrices of the target discretization. This version requires solution of nonlinear systems since the bounds for local maximum principles are defined using the local extrema of the unknown solution rather than those of a low order predictor.

Most parts of this chapter deal with algebraic properties of (linear) systems of equations. Those systems are generated by the finite element discretization of the hyperbolic model equations (2.28) and (2.30) using (multi-)linear Lagrange basis functions. Therefore, the degrees of freedom are the values of the numerical solution at the nodes of the mesh. Moreover, the solution is bounded by its nodal values. Consequently, discrete maximum principles for the solution vector carry over to the continuous finite element interpolant (cf. (3.9)).

4.2 Fundamentals

In this section, we define the key ingredients of a bound-preserving low order method. To design such a method on the basis of the continuous Galerkin approximation, we identify sufficient algebraic requirements for a linear scheme to satisfy the corresponding inequality constraints. The discrete maximum principles to be enforced are consistent with the results of Section 2.1.3, where the continuous scalar advection-reaction equation is analyzed. At this point, the way in which the desired matrix properties are achieved is not important. Practical approaches to manipulating finite element matrices in the AFC framework are presented in Sections 4.3–4.5.

4.2.1 Terms and definitions

Before presenting the design criteria that provide control mechanisms for the range of discrete solution values in numerical methods, we introduce the basic terminology for verification and enforcement of desired algebraic properties.

Definition 4.1. A matrix $\mathcal{A} \in \mathbb{R}^{N \times N}$ is called a *Z-matrix* if

$$a_{ij} \leqslant 0 \qquad \forall i, j \in \{1, \ldots, N\}, j \neq i. \tag{4.1}$$

Definition 4.2. A matrix $\mathcal{A} \in \mathbb{R}^{N \times N}$ is called an *L-matrix* if

$$a_{ii} > 0, \quad a_{ij} \leqslant 0 \qquad \forall i, j \in \{1, \ldots, N\}, j \neq i. \tag{4.2}$$

Definition 4.3. An invertible matrix $\mathcal{A} \in \mathbb{R}^{N \times N}$ is called *monotone* or *inverse-positive* if

$$(\mathcal{A}^{-1})_{ij} \geqslant 0 \qquad \forall i, j \in \{1, \ldots, N\}. \tag{4.3}$$

Definition 4.4 ([Var09, Definition 3.22]). A monotone Z-matrix is called *M-matrix*.

Definition 4.5 ([Axe94, Definitions 4.2 and 4.4]).

- A matrix $\mathcal{A} \in \mathbb{R}^{N \times N}$ is called *reducible* if there exists $I \subsetneq \{1, \ldots, N\}$, $I \neq \emptyset$, such that

$$a_{ij} = 0 \qquad \forall i \in I, j \in \{1, \ldots, N\} \setminus I. \tag{4.4}$$

Otherwise, $\mathcal{A}$ is said to be *irreducible*.

- A matrix $\mathcal{A} \in \mathbb{R}^{N \times N}$ is called *irreducibly diagonally dominant* if $\mathcal{A}$ is irreducible and

$$\forall i \in \{1, \ldots, N\}: \qquad |a_{ii}| \geqslant \sum_{j \neq i} |a_{ij}|, \tag{4.5a}$$

$$\exists i \in \{1, \ldots, N\}: \qquad |a_{ii}| > \sum_{j \neq i} |a_{ij}|. \tag{4.5b}$$

- A matrix $\mathcal{A} \in \mathbb{R}^{N \times N}$ is called *strictly diagonally dominant* if

$$|a_{ii}| > \sum_{j \neq i} |a_{ij}| \qquad \forall i \in \{1, \ldots, N\}. \tag{4.6}$$

Lemma 4.6. *A strictly diagonally dominant or irreducibly diagonally dominant L-matrix is an M-matrix.*

Proof. The statement is a direct consequence of [Var09, Theorem 3.27]. $\square$

Definition 4.7. A matrix is called *nonpositive* (*nonnegative*) if

$$a_{ij} \leqslant 0 \quad (a_{ij} \geqslant 0) \qquad \forall i, j \in \{1, \ldots, N\}. \tag{4.7}$$

Definition 4.8 ([CR73a]). A Z-matrix with nonnegative row sums is said to be of *nonnegative type*.

Remark 4.9. A regular matrix of nonnegative type is an L-matrix. This is the case because all diagonal entries must be nonnegative to provide the nonnegative row sum property. Suppose that one diagonal entry vanishes. Then the nonpositive off-diagonal entries of the corresponding row must vanish as well for the row sum to be nonnegative. This contradicts the assumption of a non-singular matrix and implies the validity of the claim.

In what follows, we will frequently require the above properties only for the reduced submatrix with row indices corresponding to non-Dirichlet node numbers. For that purpose, we introduce the following convention.

Remark 4.10. The submatrix $\mathcal{A}' \in \mathbb{R}^{M \times N}$ is defined as the restriction of $\mathcal{A} \in \mathbb{R}^{N \times N}$ to its first M rows

$$a'_{ij} := a_{ij} \qquad \forall i \in \{1, \ldots, M\}, j \in \{1, \ldots, N\}. \tag{4.8}$$

Similarly, we define the restriction $u' \in \mathbb{R}^M$ of $u \in \mathbb{R}^N$ by

$$u'_i := u_i \qquad \forall i \in \{1, \ldots, M\}. \tag{4.9}$$

In this chapter, we discretize steady and unsteady advection-reaction equations using continuous (multi-)linear finite element approximations. Due to the compact support property of the corresponding Lagrange basis functions, only nodes belonging to the stencil $\mathcal{N}_i$ produce nonvanishing entries in the i-th row of a general global matrix $\mathcal{V}$. That is, we have

$$v_{ij} = 0 \qquad \forall i, j \in \{1, \ldots, N\}, j \notin \mathcal{N}_i.$$

This compact sparsity pattern is common to all discrete operators to be considered below.

4.2.2 Steady problem

This section is devoted to the definition of sufficient conditions for a linear numerical scheme to reproduce the properties of the exact solution to the steady advection-reaction equation (2.28), as defined in Section 2.1.3.1. Using these conditions, we will verify and enforce discrete maximum principles for (subsets of) degrees of freedom that define a continuous finite element approximation. For (multi-)linear Lagrange basis functions, each degree of freedom corresponds to the value of the approximate solution at the corresponding nodal point.

To begin with, we formulate a commonly employed local discrete maximum principle.

Definition 4.11. The *local discrete maximum principle (local DMP)* holds for $u \in \mathbb{R}^N$ if

$$\min_{j \in \mathcal{N}_i \setminus \{i\}} u_j \leqslant u_i \leqslant \max_{j \in \mathcal{N}_i \setminus \{i\}} u_j \qquad \forall i \in \{1, \ldots, M\}. \qquad (4.10)$$

A numerical scheme or discrete problem satisfies the local DMP if it guarantees the validity of (4.10).

The above definition imposes upper and lower bounds only on degrees of freedom with indices less than or equal to M. These components of the solution vector are associated with function values at non-Dirichlet nodes. The remaining degrees of freedom are determined by the prescribed boundary conditions.

Let $u \in \mathbb{R}^N$ be the vector of nodal values corresponding to the solution of the linear system

$$\mathcal{A}u = g \qquad \Longleftrightarrow \qquad \sum_{j=1}^{N} a_{ij} u_j = g_i \quad \forall i \in \{1, \ldots, N\}, \qquad (4.11)$$

where $\mathcal{A} \in \mathbb{R}^{N \times N}$ is an invertible matrix and $g \in \mathbb{R}^N$ is a given right hand side vector. In the following lemma, we present sufficient conditions for a general discrete problem of this form to provide the local DMP property.

Lemma 4.12. *If $\mathcal{A}'$ is an L-matrix, the solution of (4.11) satisfies*

$$u_i \leqslant \max_{j \in \mathcal{N}_i \setminus \{i\}} u_j \qquad \forall i \in \left\{ j \in \{1, \ldots, N\} \,\middle|\, g_j \leqslant 0 \wedge \sum_{l=1}^{N} a_{jl} = 0 \right\}, \quad (4.12a)$$

$$u_i \geqslant \min_{j \in \mathcal{N}_i \setminus \{i\}} u_j \qquad \forall i \in \left\{ j \in \{1, \ldots, N\} \,\middle|\, g_j \geqslant 0 \wedge \sum_{l=1}^{N} a_{jl} = 0 \right\}, \quad (4.12b)$$

$$u_i \leqslant \max\left(0, \max_{j \in \mathcal{N}_i \setminus \{i\}} u_j\right) \qquad \forall i \in \left\{ j \in \{1, \ldots, N\} \,\middle|\, g_j \leqslant 0 \wedge \sum_{l=1}^{N} a_{jl} \geqslant 0 \right\}, \quad (4.12c)$$

$$u_i \geqslant \min\left(0, \min_{j \in \mathcal{N}_i \setminus \{i\}} u_j\right) \qquad \forall i \in \left\{ j \in \{1, \ldots, N\} \,\middle|\, g_j \geqslant 0 \wedge \sum_{l=1}^{N} a_{jl} \geqslant 0 \right\}. \quad (4.12d)$$

This result is inspired by [Bar+16, Lemma 9.1], where it is shown that the Z-matrix property of a matrix with vanishing/nonnegative row sums is sufficient for the local DMP property to hold globally.

Proof. For (4.12a), let $i \in \{1, \ldots, N\}$ be such that $g_i \leqslant 0$ and $\sum_l a_{il} = 0$. Then we have

$$a_{ii} u_i = g_i + \sum_{j \neq i} (-a_{ij}) u_j = g_i + \sum_{j \in \mathcal{N}_i \setminus \{i\}} (-a_{ij}) u_j$$

$$\leqslant \left(\sum_{j \in \mathcal{N}_i \setminus \{i\}} (-a_{ij}) \right) \max_{j \in \mathcal{N}_i \setminus \{i\}} u_j = a_{ii} \max_{j \in \mathcal{N}_i \setminus \{i\}} u_j.$$

In the case of (4.12c), the fact that $\sum_l a_{il} \geqslant 0$ implies

$$a_{ii} u_i = g_i + \sum_{j \neq i} (-a_{ij}) u_j = g_i + \sum_{j \in \mathcal{N}_i \setminus \{i\}} (-a_{ij}) u_j$$

$$\leqslant \left(\sum_{j \in \mathcal{N}_i \setminus \{i\}} (-a_{ij}) \right) \max_{j \in \mathcal{N}_i \setminus \{i\}} u_j$$

$$\leqslant \left(\sum_{j \in \mathcal{N}_i \setminus \{i\}} (-a_{ij}) \right) \max\left(0, \max_{j \in \mathcal{N}_i \setminus \{i\}} u_j\right)$$

$$\leqslant a_{ii} \max\left(0, \max_{j \in \mathcal{N}_i \setminus \{i\}} u_j\right).$$

The lower bounds (4.12b) and (4.12d) can be shown similarly. $\qquad\square$

Lemma 4.12 implies that the local DMP property holds if (4.12a) and (4.12b) are valid for all interior degrees of freedom.

Theorem 4.13. *Let $\mathcal{A}'$ be an L-matrix satisfying the zero row sum condition and assume $g' = 0$. Then the solution of (4.11) possesses the local DMP property.*

Proof. The statement is a direct consequence of Lemma 4.12.   □

This property imposes strong restrictions on the system matrix $\mathcal{A}$ and the right hand side g. Weaker assumptions suffice to guarantee the nonnegativity of the solution.

Theorem 4.14. *The solution of (4.11) is nonnegative if $\mathcal{A}$ is monotone and the entries of g are nonnegative.*

Proof. The statement follows directly from the definition of a monotone matrix.   □

In particular, Lemma 4.6 provides a handy algebraic criterion for proving the monotonicity of $\mathcal{A}$ which implies positivity preservation for $g \geqslant 0$. The nonnegativity of g is guaranteed for a nonnegative source term f if numerical integration is performed using *positive quadrature rules*, that is, quadrature rules with positive weights [MR73].

The local DMP property does not ensure monotonicity of the finite element solution: Maxima can still occur in the interior of the domain if two neighboring degrees of freedom coincide. A more useful estimate for the range of possible nodal values is given by global maximum principles that are formulated as follows.

Definition 4.15. The vector $u \in \mathbb{R}^N$ has the *global discrete maximum principle (DMP)* property if

$$\min_{j \in \{M+1,\ldots,N\}} u_j \leqslant u_i \leqslant \max_{j \in \{M+1,\ldots,N\}} u_j \qquad \forall i \in \{1,\ldots,M\}. \qquad (4.13)$$

Consider a finite element approximation using linear Lagrange basis functions. If the vector of degrees of freedom enjoys the global DMP property, global extrema can only be attained at Dirichlet nodes, where the function values are prescribed in a weak or strong manner.

As in the case of the local DMP, we prove a number of global estimates before formulating sufficient conditions for the global DMP property to hold.

In the next lemma, we estimate the maxima/minima of degrees of freedom whose node numbers belong to an index set $I \subseteq \{1,\ldots,N\}$. For this purpose, we assume that the restriction of the system matrix $\mathcal{A}$ to rows with indices in I is of nonnegative type. Furthermore, we define $J \subseteq \{1, \ldots, N\}$ as the set of indices, for which we cannot prove the given DMP property (e.g., because the right hand side has the wrong sign). Then the degrees of freedom with indices in I are bounded above/below by the maximum/minimum of nodal values with indices in $J \cap I$ and the function values at nodes in a layer around I (denoted by ∂I).

Lemma 4.16. *Consider the following index sets*

$$J_1 := \left\{ j \in \{1, \ldots, N\} \,\middle|\, g_j > 0 \vee \sum_{l=1}^{N} a_{jl} > 0 \right\},$$

$$J_2 := \left\{ j \in \{1, \ldots, N\} \,\middle|\, g_j < 0 \vee \sum_{l=1}^{N} a_{jl} > 0 \right\},$$

$$J_3 := \left\{ j \in \{1, \ldots, N\} \,\middle|\, g_j > 0 \right\},$$

$$J_4 := \left\{ j \in \{1, \ldots, N\} \,\middle|\, g_j < 0 \right\}.$$

Let $\mathcal{A} \in \mathbb{R}^{N \times N}$ be a non-singular matrix and $I \subseteq \{1, \ldots, N\}$ be such that $(a_{ij})_{i \in I, j \in \{1, \ldots, N\}}$ is (i) a Z-matrix with (ii) nonnegative row sums. Define $\bar{I} := \bigcup_{j \in I} \mathcal{N}_j$ and $\partial I := \bar{I} \setminus I$. Then the degrees of freedom $(u_i)_{i \in I}$ of the solution to (4.11) are bounded as follows:

$$\max_{i \in I} u_i \leqslant \max_{i \in (J \cap I) \cup \partial I} u_i \qquad J := J_1, \qquad (4.14\text{a})$$

$$\min_{i \in I} u_i \geqslant \min_{i \in (J \cap I) \cup \partial I} u_i \qquad J := J_2, \qquad (4.14\text{b})$$

$$\max_{i \in I} u_i \leqslant \max\left(0, \max_{i \in (J \cap I) \cup \partial I} u_i\right) \qquad J := J_3, \qquad (4.14\text{c})$$

$$\min_{i \in I} u_i \geqslant \min\left(0, \min_{i \in (J \cap I) \cup \partial I} u_i\right) \qquad J := J_4. \qquad (4.14\text{d})$$

If the invertible system matrix $\mathcal{A}$ is (iii) a Z-matrix with (iv) nonnegative row sums and

$$\sum_{j=1}^{N} a_{ij} = 0 \quad \implies \quad g_i = 0 \qquad \forall i \in \{1, \ldots, N\}, \qquad (4.15)$$

then the nodal values satisfy the generalized discrete maximum/minimum principles

$$\max_{i\in\{1,...,N\}} u_i \leqslant \max_{i\in J}\big(\sum_{j=1}^{N} a_{ij}\big)^{-1} g_i \qquad J := J_1, \tag{4.16a}$$

$$\min_{i\in\{1,...,N\}} u_i \geqslant \min_{i\in J}\big(\sum_{j=1}^{N} a_{ij}\big)^{-1} g_i \qquad J := J_2. \tag{4.16b}$$

Proof. We prove (4.14a) by contradiction and follow the idea presented in the proof of [Kno10, Theorem 5.2; Bar+18, Theorem 3]. Let us assume that all assumptions regarding the system matrix $\mathcal{A}$ are true and (4.14a) is violated, i.e., there exists an index $i \in I$ such that

$$u_i = \max_{j\in I} u_j > \max_{j\in(J\cap I)\cup\partial I} u_j. \tag{4.17}$$

Then we have $i \in I \setminus J$. Furthermore, let the set $I_* \subseteq I \setminus J$ be defined by

$$I_* := \{l \in I \setminus J \mid u_l = u_i\} \subseteq I \setminus J.$$

Then we can show that

$$\exists\iota \in I_* : \quad \sum_{j\in I_*} a_{\iota j} > 0. \tag{4.18}$$

Suppose that (4.18) does not hold. Then

$$\sum_{j\in I_*} a_{lj} = 0 \qquad \forall l \in I_*, \tag{4.19a}$$

$$\implies \qquad a_{lj} = 0 \qquad \forall l \in I_*, j \notin I_* \tag{4.19b}$$

because $(a_{lj})_{l\in I, j\in\{1,...,N\}}$ is of nonnegative type. Property (4.19a) implies the singularity of $(a_{lj})_{l,j\in I_*}$ and, according to (4.19b), $\mathcal{A}$ must be singular, too, in contradiction to the assumption of an invertible matrix $\mathcal{A}$. This shows that (4.18) holds.

By (4.11) and (4.18), we have

$$\big(\sum_{j\in I_*} a_{\iota j}\big)u_i = \sum_{j\in I_*} a_{\iota j}u_j = g_\iota + \sum_{j\notin I_*}(-a_{\iota j})u_j = g_\iota + \sum_{j\in\bar{I}\setminus I_*}(-a_{\iota j})u_j$$

$$\leqslant \big(-\sum_{j\in\bar{I}\setminus I_*} a_{\iota j}\big)\max_{j\in\bar{I}\setminus I_*} u_j = \big(\sum_{j\in I_*} a_{\iota j}\big)\max_{j\in\bar{I}\setminus I_*} u_j \tag{4.20}$$

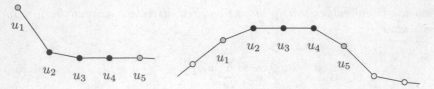

(a) Local and generalized DMPs hold. **(b)** Generalized DMP holds only if $2 \in J$, $3 \in J$, or $4 \in J$.

Figure 4.1: Illustration of local and generalized discrete maximum principles for steady problems: The local DMP of Definition 4.11 holds for every black degree of freedom. The generalized DMP (4.14a) of Lemma 4.16 is applied to J containing the given node and $I = \{2, 3, 4\}$, which implies $\bar{I} = \{1, \ldots, 5\}$, and $\partial I = \{1, 5\}$.

because $\iota \in I_* \subseteq I \setminus J$ and $g_\iota \leqslant 0$. Recalling (4.18), we arrive at

$$\max_{j \in (I \setminus J) \setminus I_*} u_j < u_i \leqslant \max_{j \in \bar{I} \setminus I_*} u_j = \max\Big(\max_{j \in (I \setminus J) \setminus I_*} u_j, \max_{j \in (J \cap I) \cup \partial I} u_j \Big)$$

due to the fact that $(I \setminus J) \setminus I_* \cup (J \cap I) \cup \partial I = (I \setminus I_*) \cup \partial I = \bar{I} \setminus I_*$. This is contradictory to assumption (4.17) and shows the upper bound (4.14a).

The proof of (4.14c) follows the same idea. Only the last identity of (4.20) needs to be replaced by an inequality with a nonnegative maximum.

To prove (4.16a), let $i \in \{1, \ldots, N\}$ be the index such that $u_i = \max_{j \in \{1, \ldots, N\}} u_j$. Without loss of generality, we assume that $\sum_j a_{ij} > 0$ by (4.14a) with $I = \{1, \ldots, N\} \setminus J$ and hypothesis (4.15). Consequently,

$$g_i = \sum_{j=1}^{N} a_{ij} u_j \geqslant \Big(\sum_{j=1}^{N} a_{ij} \Big) u_i \qquad \Longrightarrow \qquad u_i \leqslant g_i \Big(\sum_{j=1}^{N} a_{ij} \Big)^{-1},$$

which proves the statement.

The results for lower bounds can be proved similarly. $\square$

Figure 4.1 illustrates the difference between local DMPs and the results of Lemma 4.16, which are clearly more significant. In fact, the results of Lemma 4.12 are included in Lemma 4.16 if $I = \{i\}$ is considered. Furthermore, the latter lemma is applicable to both weakly and strongly enforced boundary conditions and yields global bounds depending on the components of $\mathcal{A}$ and g but not on the unknown solution of the linear system.

Theorem 4.17. *The solution of (4.11) possesses the global DMP property under the assumptions of Theorem 4.13 provided that $\mathcal{A}$ is non-singular.*

Proof. This result is a direct consequence of Lemma 4.16 using $I = \{1, \ldots, M\}$. □

4.2.3 Unsteady problem

After defining discrete maximum principles for steady problems and introducing the powerful Lemma 4.16, we perform similar analysis for unsteady problems in this section. As a first step towards that end, we consider the semi-discrete problem and present sufficient conditions for boundedness of numerical solutions in terms of the initial values. Next, we discretize in time and formulate local discrete maximum principles for the fully discretized problem. For a two-level time discretization, the local DMP property ensures boundedness by the degrees of freedom at neighboring nodes and at the old time level. The main result of our theoretical study is the time dependent counterpart of Lemma 4.16, namely Lemma 4.24, which implies global maximum principles in particular.

This section is mainly based on the work [Kuz12a]. It generalizes several results and extends them to the case of weakly imposed boundary conditions.

We begin with the definition of a property which can be interpreted as a semi-discrete global maximum principle for initial value problems. It prevents the growth of local extrema and goes back to Jameson [Jam93; Jam95] who introduced it in the context of finite volume methods.

Definition 4.18. Let $Z\big(t, u(t)\big)$ be a continuous function satisfying the Lipschitz condition

$$\big\|Z(t, v) - Z(t, \bar{v})\big\| \leqslant L\|v - \bar{v}\| \qquad \forall t \in (0, T),\ v, \bar{v} \in \mathbb{R}^N \qquad (4.21)$$

for some $L > 0$. The solution $\big(u_i(t)\big)_{i=1}^{N}$ of the semi-discrete problem

$$\frac{\mathrm{d}u(t)}{\mathrm{d}t} = Z\big(t, u(t)\big) \qquad \forall t \in (0, T) \qquad (4.22)$$

is called *local extremum diminishing (LED)* if

$$u_i = \max_{j \in \mathcal{N}_i} u_j \quad \Longrightarrow \quad \mathrm{d}_t u_i \leqslant 0 \qquad \forall i \in \{1, \ldots, M\}, \qquad (4.23\mathrm{a})$$

$$u_i = \min_{j \in \mathcal{N}_i} u_j \quad \Longrightarrow \quad \mathrm{d}_t u_i \geqslant 0 \qquad \forall i \in \{1, \ldots, M\}. \qquad (4.23\mathrm{b})$$

The LED property guarantees that no new local extrema can be formed and existing extrema cannot grow [Kuz12a]. A sufficient condition for a semi-discrete scheme to be LED is given by the following theorem.

Theorem 4.19 ([Jam95; Kuz12a]). *Consider a linear semi-discrete scheme of the form*

$$\mathcal{M}\frac{du}{dt} + \mathcal{Q}u = g, \qquad (4.24)$$

where $\mathcal{M}, \mathcal{Q} \in \mathbb{R}^{N \times N}$ *and* $g \in \mathbb{R}^N$. *The solution* $(u_i(t))_{i=1}^{N}$ *has the LED property if*

$$m_{ij} = 0, \quad q_{ij} \leqslant 0 \qquad \forall i \in \{1, \dots, M\}, \, j \in \{1, \dots, N\}, \, j \neq i, \qquad (4.25a)$$

$$m_{ii} > 0, \quad \sum_{l=1}^{N} q_{il} = 0, \quad g_i = 0 \qquad \forall i \in \{1, \dots, M\}. \qquad (4.25b)$$

Proof. Let us assume that u_i, $i \in \{1, \dots, M\}$, is a local maximum, that is, $u_i \geqslant u_j$ for all $j \in \mathcal{N}_i$. Then, according to (4.25), the time derivative of u_i is nonpositive due to

$$\frac{du_i}{dt} = -m_{ii}^{-1} \sum_{j=1}^{N} q_{ij} u_j = m_{ii}^{-1} \sum_{j \neq i} q_{ij}(u_i - u_j)$$

$$= m_{ii}^{-1} \sum_{j \in \mathcal{N}_i \setminus \{i\}} q_{ij}(u_i - u_j) \leqslant 0$$

because $q_{ij} = 0$ if $j \notin \mathcal{N}_i$ by definition of $\mathcal{N}_i$. The proof of property (4.23b) is based on similar arguments. It follows that the solution is LED due to the linearity of (4.24). $\qquad \square$

Theorem 4.19 guarantees the discrete LED property for all interior degrees of freedom. By (3.8) and (3.9), the solution $u_h(t)$ is bounded by its nodal values. Therefore, the LED criterion restricts the range of possible solution values not only at the nodal points but also in-between.

Remark 4.20. Depending on the considered PDE, the matrix $\mathcal{Q}'$ may fail to possess the zero row sum property. In this case, $g' \geqslant 0$ still guarantees that no negative function values can arise if the initial condition is nonnegative. This property is called *positivity preservation* (strictly speaking, 'nonnegativity preservation'), which means that

$$u_i(0) \geqslant 0 \quad \forall i \in \{1, \dots, N\}$$

$$\implies \quad u_i(t) \geqslant 0 \quad \forall i \in \{1, \dots, N\}, \, t \in (0, T]. \qquad (4.26)$$

Similarly to the proof of [HV03, Theorem 7.1], it suffices to show that the time derivative of vanishing degrees of freedom is nonnegative if $u_i \geqslant 0$ for all $i \in \{1, \dots, N\}$. Hence, we assume that $u_i = 0$ for some $i \in \{1, \dots, M\}$. Then the corresponding time derivative satisfies

$$\frac{du_i}{dt} = m_{ii}^{-1} g_i - m_{ii}^{-1} \sum_{j=1}^{N} q_{ij} u_j = m_{ii}^{-1} g_i + m_{ii}^{-1} \sum_{j \neq i} (-q_{ij}) u_j \geqslant 0$$

due to $g_i \geqslant 0$, $\mathcal{Q}'$ being a Z-matrix, and $u_j(t) \geqslant 0$ for all $j \in \{1, \dots, N\}$.

When it comes to the discretization in time, it is essential to guarantee that the LED and positivity preservation properties of a semi-discrete scheme carry over to its fully discrete counterpart. Since this is generally not the case, the choice of the time integrator and time increments has to be restricted by algebraic constraints. The use of the θ-scheme, as defined in Section 3.3.2, for problem (4.24) yields the algebraic system

$$\mathcal{A} u^{n+1} + \mathcal{B} u^n := (\mathcal{M} + \Delta t \theta \mathcal{Q}^{n+1}) u^{n+1} - (\mathcal{M} - \Delta t(1-\theta) \mathcal{Q}^n) u^n = \underline{g}, \quad (4.27)$$

where the right hand side is given by $\underline{g} := \Delta t \theta g^{n+1} + \Delta t(1-\theta) g^n$.

Following [Kuz12a, Definition 4], we formulate a fully discrete version of the LED criterion, in which the bounds are defined using degrees of freedom at the current and previous time level.

Definition 4.21. A vector $u^{n+1} \in \mathbb{R}^N$ possesses the *local discrete maximum principle (local DMP)* property with respect to $u^n \in \mathbb{R}^N$ if

$$u_i^{n+1} \leqslant \max\left(\max_{j \in \mathcal{N}_i} u_j^n, \max_{j \in \mathcal{N}_i \setminus \{i\}} u_j^{n+1}\right) \qquad \forall i \in \{1, \dots, M\}, \qquad (4.28a)$$

$$u_i^{n+1} \geqslant \min\left(\min_{j \in \mathcal{N}_i} u_j^n, \min_{j \in \mathcal{N}_i \setminus \{i\}} u_j^{n+1}\right) \qquad \forall i \in \{1, \dots, M\}. \qquad (4.28b)$$

For a concise formulation of the next results, we introduce the abbreviation

$$\underline{p}_i := \sum_{j=1}^{N} (a_{ij} + b_{ij}) \qquad \forall i \in \{1, \dots, N\}.$$

Theorem 4.22. *The solution $u^{n+1} \in \mathbb{R}^N$ of the general system*

$$\mathcal{A} u^{n+1} + \mathcal{B} u^n = \underline{g} \qquad (4.29)$$

possesses the local DMP property if

$$a_{ij} \leqslant 0, \quad b_{ij} \leqslant 0 \qquad \forall i \in \{1, \dots, M\}, j \in \{1, \dots, N\}, j \neq i, \qquad (4.30a)$$

$$a_{ii} > 0, \quad b_{ii} \leqslant 0, \quad \underline{g}_i = 0, \quad \underline{p}_i = 0 \qquad \forall i \in \{1, \dots, M\}. \qquad (4.30b)$$

Proof. To show that (4.28a) holds, we denote the upper bound on the right hand side of this inequality by c_i. Then the i-th row of (4.29), $i \in \{1, \ldots, M\}$, can be written and estimated as follows:

$$a_{ii}u_i^{n+1} = -\sum_{j \neq i} a_{ij}u_j^{n+1} - \sum_{j=1}^{N} b_{ij}u_j^n + \sum_{j=1}^{N} a_{ij}c_i + \sum_{j=1}^{N} b_{ij}c_i$$

$$= a_{ii}c_i + \sum_{j \neq i}(-a_{ij})(u_j^{n+1} - c_i) + \sum_{j=1}^{N}(-b_{ij})(u_j^n - c_i) \leqslant a_{ii}c_i.$$

This shows (4.28a). The proof of (4.28b) is similar.   □

As in the case of Definition 4.11, where the local DMP is defined for stationary problems, this property does not imply global boundedness of the function values (cf. Fig. 4.2). However, it constrains changes of degrees of freedom because for each new local extremum there must exist at least one degree of freedom attaining the same value. Arbitrary growth of function values can be avoided using the following design criterion, cf. [Kuz12a, Definition 6].

Definition 4.23. A vector $u^{n+1} \in \mathbb{R}^N$ possesses the *global discrete maximum principle (global DMP)* property with respect to $u^n \in \mathbb{R}^N$ if

$$u_i^{n+1} \leqslant \max\Big(\max_{j \in \{1,\ldots,N\}} u_j^n, \max_{j \in \{M+1,\ldots,N\}} u_j^{n+1}\Big) \quad \forall i \in \{1,\ldots,M\}, \quad (4.31\text{a})$$

$$u_i^{n+1} \geqslant \min\Big(\min_{j \in \{1,\ldots,N\}} u_j^n, \min_{j \in \{M+1,\ldots,N\}} u_j^{n+1}\Big) \quad \forall i \in \{1,\ldots,M\}. \quad (4.31\text{b})$$

The global DMP guarantees that all degrees of freedom are bounded by the extrema of the nodal values at the initial time step and those with indices greater than M. In the case of strongly imposed Dirichlet boundary conditions and pointwise interpolation of initial data, the bounds are readily computable and lie in the range determined by the maxima and minima of the exact initial and boundary data. The two-sided estimates of the global DMP are also applicable if the boundary conditions are imposed weakly. However, the nodal values with indices greater than M are unknown in this case. The analysis of this situation is included in the following time dependent version of Lemma 4.16.

Lemma 4.24. *Consider the following index sets*

$$J_1 := \{ j \in \{1, \ldots, N\} \mid \underline{g}_j > 0 \vee \underline{p}_j \neq 0 \},$$
$$J_2 := \{ j \in \{1, \ldots, N\} \mid \underline{g}_j < 0 \vee \underline{p}_j \neq 0 \},$$
$$J_3 := \{ j \in \{1, \ldots, N\} \mid \underline{g}_j > 0 \vee \underline{p}_j < 0 \},$$
$$J_4 := \{ j \in \{1, \ldots, N\} \mid \underline{g}_j < 0 \vee \underline{p}_j < 0 \}.$$

Let $I \subseteq \{1, \ldots, N\}$ *be such that (i)* $(a_{ij})_{i \in I, j \in \{1, \ldots, N\}}$ *is a strictly diagonally dominant L-matrix and (ii)* $(b_{ij})_{i \in I, j \in \{1, \ldots, N\}}$ *is nonpositive. Define* $\bar{I} := \bigcup_{j \in I} \mathcal{N}_j$ *and* $\partial I := \bar{I} \setminus I$. *Then the degrees of freedom* $(u_i^{n+1})_{i \in I}$ *of the solution to (4.29) are bounded as follows:*

$$\max_{i \in I} u_i^{n+1} \leqslant \max \big(\max_{i \in \bar{I}} u_i^n, \max_{i \in (J \cap I) \cup \partial I} u_i^{n+1} \big) \qquad J := J_1, \qquad (4.32a)$$

$$\min_{i \in I} u_i^{n+1} \geqslant \min \big(\min_{i \in \bar{I}} u_i^n, \min_{i \in (J \cap I) \cup \partial I} u_i^{n+1} \big) \qquad J := J_2, \qquad (4.32b)$$

$$\max_{i \in I} u_i^{n+1} \leqslant \max \big(0, \max_{i \in \bar{I}} u_i^n, \max_{i \in (J \cap I) \cup \partial I} u_i^{n+1} \big) \qquad J := J_3, \qquad (4.32c)$$

$$\min_{i \in I} u_i^{n+1} \geqslant \min \big(0, \min_{i \in \bar{I}} u_i^n, \min_{i \in (J \cap I) \cup \partial I} u_i^{n+1} \big) \qquad J := J_4. \qquad (4.32d)$$

If (iii) $\mathcal{A}$ *is monotone, (iv)* $\mathcal{B}$ *is nonpositive, (v)* $\underline{p}$ *is nonnegative, and*

$$\underline{p}_i = 0 \quad \implies \quad \underline{g}_i = 0 \qquad \forall i \in \{1, \ldots, N\}, \qquad (4.33)$$

then the nodal values satisfy the generalized maximum/minimum principles

$$\max_{i \in \{1, \ldots, N\}} u_i^{n+1} \leqslant \max \big(\max_{i \in \{1, \ldots, N\}} u_i^n, \max_{i \in J} \underline{g}_i \underline{p}_i^{-1} \big) \qquad J := J_1, \qquad (4.34a)$$

$$\min_{i \in \{1, \ldots, N\}} u_i^{n+1} \geqslant \min \big(\min_{i \in \{1, \ldots, N\}} u_i^n, \min_{i \in J} \underline{g}_i \underline{p}_i^{-1} \big) \qquad J := J_2. \qquad (4.34b)$$

Proof. The proofs are inspired by [Kuz12a, Theorem 4].

To show (4.32a), we denote the right hand side of this inequality by c and consider $I \subseteq \{1, \ldots, N\}$ such that the requirements of the lemma are satisfied. Invoking (4.29), we find that

$$\sum_{j \in I \setminus J_1} a_{ij}(u_j^{n+1} - c) = \underbrace{\underline{g}_i}_{\leqslant 0} + \sum_{j \in \bar{I}} \underbrace{(-b_{ij})(u_j^n - c)}_{\leqslant 0}$$
$$+ \sum_{j \in (J_1 \cap I) \cup \partial I} \underbrace{(-a_{ij})(u_j^{n+1} - c)}_{\leqslant 0} \leqslant 0 \quad (4.35)$$

holds for every index $i \in I \setminus J_1$. This is the case because $\sum_l a_{il} + b_{il} = 0$ and $a_{ij}, b_{ii}, b_{ij} \leqslant 0$ for all $i \in I \setminus J_1$, $j \in \{1, \ldots, N\}$, $j \neq i$. The system matrix of (4.35) is given by $(a_{ij})_{i,j \in I \setminus J_1}$, which is a strictly diagonally dominant L-matrix inheriting this property from $(a_{ij})_{i \in I, j \in \{1,\ldots,N\}}$. In view of Lemma 4.6, this leads to

$$u_j^{n+1} - c \leqslant 0 \quad \forall j \in I \setminus J_1 \qquad \Longrightarrow \qquad u_j^{n+1} \leqslant c \quad \forall j \in I$$

and proves the upper bound.

For (4.32c), we have

$$\underbrace{g_i}_{\leqslant 0} + \sum_{j \in \bar{I}} \underbrace{(-b_{ij})(u_j^n - c)}_{\leqslant 0} + \sum_{j \in (J_3 \cap I) \cup \partial I} \underbrace{(-a_{ij})(u_j^{n+1} - c)}_{\leqslant 0}$$
$$+ \sum_{j \in \bar{I}} \underbrace{(a_{ij} + b_{ij})(-c)}_{\leqslant 0} \leqslant 0 \qquad \forall i \in I \setminus J_3,$$

where $c \geqslant 0$ is the right hand side of (4.32c). Using this auxiliary result, the proof is completed as above.

To show (4.34a), we first find that

$$g_i - \sum_{j=1}^{N}(a_{ij} + b_{ij})c = g_i - p_i c \leqslant 0 \qquad \forall i \in \{1, \ldots, N\},$$

where c is again the bound of the inequality at hand. Then u^{n+1} solves the problem

$$\sum_{j=1}^{N} a_{ij}(u_j^{n+1} - c) = g_i - \sum_{j=1}^{N}(a_{ij} + b_{ij})c + \sum_{j=1}^{N} \underbrace{(-b_{ij})(u_j^n - c)}_{\leqslant 0} \leqslant 0$$
$$\forall i \in \{1, \ldots, N\}. \quad (4.36)$$

By the monotonicity of $\mathcal{A}$, this results in

$$u_i^{n+1} \leqslant c \qquad \forall i \in \{1, \ldots, N\},$$

which completes the proof of (4.34a).

The lower bounds of (4.32) and (4.34) can be shown similarly. $\square$

As in the steady case, discrete maximum principles formulated in Lemma 4.24 for local definitions of $I \subset \{1, \ldots, N\}$ are more significant and restrictive than the ones of Theorem 4.22. An illustration of this is shown in Fig. 4.2.

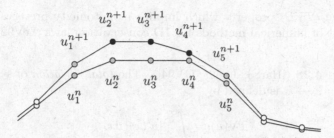

Figure 4.2: Illustration of local and generalized discrete maximum principles for unsteady problems: While the local DMP of Definition 4.21 holds for every black degree of freedom, the generalized counterpart (4.32a) of Lemma 4.24 for $I = \{2, 3, 4\}$ (implying $\bar{I} = \{1, \ldots, 5\}$, and $\partial I = \{1, 5\}$) holds only if $2 \in J$ or $3 \in J$.

Theorem 4.25. *The solution of (4.29) possesses the global DMP property if $\mathcal{A}'$ is a strictly diagonally dominant L-matrix and the assumptions of Theorem 4.22 are satisfied.*

Proof. The statement follows directly from Lemma 4.24.   □

Similarly to Definition 4.23, sufficient conditions can also be formulated for the nonnegativity of function values.

Definition 4.26. The solution u^{n+1} of the fully discretized problem is *positivity preserving* if

$$\min\left(\min_{j \in \{1, \ldots, N\}} u_j^n, \min_{j \in \{M+1, \ldots, N\}} u_j^{n+1} \right) \geqslant 0$$

$$\implies \quad u_i^{n+1} \geqslant 0 \quad \forall i \in \{1, \ldots, M\}. \quad (4.37)$$

Theorem 4.27. *The solution u^{n+1} of (4.29) is positivity preserving if $\mathcal{A}'$ is a strictly diagonally dominant L-matrix, $\mathcal{B}'$ is nonpositive, and $\underline{g}_i \geqslant 0$ for all $i \in \{1, \ldots, M\}$.*

Proof. The proof is similar to the proof of Lemma 4.24.   □

4.2.3.1 Order barrier

In what follows, we show that a fully discrete scheme in one space dimension is at most first order accurate if it represents a time-discrete counterpart of a linear LED method. For this purpose, we first define the *total variation*

diminishing (TVD) property which implies monotonicity preservation in
the context of numerical methods for 1D conservation laws [LeV92, Theorem 15.3].

Definition 4.28 ([Har83; Jam95; QV94]). The *total variation* of a smooth
function $u : \mathbb{R} \to \mathbb{R}$ is defined by

$$\mathrm{TV}(u) := \int_{-\infty}^{\infty} |u'(x)| \, \mathrm{d}x. \qquad (4.38)$$

The corresponding formula for the discrete total variation of a grid function
$u_h = (u_i)_{i \in \mathbb{Z}}$ reads

$$\mathrm{TV}(u_h) := \sum_{i \in \mathbb{Z}} |u_{i+1} - u_i|. \qquad (4.39)$$

Definition (4.39) corresponds to a piecewise linear approximation of the
gradient.

Definition 4.29 ([Har83, Eq. (2.2)]). A semi-discrete numerical solution
$u_h(t)$ is said to be *total variation diminishing (TVD)* if

$$\mathrm{TV}\big(u_h(t_1)\big) \leqslant \mathrm{TV}\big(u_h(t_2)\big) \qquad \forall t_1 \in [0, t_2]. \qquad (4.40)$$

Each (semi-discrete) LED scheme is TVD in one space dimension [Jam95].

We now follow [Wes01] to prove the famous Godunov theorem originally
published in [God59] and show that each *linear* fully discretized TVD
scheme is at most first order accurate. For this purpose, we consider the
time dependent linear advection equation in the infinite domain $\Omega = \mathbb{R}$ with
a constant velocity field $v > 0$. The initial value problem to be solved is
given by

$$\frac{\partial u}{\partial t} + v \frac{\partial u}{\partial x} = 0 \qquad \text{in } \mathbb{R} \times (0, T), \qquad (4.41\mathrm{a})$$

$$u(\cdot, 0) = u_0 \qquad \text{on } \mathbb{R}, \qquad (4.41\mathrm{b})$$

where $u_0 : \mathbb{R} \to \mathbb{R}$ is the given initial condition. The exact solution

$$u(x, t) = u_0(x - vt) \qquad \forall x \in \mathbb{R}, \, t \in [0, T], \qquad (4.42)$$

is non-decreasing/non-increasing/monotone if the initial condition is non-decreasing/non-increasing/monotone.

We discretize (4.41) using an arbitrary finite difference scheme with constant mesh size $h > 0$ and a two-level time-stepping procedure with constant time increment $\Delta t > 0$, $\frac{T}{\Delta t} \in \mathbb{N}$. Let $u_i^n \approx u(x_i, t^n)$ denote the value of the approximate solution u_h at grid point $x_i = ih$ and time level $t^n = n\Delta t$. Then the so defined degrees of freedom satisfy

$$\sum_{j \in \mathbb{Z}} a_j u_{i+j}^{n+1} + \sum_{j \in \mathbb{Z}} b_j u_{i+j}^n = 0 \qquad \forall i \in \mathbb{Z},$$

where $a_j, b_j \in \mathbb{R}$, $j \in \mathbb{Z}$, are the coefficients of the given numerical scheme. Due to the linearity of the mapping, this solution update can be written as [Wes01, Eq. (9.4)]

$$u_i^{n+1} = \sum_{j \in \mathbb{Z}} c_j u_{i+j}^n \qquad \forall i \in \mathbb{Z}, \tag{4.43}$$

where $c_j \in \mathbb{R}$ for all $j \in \mathbb{Z}$.

As in the case of a continuous function, we define the *monotonicity* of a discrete solution.

Definition 4.30 ([Wes01]). A sequence $u_h^n = (u_i^n)_{i \in \mathbb{Z}}$ is called *monotone* if, for fixed n, it is either non-decreasing or non-increasing as i increases.

This definition is consistent with the concept of monotonicity for continuous functions because their nodal values satisfy Definition 4.30.

As already mentioned, the exact solution given by (4.42) preserves the monotonicity of the initial condition at any time. A numerical method is called *monotonicity preserving* if the discrete solutions remain monotone as well.

Definition 4.31 ([Wes01, Definition 9.2.1]). A numerical scheme is called *monotonicity preserving* if for every non-decreasing (non-increasing) initial condition $u_h^0 = (u_i^0)_{i \in \mathbb{Z}}$ the numerical solution $u_h^n = (u_i^n)_{i \in \mathbb{Z}}$, $n \in \{1, \dots, \frac{T}{\Delta t}\}$, is non-decreasing (non-increasing) at all later time instants.

Theorem 4.32 ([Wes01, Theorem 9.2.1]). *The numerical scheme* (4.43) *is monotonicity preserving if and only if* $c_i \geqslant 0$ *for all* $i \in \mathbb{Z}$.

Proof. Let us first assume that $c_i \geqslant 0$ for all $i \in \mathbb{Z}$, and, without loss of generality, that $u_h^n = (u_i^n)_{i \in \mathbb{Z}}$ is non-decreasing. Then we have

$$u_i^{n+1} - u_{i-1}^{n+1} = \sum_j c_j (u_{i+j}^n - u_{i+j-1}^n) \geqslant 0 \qquad \forall i \in \mathbb{Z}.$$

On the other hand, if $c_{i_0} < 0$ for some $i_0 \in \mathbb{Z}$, the non-decreasing initial condition defined by

$$u_i^0 = \begin{cases} 0 & : i \leqslant 0, \\ 1 & : i > 0 \end{cases}$$

leads to

$$u^1_{1-i_0} - u^1_{-i_0} = \sum_j c_j(u^0_{1-i_0+j} - u^0_{-i_0+j}) = \sum_j c_j \delta_{j i_0} = c_{i_0} < 0,$$

which is a contradiction to the assumption of a monotonicity preserving method.   □

A second order accurate method is supposed to evolve quadratic polynomials exactly. A linear monotonicity-preserving scheme cannot possess this property for arbitrary data as shown in the following theorem.

Theorem 4.33 (Godunov's order barrier theorem [Wes01, Theorem 9.2.2]). *Linear one-step numerical schemes for the advection equation (4.41) cannot be monotonicity preserving and second order accurate, unless the CFL number satisfies* $\tau = \frac{|v|\Delta t}{h} \in \mathbb{N}$.

Proof. The proof follows the proof of [Wes01, Theorem 9.2.2], which in turn follows [God59].

To prove this theorem by contradiction, we suppose that a linear second order accurate and monotonicity preserving scheme does exist. In view of the above assumption regarding second order methods, this scheme evolves quadratic initial conditions exactly. If the exact solution of (4.41) depending on the mesh size $h > 0$ is given by

$$u(x,t) = \left(\frac{x - vt}{h} - \tfrac{1}{2}\right)^2 - \tfrac{1}{4} \qquad \forall x \in \mathbb{R}, \, t \in [0,T],$$

the numerical approximation produced by (4.43) is exact for

$$u_i^n = u(ih, n\Delta t) = \left(i - n\tau - \tfrac{1}{2}\right)^2 - \tfrac{1}{4} \qquad \forall i \in \mathbb{Z}, \, n \in \{0, \ldots, \tfrac{T}{\Delta t}\},$$

where $\tau = \frac{v\Delta t}{h} > 0$ is the CFL number. For $n = 0$, substitution into (4.43) gives

$$\left(i - \tau - \tfrac{1}{2}\right)^2 - \tfrac{1}{4} = \sum_{j \in \mathbb{Z}} c_j u^0_{i+j} = \sum_{j \in \mathbb{Z}} c_j \left(\left(i + j - \tfrac{1}{2}\right)^2 - \tfrac{1}{4}\right) \geqslant 0 \qquad \forall i \in \mathbb{Z} \quad (4.44)$$

by the nonnegativity of u_i^0 for all $i \in \mathbb{Z}$ and the monotonicity preservation property. However, if $\tau \notin \mathbb{N}$ and i_0 is chosen such that $i_0 - 1 < \tau < i_0$, we deduce

$$u^1_{i_0} = (i_0 - \tau - \tfrac{1}{2})^2 - \tfrac{1}{4} = (i_0 - \tau)(i_0 - \tau - 1) < 0,$$

which is a contradiction to (4.44).   □

In Theorem 4.33, we have shown that linear monotonicity preserving one-step methods can be at most first order accurate. This order barrier applies to linear multistep methods as well [Wes01, Theorem 9.2.3]. We conclude this section by mentioning that each total variation diminishing method is monotonicity preserving [Wes01] and, hence, at most first order accurate in the linear case. However, nonlinear monotonicity-preserving methods may achieve second order accuracy for sufficiently smooth data. This makes the TVD criterion a powerful tool for the design of high-resolution schemes with solution-dependent coefficients in one dimension.

4.3 Low order method

In this section, we introduce a practical approach for deriving bound-preserving low order schemes that comply with the conditions of the previous section. The main idea behind the derivation of these schemes is the concept of discrete upwinding, an algebraic correction technique that modifies matrix coefficients of a given high order scheme so as to satisfy appropriate monotonicity constraints without losing the discrete conservation property (if any). As before, stationary and transient problems are analyzed separately but using the same tools.

This section is largely based on Kuzmin's presentation of the algebraic flux correction methodology for continuous Galerkin approximations [Kuz12a]. The use of sufficient conditions to determine the optimal values of artificial diffusion coefficients goes back to the work of Book, Boris, and Hain [Boo+75] who designed a low order scheme for their flux-corrected transport (FCT) algorithm in this manner in the context of an explicit finite difference scheme.

4.3.1 Steady problem

This section introduces correction techniques based on the requirements of Section 4.2.2 for finite element discretizations of two different steady state problems. The first task is the construction of a discrete diffusion operator that transforms the Galerkin discretization of the steady advection equation into a low order scheme satisfying local and global discrete maximum principles. The second problem to be addressed is the design of algebraic correction techniques for the L^2-projection of data into the finite element space. The need for bound-preserving conservative projections arises, for example, in the context of initialization or intergrid data transfer ('remapping'). A well-designed projection operator should guarantee mass conservation and boundedness by the local extrema of the input data.

4.3.1.1 Discrete upwinding

Let us return to the discretization of the advection-reaction equation in non-conservative form (3.11). For convenience, we assume that

$$f = 0, \qquad c = 0, \qquad \mathcal{A} \text{ is non-singular.} \tag{4.45}$$

The assumptions regarding f and c are made to demonstrate the ability of the method to satisfy DMPs and are related to the requirements of Theorem 2.13. One-sided a priori estimates can be obtained, for example, if $c = 0$ and $f \leqslant 0$. Positivity preservation is guaranteed for $f \geqslant 0$ and $u_{in} \geqslant 0$.

The invertibility of the system matrix $\mathcal{A}$ can be guaranteed, for example, by the coercivity condition (2.14) as shown in Remark 3.7. In the unlikely event that the standard Galerkin discretization produces a singular matrix if (2.14) is violated, additional stabilization terms need to be included. An in-depth discussion of this topic is beyond the scope of this work and would burden the reader with a lengthy presentation of material which can be found in many publications on stabilized finite element methods for hyperbolic problems.

By definition (3.11b) of the system matrix $\mathcal{A}$, the partition of unity property of the Lagrange basis functions $(\varphi_i)_{i=1}^N$, and assumption (4.45), we have

$$\sum_{j=1}^N a_{ij} = 0 \qquad \forall i \in \{1, \ldots, M\}, \tag{4.46a}$$

$$\sum_{j=1}^N a_{ij} > 0 \qquad \forall i \in \{M+1, \ldots, N\}. \tag{4.46b}$$

If the velocity field is solenoidal, then

$$\sum_{i=1}^N a_{ij} = 0 \qquad \forall j \in \{1, \ldots, N\} \text{ s.t. } \mathbf{x}_j \notin \Gamma_{out}. \tag{4.47}$$

Indeed, integration by parts yields

$$
\begin{aligned}
a(\varphi_j, \varphi_i) &= \int_\Omega \mathbf{v} \cdot \operatorname{grad}(\varphi_j)\varphi_i + \int_{\partial\Omega} (\mathbf{v} \cdot \mathbf{n})_- \varphi_j \varphi_i \\
&= \int_\Omega \operatorname{div}(\mathbf{v}\varphi_j)\varphi_i - \int_{\Gamma_{\text{in}}} (\mathbf{v} \cdot \mathbf{n})\varphi_j \varphi_i \qquad (4.48) \\
&= -\int_\Omega \varphi_j \mathbf{v} \cdot \operatorname{grad}(\varphi_i) + \int_{\Gamma_{\text{out}}} (\mathbf{v} \cdot \mathbf{n})\varphi_j \varphi_i
\end{aligned}
$$

and $\mathcal{A}$ is skew symmetric for degrees of freedom in the interior of the domain. Hence, $\mathcal{A}'$ does not possess the L-matrix property which would be sufficient for the validity of DMPs as shown in the proofs of Theorems 4.13 and 4.17. To achieve the desired matrix properties, the discrete problem needs to be modified in an appropriate manner.

In this work, we focus on an algebraic transformation, which modifies the system matrix by adding a diffusive perturbation such that
1. the resulting system is as close as possible to the original system;
2. all requirements for DMPs or at least positivity preservation are satisfied;
3. the modified Galerkin discretization remains consistent and conservative.
The latter requirement is satisfied if row and column sums of the system matrix do not change (cf. (4.46) and (4.47)). This observation motivates the following definition of an admissible diffusion operator.

Definition 4.34 ([KT02]). A symmetric matrix $\mathcal{D} = (d_{ij})_{i,j=1}^N$ is called a *discrete diffusion operator* if $\mathcal{D}$ has zero row and column sums.

Regardless of the exact definition of the matrix entries, adding/subtracting $\mathcal{D}$ to/from a matrix $\mathcal{A}$ does not change the row and column sums and, hence, satisfies the corresponding design criterion. Therefore, the so far undefined entries d_{ij} can be used to guarantee the Z-matrix property of the resulting matrix.

Theorem 4.35 ([Kuz12a, Eq. (37)]). *Let $\mathcal{A} \in \mathbb{R}^{N \times N}$ be arbitrary. The matrix $\mathcal{D} = \mathcal{D}(\mathcal{A}) = (d_{ij})_{i,j=1}^N$ defined by*

$$
d_{ij} := \begin{cases} \max(a_{ij}, 0, a_{ji}) & : j \neq i, \\ -\sum_{l \neq i} d_{il} & : j = i \end{cases} \qquad \forall i, j \in \{1, \dots, N\} \qquad (4.49)
$$

is a discrete diffusion operator such that $\mathcal{A} - \mathcal{D}$ is a Z-matrix whose row and column sums coincide with those of $\mathcal{A}$.

Moreover, the Frobenius norm of $\mathcal{D}$ is smaller than that of any other discrete diffusion operator that provides the above properties and has no negative off-diagonal entries. While Definition 4.34 implies that $\mathcal{D}$ is symmetric, 'non-symmetric discrete diffusion operators' are also admissible and may produce less diffusive bound-preserving schemes as in the context of high order Bernstein finite element discretizations [Loh+17]. In this work, we restrict ourselves to linear finite elements, for which the above definition is sufficient.

Lemma 4.36. *The artificial diffusion matrix* $\mathcal{D} \in \mathbb{R}^{N \times N}$ *is negative semidefinite for each* $\mathcal{A} \in \mathbb{R}^{N \times N}$.

Proof. For every $v \in \mathbb{R}^N$, we have

$$\sum_{i,j=1}^{N} v_i d_{ij} v_j = \sum_{i=1}^{N} \left(-\sum_{j \neq i} d_{ij} \right) v_i^2 + \sum_{i,j,\, i \neq j} d_{ij} v_i v_j$$

$$= \sum_{i,j,\, j \neq i} d_{ij} \left(v_i v_j - \tfrac{1}{2} v_i^2 - \tfrac{1}{2} v_j^2 \right) = -\sum_{i,j,\, j \neq i} \frac{d_{ij}}{2} (v_i - v_j)^2 \leqslant 0$$

as asserted in the theorem. □

Let $\mathcal{A}$ and g be the matrix and right hand side of (3.11). Using the artificial diffusion operator $\mathcal{D} = \mathcal{D}(\mathcal{A})$ defined by (4.49) to enforce the Z-matrix property, we obtain the perturbed system

$$(\mathcal{A} - \mathcal{D})u = g. \tag{4.50}$$

In view of (4.45), we assume that

$$\mathcal{A} - \mathcal{D} \text{ is non-singular.} \tag{4.51}$$

Remark 4.37. Similarly to Remark 3.7, the coercivity condition (2.14) guarantees the positive definiteness and invertibility of the system matrix $\mathcal{A} - \mathcal{D}$. The fact that

$$\sum_{i,j=1}^{N} u_i (a_{ij}' - d_{ij}) u_j > 0 \qquad \forall u \in \mathbb{R}^N \setminus \{0\} \tag{4.52}$$

is due to Remark 3.7 and Lemma 4.36. Moreover, $\mathcal{A} - \mathcal{D}$ is an L-matrix of nonnegative type by Remark 4.9, Theorem 4.35, (4.46), and (4.52).

Then the following DMP properties hold for the solution of (4.50).

Theorem 4.38. *The solution of* (4.50) *possesses the local and global DMP property.*

Proof. The assumption that $f = 0$ implies that $g' = 0$ as well. Due to the vanishing row sums of $\mathcal{A}'$ (cf. (4.46a)) and the fact that $\mathcal{A} - \mathcal{D}$ is an invertible Z-matrix, $\mathcal{A}' - \mathcal{D}'$ is an L-matrix which inherits the zero row sum property. Then the use of Theorems 4.13 and 4.17 completes the proof. $\square$

Global boundedness by the maxima and minima of weakly imposed boundary data can be shown as follows.

Theorem 4.39. *The solution of* (4.50) *is bounded by the extrema of the boundary data* u_{in}.

Proof. By Theorem 4.35, $\mathcal{A} - \mathcal{D}$ is a Z-matrix. Furthermore, as shown in the proof of Theorem 4.38, $\mathcal{A}' - \mathcal{D}'$ is an L-matrix with vanishing row sums. Then, according to Lemma 4.16 and (3.9), the solution u_h with degrees of freedom satisfying (4.50) is bounded by

$$u_h(\mathbf{x}) \geqslant \min_{i \in \{M+1,\dots,N\}} g_i \Big(\sum_{j=1}^{N} a_{ij} - d_{ij}\Big)^{-1} \quad \forall \mathbf{x} \in \Omega,$$

$$u_h(\mathbf{x}) \leqslant \max_{i \in \{M+1,\dots,N\}} g_i \Big(\sum_{j=1}^{N} a_{ij} - d_{ij}\Big)^{-1} \quad \forall \mathbf{x} \in \Omega$$

if $g_i = 0$ for all $i \in \{1,\dots,M\}$ and $\sum_j (a_{ij} - d_{ij}) = \sum_j a_{ij} > 0$ for all $i \in \{M+1,\dots,N\}$. The former condition is satisfied since $f = 0$. The latter one holds due to

$$\sum_{j=1}^{N} a_{ij} = \int_{\Omega} \mathbf{v} \cdot \mathrm{grad}\Big(\sum_{j=1}^{N} \varphi_j\Big)\varphi_i + \int_{\partial\Omega} (\mathbf{v} \cdot \mathbf{n})_- \Big(\sum_{j=1}^{N} \varphi_j\Big)\varphi_i$$

$$= \int_{\partial\Omega} (\mathbf{v} \cdot \mathbf{n})_- \varphi_i > 0 \quad \forall i \in \{M+1,\dots,N\} \quad (4.53)$$

because $\sum_j \varphi_j = 1$. By definition of g, we have

$$g_i = \int_{\partial\Omega} (\mathbf{v} \cdot \mathbf{n})_- u_{\mathrm{in}} \varphi_i \leqslant \Big(\max_{\mathbf{s} \in \Gamma_{\mathrm{in}}} u_{\mathrm{in}}(\mathbf{s})\Big) \int_{\partial\Omega} (\mathbf{v} \cdot \mathbf{n})_- \varphi_i \quad \forall i \in \{M+1,\dots,N\},$$
$$(4.54)$$

which completes the proof. $\square$

To demonstrate the influence of the discrete diffusion operator defined by Theorem 4.35, we consider a simple one dimensional test problem. In contrast to assumption (4.45), we use a nontrivial right hand side $f \geqslant 0$ to define a boundary value problem with a non-constant exact solution. However, this definition of f implies that the solution is not bounded above by the inflow boundary data.

Example 4.40. Let us consider the one dimensional stationary advection equation with the constant velocity $v > 0$, the inflow boundary data $u_0 \in \mathbb{R}$, and a source term f to be defined later

$$v\frac{\partial u}{\partial x} = f \qquad \text{in } \Omega = (0,1), \tag{4.55a}$$

$$u(0) = u_{\text{in}}. \tag{4.55b}$$

We discretize this problem using a uniform mesh with mesh size $h = (N-1)^{-1}$, $N \geqslant 2$. The coordinates of nodal points are given by $x_1 = 0$, $x_2 = h$, ..., $x_N = 1$. This node numbering violates our convention that non-Dirichlet nodes are numbered first. However, this convention was adopted for presentation purposes and does not need to be followed in a practical implementation of the method.

The system matrix of the discrete problem with weakly imposed boundary conditions (cf. (3.11)) is

$$\mathcal{A} = \frac{1}{2}\begin{pmatrix} v & v & & & \\ -v & 0 & v & & \\ & \ddots & \ddots & \ddots & \\ & & -v & 0 & v \\ & & & -v & v \end{pmatrix} \in \mathbb{R}^{N \times N}.$$

Note that in one dimension the (inflow) boundary integrals occurring in (3.11) reduce to simple evaluations of the integrand on the boundary.

The artificial diffusion matrix corresponding to $\mathcal{A}$ is given by

$$\mathcal{D} = \frac{1}{2}\begin{pmatrix} -v & v & & & \\ v & -2v & v & & \\ & \ddots & \ddots & \ddots & \\ & & v & -2v & v \\ & & & v & -v \end{pmatrix} \in \mathbb{R}^{N \times N}.$$

Thus, the low order counterpart of the Galerkin matrix $\mathcal{A}$ reads

$$\mathcal{A} - \mathcal{D} = \begin{pmatrix} v & 0 & & & \\ -v & v & 0 & & \\ & \ddots & \ddots & \ddots & \\ & & -v & v & 0 \\ & & & -v & v \end{pmatrix}$$

and has zero row sums for all rows with indices greater than 1. Furthermore, it is easy to verify that this matrix satisfies the requirements for local and global DMPs. However, the presence of a nonvanishing source term f makes it possible to obtain one-sided estimates at best.

The modified system matrix coincides with that of the first order upwind finite difference scheme at each node that is not located on the inflow boundary. For that reason, the presented algebraic correction technique is called *discrete upwinding* [KM05]. However, already in two dimensions, the resulting scheme are generally not of upwind type in a literal sense because some degrees of freedom can depend on the solution values at downwind nodes [Kno17].

Interestingly enough, the artificial diffusion matrix $\mathcal{D}$ corresponds to the finite element discretization of the scaled Laplace operator $\frac{1}{2}|v|h\Delta$.

In Fig. 4.3, the exact, Galerkin, and low order solutions of problem (4.55) are shown for $v = 1$ and $N = 11$. In this test, we use $u_{\mathrm{in}} = 0$ and $f(x) = 10x^9$. The corresponding exact solution is given by $u(x) = x^{10}$.

The introduction of the artificial diffusion matrix $\mathcal{D}$ stabilizes the low order solution so that all unphysical overshoots and undershoots disappear. In particular, all nodal values become nonnegative in accordance with the principle of positivity preservation for $f \geqslant 0$ and $u_{\mathrm{in}} = 0$.

Remark 4.41. In Section 4.5.1.3, we will derive error estimates for the method that will be considered in Section 4.5.1. This technique includes discrete upwinding as a special case and the rate of convergence $\frac{1}{2}$ can be shown for the error measured in $\|\cdot\|_{L^2(\Omega)}$ if the coercivity condition (2.14) holds.

4.3.1.2 Mass lumping

The mass matrix defined by (3.19) was introduced in the context of the spatial discretization of the time derivative in unsteady advection-reaction equations. The Galerkin discretization of the reactive term cu produces a multiple of the mass matrix if the reaction rate c is constant. Using

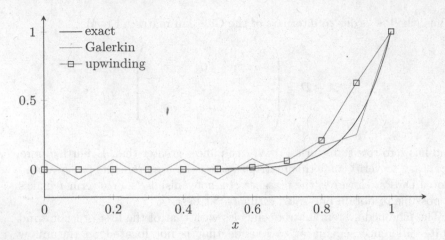

Figure 4.3: Galerkin vs. discrete upwind solution (cf. Example 4.40).

$c \equiv 1$ and $\mathbf{v} \equiv \mathbf{0}$, we consider a pure reaction equation corresponding to a projection of the right hand side f into the finite element space V_h (see below). The system matrix of the resulting discrete problem coincides with the mass matrix. No boundary conditions need to be imposed since $\Gamma_{\mathrm{in}} = \emptyset$ for vanishing velocity fields (cf. Section 2.1.2.1).

Definition 4.42 ([LT08, Eq. (5.38)]). The *(consistent) L^2-projection* operator $P_h : L^2(\Omega) \to V_h$ is defined by

$$\int_\Omega P_h u \varphi_h = \int_\Omega u \varphi_h \qquad \forall \varphi_h \in W_h = V_h. \tag{4.56}$$

The solution $P_h u \in V_h$ of the discrete problem enjoys the best approximation property w.r.t. $\| \cdot \|_{L^2(\Omega)}$ [LT08].

A typical application of the L^2-projection is the generation of discrete initial data for finite element discretizations of time dependent problems. In this context, a possibly discontinuous function $u_0 \in L^2(\Omega)$ needs to be projected into V_h while preserving desired properties like the range of function values and the total mass. In contrast to the pointwise interpolation operator which requires continuity of the input function and is generally non-conservative, the Galerkin L^2-projection is globally conservative. Indeed, substituting the admissible test function $\varphi_h \equiv 1 \in V_h$ into (4.56), we obtain the integral identity

$$\int_\Omega P_h u = \int_\Omega u, \qquad (4.57)$$

which guarantees conservation of mass in addition to the best approximation property.

As we will see in Example 4.47, the L^2-projection tends to produce oscillatory solutions. If the exact solution is not sufficiently smooth, spurious overshoots and undershoots may arise leading to projections that violate bounds given by the extrema of the input data. The theory presented in Section 4.2.2 reveals that we can solve this problem by making the mass matrix monotone (cf. Theorem 4.14). In view of Lemma 4.6, a sufficient and easily applicable condition that guarantees monotonicity is strict diagonal dominance combined with the L-matrix property. Modified mass matrices satisfying this criterion may be used for projection purposes as long as the involved approximations are consistent and conservative. That is, a bound-preserving approximate L^2-projection should be exact at least for constant data and it should conserve the total mass of the function to be projected (cf. (4.57)).

The discrete upwinding strategy of the previous section represents a general approach to construct discrete diffusion operators that convert an arbitrary matrix into a Z-matrix. Hence, this algebraic post-processing technique is readily applicable to the mass matrix as well. In the case of (multi-)linear basis functions, the resulting low order approximation is a diagonal matrix with positive diagonal entries because all entries of the consistent mass matrix are nonnegative.

Definition 4.43 ([Kuz12a, Eq. (35)]). The *lumped mass matrix* $\mathcal{M}_L = (m_i \delta_{ij})_{i,j=1}^N \in \mathbb{R}^{N \times N}$ is defined as the diagonal matrix, whose diagonal entries are given by the row sums of the consistent mass matrix $\mathcal{M} = (m_{ij})_{i,j=1}^N$. That is,

$$m_i = \sum_{j=1}^N m_{ij} = \sum_{j=1}^N \int_\Omega \varphi_i \varphi_j = \int_\Omega \varphi_i > 0 \qquad \forall i \in \{1, \ldots, N\} \qquad (4.58)$$

due to the partition of unity property of the basis functions $(\varphi_j)_{j=1}^N$.

Deviating from the convention made in Section 1.3, diagonal entries of lumped mass matrices are denoted by m_i because the notation m_{ii} is reserved for diagonal entries of the consistent mass matrix. This exception to the double-subscript addressing rule for matrix entries reflects the fact that all nonzero entries of the lumped mass matrix can be stored in a vector $(m_i)_{i=1}^N$.

Row-sum mass lumping preserves consistency because it is equivalent to numerical calculation of m_{ij} using a low order Newton-Cotes quadrature rule [Kuz12a; Han94].

Lemma 4.44. *The mass of a finite element function u_h is defined as $\int_\Omega u_h$ and can be calculated using the entries of the consistent or lumped mass matrix as follows:*

$$\int_\Omega u_h = \sum_{i=1}^N u_i \int_\Omega \varphi_i = \sum_{i=1}^N m_i u_i = \sum_{i,j=1}^N m_{ij} u_j. \qquad (4.59)$$

Due to the last identity and (4.57), the lumped-mass version of the L^2-projection is conservative, too.

Remark 4.45. Row-sum mass lumping as presented in Definition 4.43 yields the only diagonal approximation to the mass matrix which satisfies (4.59) [Han94].

To show this statement, assume that $\tilde{\mathcal{M}} = (\tilde{m}_i \delta_{ij})_{i,j=1}^N$ is conservative, too, and that $\tilde{u}_h$ is the corresponding solution of the modified L^2-projection for an arbitrary function $u : \Omega \to \mathbb{R}$. Then, according to (4.59), $\tilde{u}_h$ should satisfy

$$\int_\Omega u \overset{!}{=} \int_\Omega \tilde{u}_h = \sum_{i,j=1}^N m_{ij} \tilde{u}_j = \sum_{i,j=1}^N m_{ij} \tilde{m}_i^{-1} \int_\Omega u\varphi_i \qquad \forall u : \Omega \to \mathbb{R}.$$

Hence,

$$1 \equiv \sum_{i=1}^N \Big(\sum_{j=1}^N m_{ij} \Big) \tilde{m}_i^{-1} \varphi_i$$

must hold due to the fundamental lemma of the calculus of variations. This is equivalent to

$$\tilde{m}_i = \sum_{j=1}^N m_{ij} \qquad \forall i \in \{1, \ldots, N\}$$

by virtue of the unit sum property of the basis functions $\varphi_1, \ldots, \varphi_N$. Therefore, $\tilde{\mathcal{M}} = \mathcal{M}_\mathrm{L}$ is the only conservative and diagonal approximation of the mass matrix.

Furthermore, the requirements of Theorem 4.14 are satisfied, which guarantees the nonnegativity of the solution if the right hand side vector g is nonnegative. Indeed, the function values are bounded by the extrema of the function to be projected.

Theorem 4.46. *The lumped L^2-projection $u_h \in V_h$ is locally bounded by the extrema of the input data $u \in C^0(\bar{\Omega})$ as follows:*

$$\min_{\tilde{\mathbf{x}} \in \varpi^e} u(\tilde{\mathbf{x}}) \leqslant u_h(\mathbf{x}) \leqslant \max_{\tilde{\mathbf{x}} \in \varpi^e} u(\tilde{\mathbf{x}}) \qquad \forall \mathbf{x} \in K^e, \, e \in \{1, \dots, E\}, \qquad (4.60)$$

where ϖ^e is the patch of elements consisting of K^e and its common-vertex neighbors.

Proof. The degree of freedom u_i, $i \in \{1, \dots, N\}$, is bounded above by

$$u_i = m_i^{-1} g_i = m_i^{-1} \int_{\varpi_i} u \varphi_i \leqslant m_i^{-1} \left(\int_{\varpi_i} \varphi_i \right) \left(\max_{\mathbf{x} \in \varpi_i} u(\mathbf{x}) \right) = \max_{\mathbf{x} \in \varpi_i} u(\mathbf{x})$$
$$(4.61)$$

due to the nonnegativity of the basis functions and the fact that their local support is given by (3.7). Invoking (3.9), we obtain the upper bound of (4.60). The proof for the lower bound is similar. □

We conclude this section with a one dimensional example which illustrates the numerical behavior of the lumped L^2-projection.

Example 4.47. Let us consider the one dimensional L^2-projection (4.56) of a function u defined on the interval $\Omega = (0,1)$. As in Example 4.40, we use the uniform mesh with mesh size $h = (N-1)^{-1}$, $N \geqslant 2$. The consistent and lumped mass matrices of the corresponding linear finite element space V_h read

$$\mathcal{M} = \frac{h}{6} \begin{pmatrix} 2 & 1 & & & \\ 1 & 4 & 1 & & \\ & \ddots & \ddots & \ddots & \\ & & 1 & 4 & 1 \\ & & & 1 & 2 \end{pmatrix}, \quad \mathcal{M}_{\mathrm{L}} = \frac{h}{2} \begin{pmatrix} 1 & & & & \\ & 2 & & & \\ & & \ddots & & \\ & & & 2 & \\ & & & & 1 \end{pmatrix}.$$

In Fig. 4.4, we show the consistent and lumped L^2-projections of the function $u(x) = x^{50}$ for $N = 11$. In this example, the components of the right hand side vector g are determined using exact integration to guarantee their nonnegativity. In more realistic applications, sign-preserving numerical integration can be performed using positive quadrature rules [MR73].

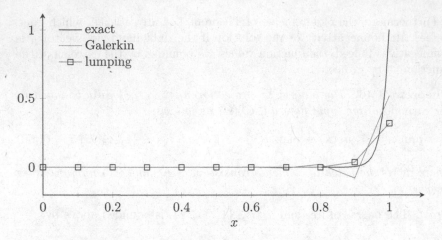

Figure 4.4: Consistent vs. lumped L^2-projection (cf. Example 4.47).

The oscillatory 'Galerkin' solution in Fig. 4.4 was obtained using the consistent L^2-projection. Its lumped counterpart yields a low order approximation (denoted by 'lumping') which is positivity-preserving and bounded by the extrema of the exact solution (cf. Theorem 4.46). However, mass lumping results in strong flattening of the steep gradient near the boundary $x = 1$.

4.3.2 Unsteady problem

In the previous section, we derived algebraic correction techniques for enforcing discrete maximum principles in the context of finite element approximations to stationary problems. Now we extend the discrete upwinding methodology to time dependent problems using the design criteria formulated in Section 4.2.3 (Theorems 4.19, 4.22, and 4.25). Furthermore, we discuss the impact of a solenoidal velocity field (cf. (4.47)) on the mass of time dependent solutions.

As a model problem, we consider the time dependent counterpart of the advection equation considered in Section 4.3.1.1. That is, we consider (2.30) and again assume that (cf. (4.45))

$$f = 0, \qquad c = 0. \tag{4.62}$$

To stabilize the matrix form (3.18) of the Galerkin discretization (3.16a), we modify the mass and convection matrices as for the stationary problem in Section 4.3.1. This yields

$$\mathcal{M}_L \frac{du}{dt} - (\mathcal{K} + \mathcal{D})u = g \qquad \forall i \in \{1, \ldots, N\}, \tag{4.63}$$

where $\mathcal{K} = -\mathcal{A}$ and $\mathcal{D}$ is the artificial diffusion operator defined in terms of $-\mathcal{K}$ such that all off-diagonal matrix entries of $\mathcal{K} + \mathcal{D}$ are nonnegative. The only difference compared to the low order scheme of Section 4.3.1.1 is the presence of the lumped mass matrix multiplied by the vector of nodal time derivatives.

By definition of the matrices $\mathcal{M}_L$ and $\mathcal{D}$, the modified Galerkin scheme (4.63) remains conservative and satisfies the LED criterion.

Theorem 4.48. *The solution of* (4.63) *is LED.*

Proof. This is a consequence of Theorem 4.19 because $f = 0$ implies $g' = 0$, $\mathcal{M}_L$ is a diagonal matrix with positive diagonal entries, and $-(\mathcal{K}' + \mathcal{D}')$ is a Z-matrix with vanishing row sums by Theorem 4.35 and (4.46a). $\qquad\square$

Lemma 4.49. *In the case of a solenoidal velocity field, the evolution of the total mass is governed by*

$$\frac{d}{dt} \int_\Omega u_h - \int_\Omega \frac{\partial u_h}{\partial t} = -\int_{\Gamma_{in}} (\mathbf{v} \cdot \mathbf{n}) u_{in} \quad \int_{\Gamma_{out}} (\mathbf{v} \cdot \mathbf{n}) u_h. \tag{4.64}$$

Proof. According to (4.59) and (4.63), the total mass of $\partial_t u_h$ is given by

$$\int_\Omega \frac{\partial u_h}{\partial t} = \sum_{i=1}^N m_i \frac{du_i}{dt} = \sum_{i=1}^N g_i + \sum_{i,j=1}^N (k_{ij} + d_{ij}) u_j$$

$$= \sum_{i=1}^N g_i + \sum_{i,j=1}^N k_{ij} u_j = \int_{\partial\Omega} (\mathbf{v} \cdot \mathbf{n})_- u_{in} - \sum_{j=1}^N \int_{\Gamma_{out}} (\mathbf{v} \cdot \mathbf{n}) \varphi_j u_j$$

$$= -\int_{\Gamma_{in}} (\mathbf{v} \cdot \mathbf{n}) u_{in} - \int_{\Gamma_{out}} (\mathbf{v} \cdot \mathbf{n}) u_h$$

by virtue of (3.11c) and (4.48). $\qquad\square$

The application of the θ-scheme to the semi-discrete problem (4.63) with weakly imposed boundary conditions leads to

$$\mathcal{A}u^{n+1} + \mathcal{B}u^n = \underline{g}, \tag{4.65a}$$

$$\mathcal{A} := \mathcal{M} - \Delta t\theta\mathcal{K}^{n+1}, \qquad \underline{g} := \Delta t\big(\theta g^{n+1} + (1-\theta)g^n\big),$$
$$\mathcal{B} := -\mathcal{M} - \Delta t(1-\theta)\mathcal{K}^n. \tag{4.65b}$$

In the case of strongly imposed boundary conditions, the corresponding entries of $\mathcal{A}$, $\mathcal{B}$, and $\underline{g}$ are modified following the methodology presented in Section 3.2.3.

Lemma 4.50. *The system matrix $\mathcal{A}$ defined in (4.65) is a strictly diagonally dominant L-matrix and, hence, a non-singular M-matrix.*

Proof. The row sums of $\mathcal{K}+\mathcal{D}$ are nonpositive by (4.46) while all off-diagonal entries are nonnegative. By definition of $\mathcal{M}_\mathrm{L}$, all off-diagonal entries of $\mathcal{A}$ are nonpositive and the diagonal entries are positive. Hence we have

$$|a_{ii}| - \sum_{j\neq i}|a_{ij}| = a_{ii} + \sum_{j\neq i}a_{ij} = m_i - \Delta t\theta\sum_{j=1}^{N}(k_{ij}^{n+1} + d_{ij}^{n+1})$$

$$= m_i + \Delta t\theta\int_{\partial\Omega}(\mathbf{v}\cdot\mathbf{n})_-\varphi_i > 0 \qquad \forall i \in \{1,\ldots,N\}$$

by virtue of (4.53) and the zero row sum property of $\mathcal{D}^{n+1}$, which completes the proof. $\qquad\square$

Next, we derive conditions under which the fully discrete scheme is guaranteed to satisfy the local DMPs. These conditions impose an upper bound on the time increment $\Delta t > 0$ for $\theta \in [0,1)$. For the backward Euler method ($\theta = 1$), arbitrary time steps are allowed.

Theorem 4.51. *The solution u_h^{n+1} of (4.65) possesses the local DMP property if Δt is chosen so that*

$$b_{ii} = -m_i - \Delta t(1-\theta)(k_{ii}^n + d_{ii}^n) \leqslant 0 \tag{4.66}$$

holds for every $i \in \{1,\ldots,M\}$.

Proof. By Theorem 4.22, the local DMP holds if $\mathcal{A}'$ is an L-matrix, $\mathcal{B}'$ is nonpositive and the row sums of $\mathcal{A}' + \mathcal{B}'$ vanish due to $\underline{g}' = 0$. The first condition was shown in Lemma 4.50. The last one is satisfied since $(\mathcal{K}^n)'$, $(\mathcal{K}^{n+1})'$, $\mathcal{D}^n$, and $\mathcal{D}^{n+1}$ have zero row sums. Indeed, we have

$$\sum_{j=1}^{N}(a_{ij}+b_{ij}) = m_i - \Delta t\theta\sum_{j=1}^{N}(k_{ij}^{n+1}+d_{ij}^{n+1})$$

$$-m_i - \Delta t(1-\theta)\sum_{j=1}^{N}(k_{ij}^n+d_{ij}^n) = 0 \qquad \forall i \in \{1,\dots,M\}.$$

$$(4.67)$$

The artificial diffusion operator $\mathcal{D}^n$ ensures that $(-\mathcal{K}^n-\mathcal{D}^n)'$ is a Z-matrix. Hence, the off-diagonal entries of $\mathcal{B}'$ are nonpositive, as desired. It follows that the local DMP holds because the diagonal entries of $\mathcal{B}'$ are nonpositive by condition (4.66). □

By Lemma 4.50 and Theorem 4.51, all assumptions of Theorem 4.25 hold and the solution of (4.65) enjoys the global DMP property, too. Furthermore, global Dirichlet bounds can be derived as for the steady state problem in Theorem 4.39.

Theorem 4.52. *The solution of* (4.65) *is bounded by the extrema of the initial data u_0 and of the boundary condition u_{in} if the lumped L^2-projection is used to calculate the initial condition $u^0 \in \mathbb{R}^N$ and if Δt is chosen such that* (4.66) *holds for all $i \in \{1,\dots,N\}$.*

Proof. According to Theorem 4.46, the discrete initial condition u_h^0 and, hence, the corresponding vector of nodal values u^0 are bounded by the maxima and minima of the exact initial condition u_0. To complete the proof, it suffices to show that u^1 is bounded by the extrema of u^0 and those of the boundary condition u_{in} on the interval $[t^0, t^1]$. Then the statement of the theorem follows by induction.

By Lemmas 4.24 and 4.50, the solution is bounded above by

$$\max_{i\in\{1,\dots,N\}} u_i^1 \leqslant \max\Big(\max_{i\in\{1,\dots,N\}} u_i^0, \max_{i\in\{M+1,\dots,N\}} g_i\underline{p}_i^{-1}\Big)$$

if $\mathcal{B}$ is nonpositive and

$$\underline{p}_i = \sum_{j=1}^{N}(a_{ij}+b_{ij}) = 0, \quad g_i = 0 \qquad \forall i \in \{1,\dots,M\}, \tag{4.68a}$$

$$\underline{p}_i = \sum_{j=1}^{N}(a_{ij}+b_{ij}) > 0, \qquad\qquad \forall i \in \{M+1,\dots,N\}. \tag{4.68b}$$

The validity of condition (4.66) for all $i \in \{1, \ldots, N\}$ guarantees the non-positivity of $\mathcal{B}$. Condition (4.68a) holds by (4.67) and the assumption that $f = 0$. The validity of (4.68b) can be shown as follows:

$$
\begin{aligned}
\underline{p}_i &= -\Delta t \theta \sum_{j=1}^{N} (k_{ij}^1 + d_{ij}^1) - \Delta t (1 - \theta) \sum_{j=1}^{N} (k_{ij}^0 + d_{ij}^0) \\
&= -\Delta t \theta \sum_{j=1}^{N} k_{ij}^1 - \Delta t (1 - \theta) \sum_{j=1}^{N} k_{ij}^0 \\
&= \Delta t \theta \int_{\partial \Omega} (\mathbf{v}^1 \cdot \mathbf{n})_- \varphi_i + \Delta t (1 - \theta) \int_{\partial \Omega} (\mathbf{v}^0 \cdot \mathbf{n})_- \varphi_i > 0 \\
&\hspace{7cm} \forall i \in \{M + 1, \ldots, N\}.
\end{aligned}
$$

The entries of the right hand side vector g associated with the Dirichlet nodes satisfy

$$
\begin{aligned}
\underline{g}_i &= \Delta t \theta g_i^1 + \Delta t (1 - \theta) g_i^0 \\
&\leqslant \Delta t \theta \big(\max_{\mathbf{s} \in \Gamma_{\mathrm{in}}} u_{\mathrm{in}}(\mathbf{s}, t^1) \big) \int_{\partial \Omega} (\mathbf{v}^1 \cdot \mathbf{n})_- \varphi_i \\
&\quad + \Delta t (1 - \theta) \big(\max_{\mathbf{s} \in \Gamma_{\mathrm{in}}} u_{\mathrm{in}}(\mathbf{s}, t^0) \big) \int_{\partial \Omega} (\mathbf{v}^0 \cdot \mathbf{n})_- \varphi_i \\
&\leqslant \max \Big(\max_{\mathbf{s} \in \Gamma_{\mathrm{in}}} u_{\mathrm{in}}(\mathbf{s}, t^0), \max_{\mathbf{s} \in \Gamma_{\mathrm{in}}} u_{\mathrm{in}}(\mathbf{s}, t^1) \Big) \\
&\quad \cdot \Big(\Delta t \theta \int_{\partial \Omega} (\mathbf{v}^1 \cdot \mathbf{n})_- \varphi_i + \Delta t (1 - \theta) \int_{\partial \Omega} (\mathbf{v}^0 \cdot \mathbf{n})_- \varphi_i \Big) \\
&\hspace{7cm} \forall i \in \{M + 1, \ldots, N\}
\end{aligned}
$$

according to (4.54). This completes the proof of the upper bound and a similar argument can be used to show the lower bound. $\qquad\square$

4.4 Fractional step approach

The low order methods presented in the previous section preserve physically relevant properties using algebraically defined artificial diffusion operators to enforce local and global DMPs. The resulting approximations are provably bound-preserving but very diffusive and too inaccurate for practical purposes. In this section, we remove excess numerical diffusion using a predictor-corrector approach which goes back to the classical flux-corrected transport (FCT) algorithm [BB73; Boo+75; BB76; Zal79]. First extensions of FCT to continuous finite elements and unstructured grids were proposed by Parrott and Christie [PC86] and Löhner et al. [Löh+87; Löh+88], respectively.

In FCT-like methods, the computation of a low order predictor satisfying DMPs is followed by a bound-preserving high order reconstruction. The formation of overshoots and undershoots is prevented using a limiting procedure which is fully multidimensional and readily applicable on unstructured meshes. The local bounds for the antidiffusive correction step are commonly defined in terms of the low order solution. In general, fractional step limiting techniques of this kind rely on the assumption that the difference between the low order approximation and the high order target is sufficiently small.

A detailed presentation of the underlying design principles can be found in [Kuz12a].

4.4.1 Steady problem

While the FCT methodology was originally developed and is commonly used for the numerical treatment of transient problems, it is also well-suited for constraining conservative projections of data in the process of initialization or remapping [Löh08, Section 13.6; Kuz+12, Section 10]. We introduce its basic ingredients in the unconventional setting of the L^2-projection and postpone the treatment of time dependent problems to Section 4.4.2.

Following Löhner [Löh08, Section 13.6], we use FCT to constrain the Galerkin L^2-projection of a possibly nonsmooth function in this section. To that end, we perform row sum mass lumping and apply limited antidiffusive corrections to the resulting low order approximation. Even though both steps of the so defined FCT algorithm guarantee preservation of local bounds, the nodal values of the FCT projected solution are not necessarily bounded by the data in elements containing the given node. However, the validity of local maximum principles can be shown for extended stencils. We illustrate this behavior by a one dimensional example.

4.4.1.1 Projection operator

Let $u^H = \mathcal{M}^{-1}g$ and $u^L = \mathcal{M}_L^{-1}g$ denote the vectors of degrees of freedom corresponding to the consistent (high order, superscript H) and lumped (low order, superscript L) L^2-projections of given data into the finite element space V_h. The two solutions are related by [Kuz+12, Eq. (96)]

$$\mathcal{M}_L u^H = g + \mathcal{M}_L u^H - g = \mathcal{M}_L u^L + (\mathcal{M}_L - \mathcal{M})u^H. \tag{4.69}$$

Applying the inverse of the lumped mass matrix, we consider the FCT splitting

$$u^{\mathrm{H}} = u^{\mathrm{L}} + \mathcal{M}_{\mathrm{L}}^{-1}(\mathcal{M}_{\mathrm{L}} - \mathcal{M})u^{\mathrm{H}}$$

in which the antidiffusive correction term $\mathcal{M}_{\mathrm{L}}^{-1}(\mathcal{M}_{\mathrm{L}} - \mathcal{M})u^{\mathrm{H}}$ has to be adjusted so as to prevent formation of overshoots and undershoots while preserving the mass conservation property of the consistent and lumped L^2-projections. The resulting solution u_h can be interpreted as a diffusive correction of u_h^{H} or an antidiffusive correction of u_h^{L}. The latter interpretation is commonly used in practical implementations. The structure of the predictor-corrector algorithm to be presented below is similar to that of the classical FCT schemes for convection-dominated transport equations [BB73; Zal79].

The design of FCT algorithms begins with a decomposition of the antidiffusive term $\mathcal{M}_{\mathrm{L}}^{-1}(\mathcal{M}_{\mathrm{L}} - \mathcal{M})u^{\mathrm{H}}$ into independent contributions that can be limited in a conservative manner. In this section, we decompose the mass lumping error into edge-based fluxes as follows.

Definition 4.53 ([Kuz+12, Eq. (96)]). A conservative decomposition of $(\mathcal{M}_{\mathrm{L}} - \mathcal{M})u^{\mathrm{H}}$ into *edge-based antidiffusive fluxes* is defined by

$$f_{ij} := m_{ij}(u_i^{\mathrm{H}} - u_j^{\mathrm{H}}) = -f_{ji} \qquad \forall i, j \in \{1, \ldots, N\}, j \neq i. \qquad (4.70)$$

It is easy to verify that the so defined fluxes add up to the total antidiffusive correction (4.69) which must be applied to $\mathcal{M}_{\mathrm{L}}u^{\mathrm{L}}$ to obtain $\mathcal{M}_{\mathrm{L}}u^{\mathrm{H}}$. Due to the compact support property of the basis functions (3.7), the flux f_{ij} vanishes if nodes i and j do not belong to at least one common element. In view of the skew symmetry $f_{ij} = -f_{ji}$, the multiplication of f_{ij} and f_{ji} by a common correction factor $\alpha_{ij} = \alpha_{ji}$ does not change the total mass of the solution. For instance, the addition of $m_i^{-1}f_{ij}$ to u_i^{L} and $m_j^{-1}f_{ji}$ to u_j^{L} does not change the mass defined by (4.59). Taking advantage of this conservation property, FCT algorithms adjust the magnitudes of f_{ij} and f_{ji} using the same adaptively chosen correction factor α_{ij}. This adjustment produces the limited FCT solution u_h with degrees of freedom $u = (u_i)_{i=1}^N$ defined by

$$u_i = u_i^{\mathrm{L}} + m_i^{-1} \sum_{j \neq i} \alpha_{ij} f_{ij} \qquad \forall i \in \{1, \ldots, N\}. \qquad (4.71)$$

The finite element approximation u_h has the same mass as u_h^{H} and u_h^{L} for any set of correction factors satisfying the symmetry condition $\alpha_{ij} = \alpha_{ji}$, $j \neq i$.

The trivial choice of correction factors ($\alpha_{ij} = 0$) leads to $u_h = u_h^L$. On the other hand, the high order projection target $u_h = u_h^H$ can be recovered from u_h^L by using $\alpha_{ij} = 1$. Adopting the terminology of [Boo+75], we say that an FCT algorithm is *phoenical* if u_h^H can be resurrected in this way "like a phoenix".

To obtain a bound-preserving solution while staying sufficiently close to the target u_h^H, the correction factors α_{ij} are chosen to be as large as possible without violating upper and lower bounds that guarantee a DMP property of the antidiffusive correction step. The inequality constraints to be enforced are given by

$$u_i^{\min} \overset{!}{\leqslant} u_i \overset{!}{\leqslant} u_i^{\max} \qquad \forall i \in \{1,\dots,N\}, \tag{4.72}$$

where $u_i^{\max}$ and $u_i^{\min}$, $i \in \{1,\dots,N\}$, are bounds that are defined so that

$$u_i^{\min} \leqslant u_i^L \leqslant u_i^{\max} \qquad \forall i \in \{1,\dots,N\} \tag{4.73}$$

holds. Condition (4.73) is a prerequisite for the existence of a set of admissible correction factors and, hence, the existence of a solution.

In general, (4.72) leads to a high dimensional optimization problem, which can be quite expensive to solve. Using the worst case assumptions

$$m_i u_i^{\min} \overset{!}{\leqslant} m_i u_i^L + \sum_{j \neq i} \alpha_{ij} \min(0, f_{ij}) \leqslant m_i u_i \qquad \forall i \in \{1,\dots,N\}, \tag{4.74a}$$

$$m_i u_i^{\max} \overset{!}{\geqslant} m_i u_i^L + \sum_{j \neq i} \alpha_{ij} \max(0, f_{ij}) \geqslant m_i u_i \qquad \forall i \in \{1,\dots,N\}, \tag{4.74b}$$

the global optimization problem reduces to a set of small decoupled subproblems with box constraints [Boc+12].

The following realization of Zalesak's multidimensional flux limiter [Zal79; Kuz12a] calculates the optimal correction factors for this worst case scenario.

Algorithm 4.54.

1. *Compute the sums of positive and negative unlimited fluxes into node i*

$$P_i^+ := \sum_{j \neq i} \max(0, f_{ij}), \quad P_i^- := \sum_{j \neq i} \min(0, f_{ij}) \quad \forall i \in \{1,\dots,N\};$$

$$\tag{4.75}$$

2. *Determine the distance from the low order solution to the local maximum and minimum*

$$Q_i^+ := m_i(u_i^{\max} - u_i^L), \quad Q_i^- := m_i(u_i^{\min} - u_i^L) \quad \forall i \in \{1,\dots,N\}; \tag{4.76}$$

3. *Define bounds for correction factors such that the sums of positive and negative limited fluxes into node i do not produce undershoots/overshoots*

$$R_i^+ := \min\left(1, \frac{Q_i^+}{P_i^+}\right), \quad R_i^- := \min\left(1, \frac{Q_i^-}{P_i^-}\right) \quad \forall i \in \{1, \dots, N\}; \quad (4.77)$$

4. *Calculate edge-based correction factors satisfying* (4.74) *and the symmetry condition*

$$\alpha_{ij} := \begin{cases} \min(R_i^+, R_j^-) & : f_{ij} > 0, \\ 1 & : f_{ij} = 0, \qquad \forall i, j \in \{1, \dots, N\}, \ j \neq i. \ (4.78) \\ \min(R_i^-, R_j^+) & : f_{ij} < 0. \end{cases}$$

Inequality constraints resembling local DMPs for steady problems (cf. Definition 4.11) can be formulated using the local extrema $u_i^{\max}$ and $u_i^{\min}$ of the bound-preserving lumped mass approximation in the direct neighborhood of node i.

Definition 4.55 ([Kuz12a]). The *local bounds* for FCT constraints are defined by

$$u_i^{\max} := \max_{j \in \mathcal{N}_i} u_j^{\mathrm{L}}, \quad u_i^{\min} := \min_{j \in \mathcal{N}_i} u_j^{\mathrm{L}} \quad \forall i \in \{1, \dots, N\}. \qquad (4.79)$$

While these bounds satisfy condition (4.73) and guarantee the existence of correction factors such that (4.72) holds, the use of global bounds is also feasible and possibly appropriate for some applications.

Even though each step of the FCT algorithm is locally bound-preserving per se, the final FCT solution is not necessarily bounded by the local extrema of the function to be projected. The bounds defined by the data in common-vertex neighbors as in Theorem 4.46 can generally not be preserved due to the splitting error of the two-step limiting procedure. In the case of the bounds defined by (4.79), the estimate

$$u_h(\mathbf{x}) \leqslant \max_{i \in \mathcal{N}^e} u_i \leqslant \max_{i \in \mathcal{N}^e} u_i^{\max} = \max_{j \in \mathcal{N}_i, i \in \mathcal{N}^e} u_j^{\mathrm{L}}$$

$$\leqslant \max_{\substack{\tilde{\mathbf{x}} \in K^{\bar{e}}, \bar{e} \in \mathcal{E}_j, \\ j \in \mathcal{N}_i, i \in \mathcal{N}^e}} u(\tilde{\mathbf{x}}) = \max_{\tilde{\mathbf{x}} \in \varpi_j, j \in \mathcal{N}_i, i \in \mathcal{N}^e} u(\tilde{\mathbf{x}})$$

$$\forall \mathbf{x} \in K^e, e \in \{1, \dots, E\} \quad (4.80)$$

can be established using (3.9) and (4.61). The splitting procedure extends the local domain of dependence by one layer of nodes/elements ('$j \in \mathcal{N}_i$'

in (4.80)). Hence, preservation of local bounds is guaranteed only for extended stencils including the nearest neighbors of nearest neighbors. Obviously, preservation of global bounds follows from (4.80) as well.

The possible lack of boundedness by the data at nearest neighbors is illustrated by the following example.

Example 4.56 (Extension of Example 4.47). Figure 4.5 shows different numerical approximations of the function

$$u(x) = \begin{cases} 0 & : 0 \leqslant x < \frac{1}{2}, \\ 1 & : \frac{1}{2} \leqslant x \leqslant 1 \end{cases} \qquad (4.81)$$

using $E = 9$ elements of equal length. The test configuration and legends of projected solutions are defined in Example 4.47.

It can be seen that the FCT solution exhibits an undershoot at node $x_7 = 0.\bar{6}$ and the principle of monotonicity preservation is violated. Nevertheless, local bounds based on the values of the low order solution and the input data are satisfied. As mentioned above, the lack of the local DMP property w.r.t. compact-stencil values of the input data is caused by the fact that the solution is calculated in two steps.

Remark 4.57. The worst case assumptions made in (4.74) prevent Algorithm 4.54 from delivering the optimal correction factors. That is why the FCT solution u_h does not generally coincide with the solution of the optimization problem that produces the best approximation to the target u_h^H subject to the inequality constraints (4.72). However, the solution u_h of the worst case optimization problem should be closer to u_h^H than u_h^L because parts of the antidiffusive fluxes are already built into this approximate solution. Further improvements can be achieved using an iterative version of Algorithm 4.54. In iterative FCT schemes as presented in [Kuz12a], the FCT solution becomes the 'low order' solution at the end of each limiting cycle and the antidiffusive fluxes are replaced by those that have not been added to the solution yet. Numerical experiments show that the benefit of iterative limiting is hardly worth the effort as long as the low order scheme is consistent and a large percentage of artificial diffusion can safely be removed at the first iteration.

The FCT limiter presented so far avoids the need for solving global optimization problems by assuming the worst case scenario in which all fluxes have the same sign. Additional assumptions and estimates can be

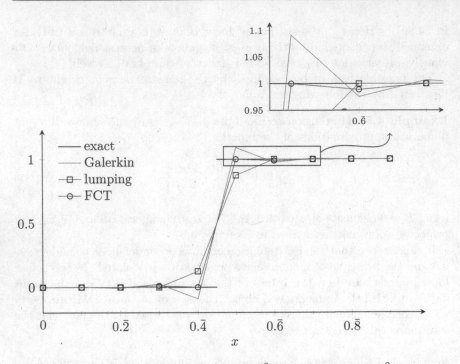

Figure 4.5: Comparison between consistent L^2-projection, lumped L^2-projection, and FCT solution using Zalesak's limiter (cf. Example 4.56).

used to make the calculation of the correction factors more efficient. For example, the decomposition (cf. (4.58))

$$m_i = \sum_{j \neq i} \tilde{m}_{ij}, \qquad \text{where} \qquad \tilde{m}_{ij} := \frac{m_{ij} m_i}{m_i - m_{ii}} \leqslant m_{ij} \quad \forall j \neq i,$$

leads to the partitioned FCT constraints (cf. [Loh17a])

$$\tilde{m}_{ij} u_i^{\min} \overset{!}{\leqslant} \tilde{m}_{ij} u_i^{\mathrm{L}} + \alpha_{ij} f_{ij} \overset{!}{\leqslant} \tilde{m}_{ij} u_i^{\max} \qquad \forall i, j \in \{1, \dots, N\}, j \neq i. \quad (4.82)$$

This limiting criterion implies (4.72) due to

$$m_i u_i^{\min} = \sum_{j \neq i} \tilde{m}_{ij} u_i^{\min} \leqslant \sum_{j \neq i} (\tilde{m}_{ij} u_i^{\mathrm{L}} + \alpha_{ij} f_{ij}) = m_i u_i \qquad \forall i \in \{1, \dots, N\},$$

$$m_i u_i^{\max} = \sum_{j \neq i} \tilde{m}_{ij} u_i^{\max} \geqslant \sum_{j \neq i} (\tilde{m}_{ij} u_i^{\mathrm{L}} + \alpha_{ij} f_{ij}) = m_i u_i \qquad \forall i \in \{1, \dots, N\}.$$

The use of the fact that (4.82) is a sufficient condition for the validity of (4.72) leads to a localized version of the edge-based FCT limiter presented in Algorithm 4.54. Instead of calculating and limiting sums of antidiffusive fluxes, the correction factors α_{ij} can be readily determined using (4.82). This simple and natural localization leads to the following algorithm which was originally proposed by the author in [Loh17a]. Closely related approaches to edge-based convex limiting were developed independently in [Gue+18].

Algorithm 4.58. *Admissible correction factors satisfying* (4.82) *can be determined by*

$$
\alpha_{ij} := \begin{cases}
\min\Big(1, \tilde{m}_{ij}(u_i^{\max} - u_i^{\mathrm{L}})|f_{ij}|^{-1}, \\
\qquad \tilde{m}_{ij}(u_j^{\mathrm{L}} - u_j^{\min})|f_{ij}|^{-1}\Big) & : f_{ij} > 0, \\
1 & : f_{ij} = 0, \\
\min\Big(1, \tilde{m}_{ij}(u_i^{\mathrm{L}} - u_i^{\min})|f_{ij}|^{-1}, \\
\qquad \tilde{m}_{ij}(u_j^{\max} - u_j^{\mathrm{L}})|f_{ij}|^{-1}\Big) & : f_{ij} < 0
\end{cases}
$$
$$\forall i, j \in \{1, \ldots, N\}, \ j \neq i. \quad (4.83)$$

If the global consistent mass matrix $\mathcal{M}$ is readily available in assembled form, the calculation of f_{ij} and α_{ij} can be performed efficiently. Another benefit of localized edge-based limiting will be exploited in Chapter 5. If the structure of a given finite element code or the nature of the problem at hand dictates the use of element-based data structures and limiting approaches, a decomposition of the antidiffusive term into element contributions may be more appropriate. An element-based localized version of the FCT-constrained L^2-projection can be readily designed using the following decomposition.

Definition 4.59. The correction term (4.69) admits a conservative decomposition into *antidiffusive element contributions* defined by

$$
f_i^e := \sum_{j=1}^{N} m_{ij}^e (u_i^{\mathrm{H}} - u_j^{\mathrm{H}}) = m_i^e u_i^{\mathrm{H}} - \sum_{j=1}^{N} m_{ij}^e u_j^{\mathrm{H}}
$$
$$\forall i \in \{1, \ldots, N\}, \ e \in \{1, \ldots, E\}, \quad (4.84)$$

where m_{ij}^e and m_i^e are the contributions of element K^e to the (lumped) mass matrix entries m_{ij} and m_i, that is,

$$m_i^e := \sum_{j=1}^{N} m_{ij}^e := \sum_{j=1}^{N} \int_{K^e} \varphi_i \varphi_j \qquad \forall i \in \{1, \dots, N\}, \, e \in \{1, \dots, E\}. \quad (4.85)$$

By definition of f_i^e, the antidiffusive element contributions satisfy

$$\sum_{i=1}^{N} f_i^e = \sum_{i,j=1}^{N} m_{ij}^e (u_i^{\mathrm{H}} - u_j^{\mathrm{H}}) = \sum_{i,j=1}^{N} m_{ij}^e u_i^{\mathrm{H}} - \sum_{i,j=1}^{N} m_{ij}^e u_j^{\mathrm{H}} = 0$$

$$\forall e \in \{1, \dots, E\}. \quad (4.86)$$

Note that in view of the compact support property, the sums reduce to those over $i, j \in \mathcal{N}^e$. The zero sum property is preserved if the same correction factor $\alpha^e \in [0, 1]$ is applied to f_i^e for all $i \in \mathcal{N}^e$, where $e \in \{1, \dots, E\}$. Therefore, the element-based FCT projection

$$u_i = u_i^{\mathrm{L}} + m_i^{-1} \sum_{e=1}^{E} \alpha^e f_i^e \qquad \forall i \in \{1, \dots, N\}$$

guarantees conservation of mass for any set of element-based correction factors. An element-based version of Zalesak's limiter for sums of antidiffusive element contributions was proposed in [Löh+87] and generalized in [KT02; KH14]. Similarly to the edge-based FCT schemes presented so far, it can be restricted to individual element contributions using the fact that $m_i = \sum_e m_i^e$ to derive the partitioned FCT constraints

$$m_i^e u_i^{\min} \overset{!}{\leqslant} m_i^e u_i^{\mathrm{L}} + \alpha^e f_i^e \overset{!}{\leqslant} m_i^e u_i^{\max} \qquad \forall i \in \{1, \dots, N\}, \, e \in \{1, \dots, E\}. \quad (4.87)$$

The corresponding localized limiter was originally proposed in [CK16]. In this work, we do not go into details of element-based FCT algorithms and refer the interested reader, e.g., to [KH14; CK16; Loh+17].

The use of the oscillatory target u^{H} in the definition of the antidiffusive terms (4.70) and (4.84) requires solution of a linear system for u^{H}. While the consistent mass matrix $\mathcal{M}$ is well-conditioned, the computation of u^{H} may be very expensive in problems more advanced than L^2-projections. In many applications of practical interest, the system matrix of the high order scheme is ill-conditioned and the high cost of its 'inversion' may become a major bottleneck to efficiency. To avoid troubles associated with the computation of u^H, the antidiffusive contributions can also be defined in terms of u^L [Kuz09]. FCT algorithms based on such linearizations are more robust and efficient, but no longer phoenical since the high order target cannot be recovered exactly even in the case when all correction factors are set to one.

4.4.1.2 Advection equation

The FCT methodology rests on the assumption that the low order solution is already a reasonable approximation of the exact solution. Local bounds are defined in terms of the low order predictor and only function values that lie in the same range are accepted by the limiting procedure. Therefore, the FCT solution becomes more accurate as the low order solution improves. However, in the case of steady hyperbolic equations, the accuracy of the low order solution is typically very poor and cannot be improved significantly using algebraic corrections of FCT type unless the reaction term plays a dominant role (as in case of the L^2-projection corresponding to the limit of a vanishing velocity field).

The application of FCT to stationary advection-dominated problems requires the use of a pseudo time stepping approach. For this purpose, the steady PDE is augmented by a pseudo time derivative and solved until the solution converges to a time independent result. The fully discrete problem to be solved at each intermediate step can be interpreted as the discretization of a steady advection reaction equation with a right hand side depending on the solution from the previous pseudo time level. If the time increment is chosen sufficiently small, the reactive part dominates and FCT produces just moderate amounts of numerical diffusion. However, convergence to steady-state solutions is difficult to achieve and the accuracy of the result depends on the artificial time step which acts as a relaxation parameter.

An in-depth description of the FCT methodology for transient problems can be found in Section 4.4.2.

4.4.1.3 Error analysis

Disappointingly, no theoretical results regarding the convergence behavior of FCT schemes have been published so far to the best knowledge of the author. The purpose of what follows is to take a first step toward filling this gap and to motivate further theoretical studies. The analysis presented in this section is restricted to linear (not multilinear) finite element functions and shows that any locally bounded finite element function converges to the exact solution if the low order approximation does and is sufficiently smooth.

We first show that the locally defined bounds employed by the FCT algorithms presented above converge to the low order solution under some assumptions regarding u_h^L.

Theorem 4.60. *Let V_h be the space of linear (not multilinear) finite elements. Assume that the finite element approximation $u_h^L \in V_h$ satisfies $|u_h^L|_{H^1(\Omega)} \leqslant C$ for some $C > 0$ independently of h. Furthermore, define $u_h^{\max}, u_h^{\min} \in V_h$ as finite element functions interpolating the nodal bounds*

$$u_h^{\max}(\mathbf{x}_i) = u_i^{\max}, \quad u_h^{\min}(\mathbf{x}_i) = u_i^{\min} \quad \forall i \in \{1, \dots, N\}, \qquad (4.88)$$

where $u^{\max} = (u_i^{\max})_{i=1}^N$ and $u^{\min} = (u_i^{\min})_{i=1}^N$ are given by Definition 4.55. Then $u_h^{\max}$ and $u_h^{\min}$ satisfy

$$\|u_h^{\max} - u_h^L\|_{L^2(\Omega)} \leqslant Ch, \qquad \|u_h^{\min} - u_h^L\|_{L^2(\Omega)} \leqslant Ch. \qquad (4.89)$$

Proof. Without loss of generality, we prove the result for $u_h^{\max}$.

The H^1-seminorm of a linear finite element approximation u_h^L on a single element K^e satisfies

$$|u_h^L|_{H^1(K^e)}^2 \geqslant C \int_{K^e} \|\nabla u_h^L\|_2^2 = C \int_{K^e} \left\||\nabla u_h^L|_{K^e}\right\|_2^2 = C|K^e| \left\||\nabla u_h^L|_{K^e}\right\|_2^2$$
$$\forall e \in \{1, \dots, E\}$$

since the gradient of u_h^L is constant on K^e. Using this result and (3.3), the difference between two neighboring degrees of freedom can be estimated by

$$|u_i^L - u_j^L| = \int_{v_{ij}} \left|\frac{\partial u_h^L}{\partial v_{ij}}\right| \leqslant \int_{v_{ij}} \left\||\nabla u_h^L|_{K^e}\right\|_2 = |v_{ij}| \left\||\nabla u_h^L|_{K^e}\right\|_2$$
$$\leqslant C|v_{ij}| |K^e|^{-\frac{1}{2}} |u_h^L|_{H^1(K^e)} \leqslant Ch^{1-\frac{d}{2}} |u_h^L|_{H^1(K^e)}$$
$$\forall i, j \in \mathcal{N}^e, e \in \{1, \dots, E\}, \quad (4.90)$$

where v_{ij} is the edge connecting nodes $\mathbf{x}_i$ and $\mathbf{x}_j$. Invoking the fact that [RW17, Lemma 4.1]

$$\sum_{i=1}^N m_i v_i^2 \leqslant C\|v_h\|_{L^2(\Omega)}^2 = C \sum_{i,j=1}^N m_{ij} v_i v_j$$

$$\leqslant C \sum_{i,j=1}^N \frac{m_{ij}}{2}(v_i^2 + v_j^2) = C \sum_{i=1}^N m_i v_i^2 \qquad \forall v_h \in V_h, \qquad (4.91)$$

we find that

$$\|u_h^{\max} - u_h^{\mathrm{L}}\|_{L^2(\Omega)}^2 \leqslant C \sum_{i=1}^{N} m_i (u_i^{\max} - u_i^{\mathrm{L}})^2 = C \sum_{i=1}^{N} m_i \big(\max_{j \in \mathcal{N}_i}(u_j^{\mathrm{L}} - u_i^{\mathrm{L}})\big)^2$$

$$\leqslant C \sum_{i=1}^{N} m_i \max_{j \in \mathcal{N}_i}(u_j^{\mathrm{L}} - u_i^{\mathrm{L}})^2 \leqslant C \sum_{e=1}^{E} \sum_{i,j \in \mathcal{N}^e} m_i(u_j^{\mathrm{L}} - u_i^{\mathrm{L}})^2$$

$$\leqslant C \sum_{e=1}^{E} \sum_{i,j \in \mathcal{N}^e} h^d h^{2-d} |u_h^{\mathrm{L}}|_{H^1(K^e)}^2 \leqslant Ch^2 |u_h^{\mathrm{L}}|_{H^1(\Omega)}^2,$$

as long as $|\mathcal{N}^e| < C$ independently of h for all $e \in \{1, \dots, E\}$. $\qquad\square$

Proposition 4.61. *Assume that all requirements of Theorem 4.60 hold and*

$$\|u - u_h^{\mathrm{L}}\|_{L^2(\Omega)} \leqslant Ch^q \tag{4.92}$$

for some $q > 0$. Then each finite element function $u_h \in V_h$ s.t. $u_h^{\min} \leqslant u_h \leqslant u_h^{\max}$ satisfies

$$\|u - u_h\|_{L^2(\Omega)} \leqslant \|u - u_h^{\mathrm{L}}\|_{L^2(\Omega)} + \|u_h - u_h^{\mathrm{L}}\|_{L^2(\Omega)} \leqslant Ch^q + Ch. \tag{4.93}$$

Proof. The statement of the proposition can be easily proved as follows:

$$\|u_h - u_h^{\mathrm{L}}\|_{L^2(\Omega)} \leqslant \big\|\max(0, u_h - u_h^{\mathrm{L}})\big\|_{L^2(\Omega)} + \big\|\min(0, u_h - u_h^{\mathrm{L}})\big\|_{L^2(\Omega)}$$

$$\leqslant \|u_h^{\max} - u_h^{\mathrm{L}}\|_{L^2(\Omega)} + \|u_h^{\min} - u_h^{\mathrm{L}}\|_{L^2(\Omega)} \leqslant Ch. \qquad\square$$

The above analysis is applicable to any FCT algorithm that guarantees the validity of (4.72) with local bounds defined by (4.79). In particular, arbitrary correction factors are admissible and the approximation u_h does not even have to be conservative, i.e., we may have $\int_\Omega u_h \neq \int_\Omega u_h^{\mathrm{L}}$. Indeed, the proof of convergence relies heavily on the definition of the squeezing functions $u_h^{\max}$ and $u_h^{\min}$ converging to the low order approximation u_h^{L}.

Remark 4.62. In Theorem 4.72 of Section 4.5.1.3, we will see that requirement (4.92) of Proposition 4.61 is satisfied for $q = \frac{1}{2}$ if the exact solution of the steady advection-reaction equation is sufficiently smooth and conditions (2.11) and (2.14) are satisfied. The same theorem guarantees that the estimates of (4.89) hold with order $\frac{1}{2}$ without the assumption that $|u_h^{\mathrm{L}}|_{H^1(\Omega)}$ is bounded independently of h (cf. Remark 4.73). Therefore, any bound-preserving correction of the low order approximation u_h^{L} does not degrade the rate of convergence.

4.4.2 Unsteady problem

In the case of the unsteady advection equation, the high order target defined by (3.21) can be recovered by adding an antidiffusive correction to the bound-preserving low order solution defined by (4.65). For limiting purposes, the correction terms must be decomposed into fluxes or element contributions. To find a suitable conservative decomposition, we consider the following representation of the difference between the high and low order solutions:

$$
\begin{aligned}
\mathcal{M}_L u^{H,n+1} &= \underline{g} + \mathcal{M}_L u^{H,n+1} - \underline{g} \\
&= (\mathcal{M}_L - \Delta t\theta(\mathcal{K}^{n+1} + \mathcal{D}^{n+1}))u^{L,n+1} \\
&\quad - (\mathcal{M}_L + \Delta t(1-\theta)(\mathcal{K}^n + \mathcal{D}^n))u^n \\
&\quad + \mathcal{M}_L u^{H,n+1} - (\mathcal{M} - \Delta t\theta\mathcal{K}^{n+1})u^{H,n+1} \\
&\quad + (\mathcal{M} + \Delta t(1-\theta)\mathcal{K}^n)u^n \\
&= \mathcal{M}_L u^{L,n+1} - \Delta t\theta(\mathcal{K}^{n+1} + \mathcal{D}^{n+1})u^{L,n+1} \\
&\quad + (\mathcal{M}_L - \mathcal{M} + \Delta t\theta\mathcal{K}^{n+1})u^{H,n+1} \\
&\quad - (\mathcal{M}_L - \mathcal{M} + \Delta t(1-\theta)\mathcal{D}^n)u^n \\
&= \mathcal{M}_L u^{L,n+1} - \Delta t\theta(\mathcal{K}^{n+1} + \mathcal{D}^{n+1})u^{L,n+1} \\
&\quad + \Delta t\theta\mathcal{K}^{n+1}u^{H,n+1} - \Delta t(1-\theta)\mathcal{D}^n u^n \\
&\quad + (\mathcal{M}_L - \mathcal{M})(u^{H,n+1} - u^n).
\end{aligned}
\tag{4.94}
$$

The addition of matrix-vector products involving discrete diffusion operators like $\mathcal{M}_L - \mathcal{M}$ and $\mathcal{D}$ does not change the total mass. Hence, any difference between the integrals of $u_h^{H,n+1}$ and $u_h^{L,n+1}$ is caused by the contribution of $\mathcal{K}^{n+1}(u^{H,n+1} - u^{L,n+1})$. Recalling that the entries of $\mathcal{K}$ are defined by

$$
\begin{aligned}
k_{ij} = -a(\varphi_j, \varphi_i) &= -\int_\Omega c\varphi_j\varphi_i - \int_\Omega \mathbf{v}\cdot\mathrm{grad}(\varphi_j)\varphi_i - \int_{\partial\Omega}(\mathbf{v}\cdot\mathbf{n})_-\varphi_j\varphi_i \\
&= -\int_\Omega c\varphi_j\varphi_i + \int_\Omega \mathrm{div}(\mathbf{v}\varphi_i)\varphi_j - \int_{\partial\Omega}\mathbf{v}\cdot\mathbf{n}\varphi_i\varphi_j + \int_{\Gamma_{in}}\mathbf{v}\cdot\mathbf{n}\varphi_j\varphi_i \\
&= \int_\Omega (\mathrm{div}(\mathbf{v}) - c)\varphi_j\varphi_i + \int_\Omega \mathbf{v}\cdot\mathrm{grad}(\varphi_i)\varphi_j - \int_{\Gamma_{out}}\mathbf{v}\cdot\mathbf{n}\varphi_i\varphi_j,
\end{aligned}
$$

we can decompose $\mathcal{K}$ into the sum of element matrices $\mathcal{K}^e = (k_{ij}^e)_{i,j=1}^N$, $e \in \{1, \ldots, E\}$, whose entries are given by

$$k_{ij}^e = \int_{K^e} (\mathrm{div}(\mathbf{v}) - c)\varphi_j \varphi_i + \int_{K^e} \mathbf{v} \cdot \mathrm{grad}(\varphi_i)\varphi_j - \int_{\partial K^e \cap \Gamma_{\mathrm{out}}} \mathbf{v} \cdot \mathbf{n}\varphi_i \varphi_j$$
$$\forall i, j \in \{1, \ldots, N\}, \, e \in \{1, \ldots, E\}.$$

In view of the fact that the Lagrange basis functions form a partition of unity, the matrix $\mathcal{K}^e$ has zero column sums if

$$\mathrm{div}(\mathbf{v}(\mathbf{x})) - c(\mathbf{x}) = 0 \qquad \forall \mathbf{x} \in K^e,$$
$$\mathbf{v}(\mathbf{s}) \cdot \mathbf{n}(\mathbf{s}) = 0 \qquad \forall \mathbf{s} \in \partial K^e \cap \Gamma_{\mathrm{out}}.$$

In this case, the vector $f^e \in \mathbb{R}^N$ of antidiffusive element contributions

$$f^e := (\mathcal{M}_{\mathrm{L}}^e - \mathcal{M}^e)(u^{\mathrm{H},n+1} - u^n) - \Delta t\theta(\mathcal{K}^{e,n+1} + \mathcal{D}^{e,n+1})u^{\mathrm{L},n+1}$$
$$+ \Delta t\theta\mathcal{K}^{e,n+1}u^{\mathrm{H},n+1} - \Delta t(1-\theta)\mathcal{D}^{e,n}u^n \in \mathbb{R}^N \qquad \forall e \in \{1, \ldots, E\}$$
$$(4.95)$$

has zero net mass, i.e., $\sum_i f_i^e = 0$. Hence, the multiplication of f^e by a correction factor α^e preserves the total mass in the context of an element-based FCT formulation. If the assumption regarding boundary fluxes is not valid, the use of the antidiffusive element contributions f^e defined by (4.95) may cause a lack of conservation. However, fluxes across Γ_{out} are physically admissible and f^e can still be used.

In the edge-based version of FCT, matrix vector products of the form $\mathcal{D}u$ and $(\mathcal{M}_{\mathrm{L}} - \mathcal{M})u$ admit a simple and natural decomposition into antidiffusive fluxes [Kuz10]. Products of the form $\mathcal{K}u$ can also be decomposed into such fluxes exploiting the equivalence of low order continuous finite elements and vertex centered finite volume schemes [Sel93; SF96]. However, the use of such decompositions for $\mathcal{K}u$ requires implementation of edge-based data structures [Löh08; Kuz10].

Phoenical FCT algorithms in which the antidiffusive terms involve only products with discrete diffusion operators $\mathcal{D}$ and $\mathcal{M}_{\mathrm{L}} - \mathcal{M}$ can be designed using deferred correction splittings [KT02; Kuz12a]. In implicit FCT schemes of this kind, limited antidiffusive fluxes are inserted into the right-hand side of the low order system instead of calculating u^{L} and correcting it explicitly.

To keep things simple, we give up the phoenical property and define the high order target $u^{\mathrm{H},n+1}$ using the linearized FCT splitting

$$\mathcal{M}_{\mathrm{L}}u^{\mathrm{H},n+1} = \mathcal{M}_{\mathrm{L}}u^{\mathrm{L},n+1} - \Delta t\theta\mathcal{D}^{n+1}u^{\mathrm{L},n+1}$$
$$- \Delta t(1-\theta)\mathcal{D}^n u^n + \Delta t(\mathcal{M}_{\mathrm{L}} - \mathcal{M})\dot{u}^{\mathrm{L}} \qquad (4.96)$$

in which the antidiffusive part is defined in terms of the low order predictor, as proposed in [Kuz09], and $\dot{u}^L := \Delta t^{-1}(u^{L,n+1} - u^n)$ approximates the time derivative of the degrees of freedom. The corresponding decomposition of $\mathcal{M}_L(u^{H,n+1} - u^{L,n+1})$ into edge-based antidiffusive fluxes is defined by

$$f_{ij} = \Delta t m_{ij}(\dot{u}_i^L - \dot{u}_j^L) - \Delta t \theta d_{ij}^{n+1}(u_j^{L,n+1} - u_i^{L,n+1})$$
$$- \Delta t(1-\theta)d_{n+1}^n(u_j^n - u_i^n) \qquad \forall i,j \in \{1,\dots,N\} \qquad (4.97)$$

and bound-preserving antidiffusive corrections of the low order solution can be performed as in Section 4.4.1. In applications to time dependent problems, the numerical solution is advanced from one step to the next using the following linearized FCT algorithm (cf. [Kuz09; Kuz12a]).

Algorithm 4.63. *Given a vector of degrees of freedom u^n at the time level t^n, update it as follows:*

1. *Calculate the low order solution $u^{L,n+1}$ defined by (4.65);*

2. *Compute the local bounds $u_i^{\max,n+1}$ and $u_i^{\min,n+1}$ using (4.79);*

3. *Compute the raw antidiffusive fluxes given by (4.97);*

4. *Use a limiting procedure like Algorithm 4.54 or Algorithm 4.58 to define the edge-based correction factors α_{ij};*

5. *Correct $u^{L,n+1}$ using (4.71) to obtain the degrees of freedom u^{n+1} corresponding to the bound-preserving FCT solution u_h^{n+1}.*

The edge-based decomposition (4.97) exploits the fact that the same time integrator is applied to the semi-discrete high and low order scheme and, hence, convection matrices do not contribute to (4.96). To avoid time step restrictions while achieving second order temporal accuracy, an FCT algorithm may be configured to use $\theta = 1$ for the low order solution and $\theta = \frac{1}{2}$ for the (linearized) high order target. However, the antidiffusive term corresponding to this combination cannot be expressed in terms of matrix-vector multiplications involving only discrete diffusion operators even in the linearized version. If the use of edge-based data structures for convective fluxes is not an option, decompositions into element contributions and element-based limiting procedures can still be used to design FCT algorithms with desired properties. Examples of such element-based FCT schemes can be found in [KH14; Loh+17].

4.5 Monolithic approach

So far, we presented a fractional step limiting strategy based on a lagged antidiffusive correction of a low order predictor. In the process of this correction, the artificial diffusion that was previously added to guarantee DMPs is partially removed leading to improved approximations that stay bounded by local extrema of the low order solution. For reasons explained above, the presented methodology is guaranteed to satisfy local DMPs only with respect to extended stencils. Although such a predictor-corrector technique can be used to solve unsteady transport problems using implicit time integrators with very large time increments, the resulting solutions are likely to be very diffusive and may exhibit unphysical artifacts.

In this section, we introduce an alternative limiting strategy which incorporates the antidiffusive correction into the residual of a nonlinear problem. This approach avoids splitting errors and convergence problems associated with the use of low order solutions in predictor-corrector algorithms. In the context of steady problems, the design of limiter functions is backed by theoretical studies leading to proofs of well-posedness and a priori error estimates. Furthermore, the validity of DMPs, as defined in Section 4.2, can be shown for the steady and unsteady advection-reaction equation.

Theoretical foundations for the development of the limiting techniques to be presented in this section were laid in [Bar+16], where rigorous analysis of the discrete upwind method and of the algebraic flux correction (AFC) scheme proposed in [Kuz07] was performed for stationary singularly perturbed advection-diffusion-reaction equations. Improved limiting approaches based on the design philosophy of [Kuz07] and [Bar+16] were recently developed, e.g., in [BB17; Kuz+17; Bar+17a; Bar+17b]. In this section, we review the recent advances and perform further analysis of AFC schemes for hyperbolic equations.

4.5.1 Steady problem

Consider a linear system of equations corresponding to the discretization of a steady boundary value problem like (3.11) and its low order counterpart given by (4.50). In contrast to the FCT methodology, the key idea behind the approach that we pursue in this section is to limit the artificial diffusion operator before adding it to the residual of the target scheme. We call limiting techniques of this kind *monolithic* because they produce bound-preserving approximations in one step without calculating and correcting auxiliary solutions. The local bounds and correction factors are defined in terms of the unknown degrees of freedom. The monolithic approach to edge-based flux correction for our steady model problem leads to the nonlinear system of equations

$$\sum_{j=1}^{N} a_{ij} u_j - \sum_{j \neq i} (1 - \alpha_{ij}) d_{ij} (u_j - u_i) = g_i \qquad \forall i \in \{1, \ldots, N\}, \qquad (4.98)$$

where $\alpha_{ij} = \alpha_{ji} \in [0, 1]$, $i, j \in \{1, \ldots, N\}$, $j \neq i$, are correction factors which can be used to adjust the amount of artificial diffusion. As in the FCT context, they should be chosen as close to 1 as possible without violating properly defined DMP constraints. However, the dependence of local bounds on the unknown solution $u \in \mathbb{R}^N$ in monolithic limiting approaches makes system (4.98) nonlinear even in the case of a linear governing equation. The coefficients of the solution-dependent matrix $\tilde{\mathcal{D}} = (\tilde{d}_{ij})_{i,j=1}^{N}$, which is supposed to correct the artificial diffusion operator $\mathcal{D}$, are defined by

$$\tilde{d}_{ij} := \begin{cases} \alpha_{ij} d_{ij} & : j \neq i, \\ -\sum_{l \neq i} \alpha_{il} d_{il} & : j = i \end{cases} \qquad \forall i, j \in \{1, \ldots, N\}. \qquad (4.99)$$

Using this matrix notation, the AFC system (4.98) can be written as

$$\hat{\mathcal{A}} u := (\mathcal{A} - \mathcal{D} + \tilde{\mathcal{D}}) u = g. \qquad (4.100)$$

As in the continuous case, we assume that the coercivity condition (2.14) is satisfied. This implies the positive definiteness of the system matrix $\mathcal{A}$ (cf. (3.12)).

4.5.1.1 Well-posedness

In contrast to the predictor-corrector approach considered in Section 4.4, the above monolithic limiting strategy requires iterative solution of a nonlinear system. Moreover, it is essential to guarantee that a solution to (4.98) exists and is unique. Therefore, we need to clarify under which assumptions this will be the case. After formulating sufficient conditions for solvability and uniqueness of the solution to the nonlinear discrete problem, we will proceed to the derivation of requirements for the validity of local and global DMPs.

First results on the solvability of (4.98) were obtained by Barrenechea, John, and Knobloch [Bar+15] in one dimension. However, the analysis presented in this paper was performed without assuming the symmetry of the correction factors. It turned out that limiting techniques violating the condition $\alpha_{ij} = \alpha_{ji}$ for $j \neq i$ may cause a lack of conservation and prevent the existence of a unique solution. Taking the symmetry condition into account, proofs of existence and a priori error estimates for advection-diffusion-reaction equations were obtained in [Bar+16]. The most important theoretical results adapted to hyperbolic problems are summarized in this section.

A cornerstone of the theoretical framework developed in [Bar+16] and a key ingredient of our proofs is the following consequence of Brouwer's fixed-point theorem, which guarantees the existence of a solution for coercive and 'sufficiently smooth' nonlinear approximations.

Lemma 4.64 ([Tem77, Lemma 1.4, Chapter II]). *Let X be a finite dimensional Hilbert space with inner product $(\cdot, \cdot)_X$ and norm $\| \cdot \|_X$. Let $\mathscr{L} : X \to X$ be a continuous mapping and $c_1 > 0$ a real number such that*

$$(\mathscr{L}x, x)_X > 0 \qquad \forall x \in X \ s.t. \ \|x\|_X = c_1. \tag{4.101}$$

Then there exists at least one $x \in X$ such that $\|x\|_X \leqslant c_1$ and $\mathscr{L}x = 0$.

Proof. The proof can be found in [Tem77, Chapter II]. □

This lemma yields a sufficient condition for the existence of a solution to (4.98).

Theorem 4.65. *The AFC system* (4.98) *possesses at least one solution $u \in \mathbb{R}^N$ with $\|u\|_2 \leqslant C_M^{-1}\|g\|_2$ for $C_M > 0$ defined by (3.12) if*

$$\mathbb{R}^N \ni v \ \mapsto \ \alpha_{ij}(v)(v_i - v_j) \in \mathbb{R} \quad \text{is continuous}$$
$$\forall i, j \in \{1, \dots, N\}, \ j \neq i. \tag{4.102}$$

Proof. The proof follows the proof of [Bar+16, Theorem 3.3].

For our AFC problem (4.98), the operator $\mathscr{L} : \mathbb{R}^N \to \mathbb{R}^N$ of Lemma 4.64 is defined by

$$\mathscr{L}v := (\mathcal{A} - \mathcal{D} + \tilde{\mathcal{D}})v - g \qquad \forall v \in \mathbb{R}^N,$$

where $\tilde{\mathcal{D}}$ depends on v by definition of the correction factors, and $(\cdot,\cdot)_X$ is the Euclidean scalar product of two vectors in $\mathbb{R}^N$. Condition (4.101) holds due to [Bar+16]

$$(\mathscr{L}v,v)_2 = \sum_{i,j=1}^N v_i a_{ij} v_j - \sum_{i,j,\,j\neq i} v_i(1-\alpha_{ij})d_{ij}(v_j - v_i) - \sum_{i=1}^N v_i g_i$$

$$\geqslant C_M\|v\|_2^2 + \sum_{i,j,\,j<i}(1-\alpha_{ij})d_{ij}(v_j-v_i)^2 - c_2\|v\|_2^2 - (4c_2)^{-1}\|g\|_2^2$$

$$\geqslant C_M\|v\|_2^2 - c_2\|v\|_2^2 - (4c_2)^{-1}\|g\|_2^2 = \frac{C_M}{2}\|v\|_2^2 - (2C_M)^{-1}\|g\|_2^2 > 0$$

$$\forall v \in \mathbb{R}^N \text{ s.t. } \|v\|_2^2 = C_M^{-2}\|g\|_2^2 + \varepsilon$$

and all $\varepsilon > 0$ by virtue of the Cauchy-Schwarz inequality, Young's inequality, and (3.12). The involved constant is given by $c_2 = \frac{C_M}{2}$. The desired result follows by Lemma 4.64 and the assumption of the theorem. $\qquad\square$

The uniqueness of the solution is not as easy to guarantee. Barrenechea, Burman, and Karakatsani [Bar+17a] focused on this issue in the context of the advection-diffusion-reaction equation and an edge-based nonlinear stabilization technique. A proof of uniqueness was obtained under the assumption that the diffusion coefficient is sufficiently large. However, the extension of this analysis to AFC discretizations of hyperbolic problems seems to be problematic.

Inspired by [Abg17, Proposition 4.3], we prove the following proposition, which can be exploited to show the uniqueness of an AFC solution under some assumptions regarding the reaction rate c and its numerical treatment.

Proposition 4.66. *Let* $\|\cdot\| : \mathbb{R}^N \to \mathbb{R}_0^+$ *be an arbitrary norm and* $\mathscr{L}_L, \mathscr{L} : \mathbb{R}^N \to \mathbb{R}^N$ *be two operators such that*

1. *There exists at least one solution* $u \in \mathbb{R}^N$ *such that* $\mathscr{L}u = 0$*;*

2. *For every* $v \in \mathbb{R}^N$ *there exists exactly one* $u \in \mathbb{R}^N$ *such that* $\mathscr{L}_L u = v$*;*

3. *There exists a constant* $c_3 > 0$ *such that*

$$c_3\|v-\bar{v}\| \leqslant \|\mathscr{L}_L v - \mathscr{L}_L \bar{v}\| \qquad \forall v,\bar{v} \in \mathbb{R}^N; \qquad (4.103)$$

4. $\mathscr{L}_L - \mathscr{L}$ *is Lipschitz continuous with Lipschitz constant* $c_4 \in (0,c_3)$*, i.e.,*

$$\left\|(\mathscr{L}_L v - \mathscr{L}v) - (\mathscr{L}_L \bar{v} - \mathscr{L}\bar{v})\right\| \leqslant c_4\|v-\bar{v}\| \qquad \forall v,\bar{v} \in \mathbb{R}^N. \quad (4.104)$$

Then the solution $u \in \mathbb{R}^N$ to $\mathscr{L}u = 0$ is unique and the defect correction

$$\mathscr{L}_{\mathrm{L}} u^{i+1} = \mathscr{L}_{\mathrm{L}} u^i - \mathscr{L} u^i \qquad \forall i = 0, 1, 2, \ldots \qquad (4.105)$$

converges to the solution u for every initial condition $u^0 \in \mathbb{R}^N$ at the rate $\frac{c_4}{c_3} < 1$, i.e.,

$$\|u^{i+1} - u\| \leqslant \tfrac{c_4}{c_3} \|u^i - u\| \leqslant \left(\tfrac{c_4}{c_3}\right)^{i+1} \|u^0 - u\|. \qquad (4.106)$$

Proof. Assume that $u, \bar{u} \in \mathbb{R}^N$ are two solutions satisfying $\mathscr{L}u = \mathscr{L}\bar{u} = 0$. Then we deduce that

$$c_3 \|u - \bar{u}\| \leqslant \|\mathscr{L}_{\mathrm{L}} u - \mathscr{L}_{\mathrm{L}} \bar{u}\| = \left\|(\mathscr{L}_{\mathrm{L}} u - \mathscr{L} u) - (\mathscr{L}_{\mathrm{L}} \bar{u} - \mathscr{L} \bar{u})\right\| \leqslant c_4 \|u - \bar{u}\|$$

and the solutions coincide due to $c_4 < c_3$.

The solution obtained at the i-th cycle of the fixed point iteration satisfies

$$\mathscr{L}_{\mathrm{L}} u^{i+1} - \mathscr{L}_{\mathrm{L}} u = (\mathscr{L}_{\mathrm{L}} u^i - \mathscr{L} u^i) - (\mathscr{L}_{\mathrm{L}} u - \mathscr{L} u)$$

due to $\mathscr{L}u = 0$. This leads to the error estimate

$$\begin{aligned} c_3 \|u^{i+1} - u\| &\leqslant \|\mathscr{L}_{\mathrm{L}} u^{i+1} - \mathscr{L}_{\mathrm{L}} u\| \\ &= \left\|(\mathscr{L}_{\mathrm{L}} u^i - \mathscr{L} u^i) - (\mathscr{L}_{\mathrm{L}} u - \mathscr{L} u)\right\| \leqslant c_4 \|u^i - u\| \end{aligned}$$

and $u^i \to u$ for $i \to \infty$ at the rate $\frac{c_4}{c_3} < 1$ for every initial condition $u^0 \in \mathbb{R}^N$. $\qquad \square$

The proposition ensures the invertibility of an operator if it is close enough to an invertible (low order) approximation and suggests an iterative procedure to calculate the solution. This result can be used to derive sufficient conditions for $\mathcal{A}$ and $\mathcal{D}$ to guarantee the existence of a unique solution to the nonlinear AFC scheme.

Lemma 4.67. *The AFC system (4.98) possesses a unique solution if*

$$\left|\alpha_{ij}(v)(v_i - v_j) - \alpha_{ij}(\bar{v})(\bar{v}_i - \bar{v}_j)\right| \leqslant L_{\mathrm{li}} \|v - \bar{v}\|_\infty$$
$$\forall i, j \in \{1, \ldots, N\}, \ j \neq i, \ v, \bar{v} \in \mathbb{R}^N \quad (4.107)$$

holds for some $L_{\mathrm{li}} > 0$ which satisfies

$$L_{\mathrm{li}} \max_{i \in \{1, \ldots, N\}} |d_{ii}| < \min_{i \in \{1, \ldots, N\}} \sum_{j=1}^{N} a_{ij}. \qquad (4.108)$$

Proof. To exploit the result of Proposition 4.66, the mappings $\mathscr{L}_L$ and $\mathscr{L}$ are defined by

$$\mathscr{L}_L v := (\mathcal{A} - \mathcal{D})v - g, \quad \mathscr{L}v := (\mathcal{A} - \mathcal{D} + \tilde{\mathcal{D}})v - g \quad \forall v \in \mathbb{R}^N.$$

The first assumption of Proposition 4.66 holds due to Theorem 4.65, where condition (4.102) follows from (4.107). The operator $\mathscr{L}_L$ is invertible due to the positive definiteness of the (low order) system matrix $\mathcal{A} - \mathcal{D}$ (cf. Remark 4.37). To verify the third assumption of Proposition 4.66, consider arbitrary vectors $v, \bar{v} \in \mathbb{R}^N$ and choose the index $l \in \{1, \dots, N\}$ such that $|v_l - \bar{v}_l| = \|v - \bar{v}\|_\infty$. Then the Z-matrix property of $\mathcal{A} - \mathcal{D}$ (cf. Theorem 4.35) yields

$$\max_{i \in \{1,\dots,N\}} \sum_{j=1}^N (a_{ij} - d_{ij})|v_j - \bar{v}_j| \geqslant \sum_{j=1}^N (a_{lj} - d_{lj})|v_j - \bar{v}_j|$$

$$\geqslant |v_l - \bar{v}_l| \sum_{j=1}^N (a_{lj} - d_{lj}) \geqslant \|v - \bar{v}\|_\infty \min_{i \in \{1,\dots,N\}} \sum_{j=1}^N a_{ij}.$$

According to the L-matrix property of $\mathcal{A} - \mathcal{D}$ (cf. Remark 4.37), this leads to

$$\|\mathscr{L}_L v - \mathscr{L}_L \bar{v}\|_\infty = \max_{i \in \{1,\dots,N\}} \left| \sum_{j=1}^N (a_{ij} - d_{ij})(v_j - \bar{v}_j) \right|$$

$$\geqslant \max_{i \in \{1,\dots,N\}} \left(|(a_{ii} - d_{ii})(v_i - \bar{v}_i)| - \sum_{j \neq i} |(a_{ij} - d_{ij})(v_j - \bar{v}_j)| \right)$$

$$= \max_{i \in \{1,\dots,N\}} \sum_{j=1}^N (a_{ij} - d_{ij})|v_j - \bar{v}_j| \geqslant \left(\min_{i \in \{1,\dots,N\}} \sum_{j=1}^N a_{ij} \right) \|v - \bar{v}\|_\infty$$

$$\forall v, \bar{v} \in \mathbb{R}^N$$

by the triangle inequality. Finally, we have

$$\left\| (\mathscr{L}_L v - \mathscr{L}v) - (\mathscr{L}_L \bar{v} - \mathscr{L}\bar{v}) \right\|_\infty = \left\| \tilde{\mathcal{D}}(v)v - \tilde{\mathcal{D}}(\bar{v})\bar{v} \right\|_\infty$$

$$= \max_{i \in \{1,\dots,N\}} \left| \sum_{j \neq i} d_{ij} \big(\alpha_{ij}(v)(v_j - v_i) - \alpha_{ij}(\bar{v})(\bar{v}_j - \bar{v}_i) \big) \right|$$

$$\leqslant \max_{i \in \{1,\dots,N\}} \left(\sum_{j \neq i} d_{ij} \right) \left(\max_{j \neq i} |\alpha_{ij}(v)(v_j - v_i) - \alpha_{ij}(\bar{v})(\bar{v}_j - \bar{v}_i)| \right)$$

$$\leqslant \left(\max_{i \in \{1,\dots,N\}} |d_{ii}| \right) L_{li} \|v - \bar{v}\|_\infty \quad \forall v, \bar{v} \in \mathbb{R}^N.$$

Proposition 4.66 and the assumptions in the statement of this lemma complete the proof. $\qquad\square$

Lemma 4.67 provides algebraic conditions for the existence of unique solutions. The following theorem translates them into assumptions regarding the contribution of the reaction term cu to the discretized advection-reaction equation.

Theorem 4.68. *Let the matrix $A \in \mathbb{R}^{N \times N}$ be defined using a lumped approximation of the reactive term, i.e.,*

$$a_{ij} := \delta_{ij} \int_\Omega c\varphi_i + \int_\Omega \mathbf{v} \cdot \mathrm{grad}(\varphi_j)\varphi_i + \int_{\partial\Omega} (\mathbf{v} \cdot \mathbf{n})_- \varphi_j\varphi_i$$

$$\forall i,j \in \{1,\ldots,N\}. \quad (4.109)$$

Additionally, suppose that the reactivity c satisfies

$$L_{\mathrm{li}} \max_{i\in\{1,\ldots,N\}} \left(\int_\Omega |\mathrm{div}(\mathbf{v}\varphi_i)| + \int_{\partial\Omega} (\mathbf{v}\cdot\mathbf{n})_+\varphi_i \right) < \left(\inf_{\mathbf{x}\in\Omega} c(\mathbf{x}) \right) \min_{i\in\{1,\ldots,N\}} m_i.$$

$$(4.110)$$

Then the AFC system (4.98) possesses a unique solution.

Proof. Due to the lumped treatment of the reactive part, $|d_{ii}|$ is bounded by

$$|d_{ii}| \leqslant \sum_{j\neq i} |a_{ij}| = \sum_{j\neq i} \left| \delta_{ij} \int_\Omega c\varphi_i + \int_\Omega \mathbf{v} \cdot \mathrm{grad}(\varphi_j)\varphi_i + \int_{\partial\Omega} (\mathbf{v}\cdot\mathbf{n})_- \varphi_j\varphi_i \right|$$

$$\leqslant \sum_{j=1}^N \left| \int_\Omega \mathbf{v} \cdot \mathrm{grad}(\varphi_j)\varphi_i + \int_{\partial\Omega} (\mathbf{v}\cdot\mathbf{n})_- \varphi_j\varphi_i \right|$$

$$= \sum_{j=1}^N \left| -\int_\Omega \varphi_j \, \mathrm{div}(\mathbf{v}\varphi_i) + \int_{\partial\Omega} (\mathbf{v}\cdot\mathbf{n})_+ \varphi_j\varphi_i \right|$$

$$\leqslant \sum_{j=1}^N \left(\int_\Omega |\varphi_j \, \mathrm{div}(\mathbf{v}\varphi_i)| + \int_{\partial\Omega} (\mathbf{v}\cdot\mathbf{n})_+ \varphi_j\varphi_i \right)$$

$$= \int_\Omega |\mathrm{div}(\mathbf{v}\varphi_i)| + \int_{\partial\Omega} (\mathbf{v}\cdot\mathbf{n})_+ \varphi_i.$$

On the other hand, we have

$$\sum_{j=1}^{N} a_{ij} = \sum_{j=1}^{N} \Big(\delta_{ij} \int_{\Omega} c\varphi_i + \int_{\Omega} \mathbf{v} \cdot \mathrm{grad}(\varphi_j)\varphi_i + \int_{\partial\Omega} (\mathbf{v} \cdot \mathbf{n})_- \varphi_j \varphi_i \Big)$$

$$= \int_{\Omega} c\varphi_i + \int_{\partial\Omega} (\mathbf{v} \cdot \mathbf{n})_- \varphi_i \geqslant \int_{\Omega} c\varphi_i \geqslant \big(\inf_{\mathbf{x} \in \Omega} c(\mathbf{x}) \big) m_i.$$

The use of Lemma 4.67 completes the proof. □

The assumption in the statement of the above theorem is very restrictive because the reactive term must behave like $c \sim h^{-1}$. However, this condition becomes realistic in the context of pseudo time stepping approaches in which $cu = \Delta t^{-1} u$ represents the implicit part of the finite difference approximation to the pseudo time derivative. If the backward Euler method is used to march the solution to a steady state, existence and uniqueness of solutions to the AFC problems associated with individual time steps are guaranteed under a CFL-like condition $\Delta t \sim h$. However, additional assumptions may be required to prove convergence to steady state solutions.

The lumped treatment of the reaction term as in the above theorem was also employed in [Bar+16] to achieve desired properties of the system matrix.

4.5.1.2 Discrete maximum principles

Assuming the existence of a unique solution to problem (4.98), we now derive sufficient conditions for the so far undefined correction factors to guarantee certain DMP properties of converged solutions. For this purpose, we recall the definition of $\hat{A}$ (cf. (4.100)). The matrix form of the nonlinear system

$$\sum_{j=1}^{N} \hat{a}_{ij} u_j := \Big(a_{ii} + \sum_{l \neq i} (1 - \alpha_{il}) d_{il} \Big) u_i + \sum_{j \neq i} \big(a_{ij} - (1 - \alpha_{ij}) d_{ij} \big) u_j = g_i$$

$$\forall i \in \{1, \dots, N\}$$

for the nodal values of u_h is given by $\hat{A} u = g$, where the row sums of $\hat{A}$ coincide with those of A and $A - D$

$$\sum_{j=1}^{N} \hat{a}_{ij} = a_{ii} + \sum_{l \neq i} (1 - \alpha_{il}) d_{il} + \sum_{j \neq i} \big(a_{ij} - (1 - \alpha_{ij}) d_{ij} \big) = \sum_{j=1}^{N} a_{ij}$$

$$\forall i \in \{1, \dots, N\}. \quad (4.111)$$

Lemma 4.69. *Assume that the correction factors satisfy*

$$\alpha_{ij}(u_i - u_j) = 0 \quad \forall i, j \in \{1, \ldots, N\},\ j \neq i,\ s.t.\ u_i = \max_{l \in \mathcal{N}_i} u_l \wedge a_{ij} > 0,$$
(4.112a)

$$\alpha_{ij}(u_i - u_j) = 0 \quad \forall i, j \in \{1, \ldots, N\},\ j \neq i,\ s.t.\ u_i = \min_{l \in \mathcal{N}_i} u_l \wedge a_{ij} > 0.$$
(4.112b)

Then the inequality constraints of Lemma 4.12 hold for the solution $u \in \mathbb{R}^N$ *of* (4.98).

Proof. Without loss of generality, we assume that $\alpha_{ij} = 0$ for $i, j \in \{1, \ldots, N\}$, $j \neq i$, if $u_i = u_j$. For every index $i \in \{1, \ldots, N\}$ corresponding to a local extremum, i.e., $u_i = \max_{l \in \mathcal{N}_i} u_l$ or $u_i = \min_{l \in \mathcal{N}_i} u_l$, the requirements set forth in (4.112) lead to

$$\hat{a}_{ij} = a_{ij} - (1 - \alpha_{ij})d_{ij} \leqslant \min(0, a_{ij}) \leqslant 0 \quad \forall j \in \{1, \ldots, N\},\ j \neq i. \quad (4.113)$$

On the other hand, the diagonal entry satisfies

$$\hat{a}_{ii} = a_{ii} + \sum_{l \neq i}(1 - \alpha_{il})d_{il} \geqslant a_{ii}. \quad (4.114)$$

In combination with (4.111), this shows that $\hat{\mathcal{A}}$ and $\mathcal{A} - \mathcal{D}$ have the same properties at a local extremum and local DMPs can be shown as in Lemma 4.12. $\qquad \square$

In addition to local DMPs, their global counterparts hold.

Lemma 4.70. *Assume that condition* (4.112) *is satisfied. Then the assertions of Lemma 4.16 without requirements (i) and (iii) hold for the solution to problem* (4.98).

Proof. To verify the statement of this lemma, we adapt the proof of Lemma 4.16 and replace $\mathcal{A}$ by $\hat{\mathcal{A}}$. We follow the analysis presented in [Loh19].

The crucial part of the adapted proof is the explanation why (4.18) still holds when $\mathcal{A}$ is replaced by $\hat{\mathcal{A}}$: We first notice that for every index $l \in I_*$ the degree of freedom u_l is a local extremum because $u_l \geqslant \max_{j \in \bar{I}} u_j$. Therefore, (4.113) and (4.114) hold for all $l \in I_*$. Then, according to the positive definiteness of $\mathcal{A}$, property (4.111), and (4.113), we have

$$0 < C_M\left(\sum_{j \in I_*} 1\right) \leqslant \sum_{l, j \in I_*} a_{lj} = \sum_{l, j \in I_*} \hat{a}_{lj} + \sum_{l \in I_*, j \notin I_*} \underbrace{(\hat{a}_{lj} - a_{lj})}_{\leqslant 0} \leqslant \sum_{l, j \in I_*} \hat{a}_{lj}$$
(4.115)

and condition (4.18) must be valid. The remainder of the proof follows that of Lemma 4.16.   □

4.5.1.3 Error analysis

Let us now show that the AFC solution with nodal values satisfying (4.98) converges to the exact solution of the advection-reaction equation with order at least $\frac{1}{2}$ for any choice of the correction factors. For this purpose, we adapt the results presented in [Bar+16, Sections 6 and 7], where the advection-diffusion-reaction equation with strongly enforced Dirichlet boundary conditions was considered. Since the main focus in this work is on hyperbolic problems, we neglect the diffusive part, consider the non-conservative form (2.28) of our model problem with weakly imposed boundary conditions, and assume that requirements (2.11) and (2.14) are satisfied.

First notice that the bilinear form $a : V \times V \to \mathbb{R}$ is coercive in the sense of (2.24) and define the problem-dependent norm $\| \cdot \|_a : V \to \mathbb{R}_0^+$ as follows:

$$a(\varphi, \varphi) \geqslant \|\varphi\|_a^2 := c_0 \|\varphi\|_{L^2(\Omega)}^2 + \tfrac{1}{2}\|\varphi\|_{L^2(|\mathbf{v}\cdot\mathbf{n}|;\partial\Omega)}^2 \qquad \forall \varphi \in V. \qquad (4.116)$$

Furthermore, the contribution of the artificial diffusion operator is represented by the nonlinear form

$$d_h(v; z, w) := - \sum_{i,j,\, j \neq i} \big(1 - \alpha_{ij}(v)\big) d_{ij}\big(z(\mathbf{x}_j) - z(\mathbf{x}_i)\big) w(\mathbf{x}_i)$$

$$\forall v, z, w \in C^0(\bar{\Omega}),$$

where, $\alpha_{ij}(v)$ are correction factors depending on the values of v at $\mathbf{x}_1, \ldots, \mathbf{x}_N$. It follows that

$$d_h(v_h; z_h, w_h) = - \sum_{i,j,\, j \neq i} \big(1 - \alpha_{ij}(v_h)\big) d_{ij}(z_j - z_i) w_i \qquad \forall v_h, z_h, w_h \in V_h$$

and the AFC system (4.98) is equivalent to

$$a(u_h, \varphi_h) + d_h(u_h; u_h, \varphi_h) = \int_\Omega f\varphi_h + \int_{\partial\Omega} (\mathbf{v} \cdot \mathbf{n})_- u_{\mathrm{in}}\varphi_h \quad \forall \varphi_h \in W_h = V_h. \tag{4.117}$$

For any $v \in C^0(\bar{\Omega})$, the mapping $d_h(v; \cdot, \cdot)$ is a symmetric bilinear form

$$
\begin{aligned}
d_h(v; z, w) &= - \sum_{i,j, j \neq i} \big(1 - \alpha_{ij}(v)\big) d_{ij} \big(z(\mathbf{x}_j) - z(\mathbf{x}_i)\big) w(\mathbf{x}_i) \\
&= - \sum_{i,j, j < i} \big(1 - \alpha_{ij}(v)\big) d_{ij} \big(z(\mathbf{x}_j) - z(\mathbf{x}_i)\big) w(\mathbf{x}_i) \\
&\quad - \sum_{i,j, j > i} \big(1 - \alpha_{ij}(v)\big) d_{ij} \big(z(\mathbf{x}_j) - z(\mathbf{x}_i)\big) w(\mathbf{x}_i) \\
&= - \sum_{i,j, j < i} \big(1 - \alpha_{ij}(v)\big) d_{ij} \big(z(\mathbf{x}_j) - z(\mathbf{x}_i)\big) \big(w(\mathbf{x}_i) - w(\mathbf{x}_j)\big) \\
&= d_h(v; w, z) \qquad \forall z, w \in C^0(\bar{\Omega})
\end{aligned}
$$

due to the symmetry of the correction factors $\alpha_{ij} = \alpha_{ji} \in [0, 1]$ and artificial diffusion coefficients $d_{ij} = d_{ji}$. Obviously, $d_h(v; z, z) \geqslant 0$ holds for all $z \in C^0(\bar{\Omega})$ and an upper bound can be obtained using the Cauchy-Schwarz inequality

$$
d_h(v; z, w)^2 \leqslant d_h(v; z, z) d_h(v; w, w) \qquad \forall v, z, w \in C^0(\bar{\Omega}). \tag{4.118}
$$

To show this, write $d_h(v; z, w)$ as $\bar{w}^\top (\tilde{\mathcal{D}} - \mathcal{D}) \bar{z}$, where $\bar{w}, \bar{z} \in \mathbb{R}^N$ are vectors with components $\bar{w}_i = w(\mathbf{x}_i)$, $\bar{z}_i = z(\mathbf{x}_i)$, $i \in \{1, \ldots, N\}$, and $\tilde{\mathcal{D}} = \tilde{\mathcal{D}}(z)$ is defined by (4.99). Then (4.118) holds due to

$$
\begin{aligned}
d_h(v; z, w)^2 &= (\bar{w}^\top (\tilde{\mathcal{D}} - \mathcal{D}) \bar{z})^2 = \big(\bar{w}^\top (\tilde{\mathcal{D}} - \mathcal{D})^{\frac{1}{2}} (\tilde{\mathcal{D}} - \mathcal{D})^{\frac{1}{2}} z\big)^2 \\
&\leqslant \big\| (\tilde{\mathcal{D}} - \mathcal{D})^{\frac{1}{2}} \bar{w} \big\|_2^2 \big\| (\tilde{\mathcal{D}} - \mathcal{D})^{\frac{1}{2}} \bar{z} \big\|_2^2 \\
&= \bar{w}^\top (\tilde{\mathcal{D}} - \mathcal{D}) \bar{w} \cdot \bar{z}^\top (\tilde{\mathcal{D}} - \mathcal{D}) \bar{z} = d_h(v; w, w) d_h(v; z, z),
\end{aligned}
$$

due to the symmetry and positive semidefiniteness of $\tilde{\mathcal{D}} - \mathcal{D}$. This motivates the definition of

$$
\|\varphi\|_h := \sqrt{\|\varphi\|_a^2 + d_h(u_h; \varphi, \varphi)} \qquad \forall \varphi \in V \cap C^0(\bar{\Omega}),
$$

which is a natural norm induced by the left hand side of (4.117).

Lemma 4.71. Let $u \in V \cap C^0(\bar{\Omega})$ be the exact solution of (2.28). Then the following estimate holds

$$
\|u - u_h\|_h \leqslant \|u - I_h u\|_a + \sup_{\varphi_h \in W_h \setminus \{0\}} \frac{a(u, \varphi_h) - a(I_h u, \varphi_h)}{\|\varphi_h\|_h} \\
+ \sqrt{d_h(u_h; I_h u, I_h u)}, \tag{4.119}
$$

where $I_h u \in V_h$ is the pointwise interpolant of u and $u_h \in V_h$ is the finite element approximation associated with the solution of (4.98). Hence, the total error is caused by a combined effect of (i) an interpolation error, (ii) a consistency error of the high order method, and (iii) an error due to the AFC stabilization.

Proof. The proof follows [Bar+16] with minor changes in the definitions of function spaces, bilinear forms, and boundary conditions.

We start with the derivation of an estimate for the difference between the numerical solution u_h and an arbitrary function $z_h \in V_h$. After accomplishing this task, we will estimate the error $\|u - u_h\|_h$ in terms of z_h using the triangle inequality and substitute the interpolant $z_h = I_h u$ of the exact solution to obtain a bound for the consistency error of the AFC system (4.117).

Let $z_h \in V_h$ be arbitrary. By (2.29) and (4.117), we have

$$a(u_h - z_h, \varphi_h) + d_h(u_h; u_h - z_h, \varphi_h) = a(u, \varphi_h) - a(z_h, \varphi_h) - d_h(u_h; z_h, \varphi_h)$$
$$\forall \varphi_h \in W_h.$$

Dividing both sides of this identity by $\|\varphi_h\|_h$, choosing $\varphi_h = u_h - z_h \in W_h$ on the left hand side, taking the supremum over $\varphi_h \in W_h$ on the right hand side, and exploiting (4.116) and (4.118), we obtain

$$
\begin{aligned}
\|u_h - z_h\|_h &= \|u_h - z_h\|_h^{-1} \left(\|u_h - z_h\|_a^2 + d_h(u_h; u_h - z_h, u_h - z_h) \right) \\
&\leq \|u_h - z_h\|_h^{-1} \left(a(u_h - z_h, u_h - z_h) + d_h(u_h; u_h - z_h, u_h - z_h) \right) \\
&\leq \sup_{\varphi_h \in W_h \setminus \{0\}} \|\varphi_h\|_h^{-1} \left(a(u, \varphi_h) - a(z_h, \varphi_h) - d_h(u_h; z_h, \varphi_h) \right) \\
&\leq \sup_{\varphi_h \in W_h \setminus \{0\}} \|\varphi_h\|_h^{-1} \Big(a(u, \varphi_h) - a(z_h, \varphi_h) \\
&\qquad\qquad\qquad\qquad + \sqrt{d_h(u_h; z_h, z_h)} \sqrt{d_h(u_h; \varphi_h, \varphi_h)} \Big) \\
&\leq \sup_{\varphi_h \in W_h \setminus \{0\}} \frac{a(u, \varphi_h) - a(z_h, \varphi_h)}{\|\varphi_h\|_h} + \sqrt{d_h(u_h; z_h, z_h)},
\end{aligned}
$$

where $u \in V \cap C^0(\bar{\Omega})$ is the exact weak solution. The triangle inequality yields the error estimate

$$
\begin{aligned}
\|u - u_h\|_h &\leq \inf_{z_h \in V_h} \|u - z_h\|_h + \|u_h - z_h\|_h \\
&\leq \inf_{z_h \in V_h} \Big(\|u - z_h\|_h + \sup_{\varphi_h \in W_h} \frac{a(u, \varphi_h) - a(z_h, \varphi_h)}{\|\varphi_h\|_h} \\
&\qquad\qquad\qquad\qquad + \sqrt{d_h(u_h; z_h, z_h)} \Big).
\end{aligned}
$$

Approximating u by its interpolant $z_h := I_h u$ as usual and exploiting the fact that $d_h(u_h; u - I_h u, u - I_h u) = 0$ since $u(\mathbf{x}_i) = I_h u(\mathbf{x}_i)$ for all $i \in \{1, \ldots, N\}$, we obtain (4.119). $\qquad\qquad\qquad\qquad\qquad\qquad\qquad\qquad\qquad\qquad\qquad$ $\square$

To prove convergence of the AFC solution u_h to u, we now need to estimate the terms on the right hand side of (4.119). This can be done without imposing any further restrictions on the correction factors $\alpha_{ij} = \alpha_{ji} \in [0, 1]$. In particular, the trivial choice $\alpha_{ij} = 0$ for all $i, j \in \{1, \ldots, N\}$, $j \neq i$, leads to an error estimate for the low order solution which is valid if the exact solution is sufficiently smooth.

Theorem 4.72. *Let $u \in H^2(\Omega)$ be the exact solution of (2.28), V_h the space of linear (not multilinear) finite elements, and $u_h \in V_h$ a numerical approximation whose nodal values solve the AFC system (4.98). Then u_h satisfies*

$$\|u - u_h\|_h \leqslant C h^{\frac{1}{2}} \tag{4.120}$$

for any choice of the correction factors $\alpha_{ij} = \alpha_{ji} \in [0, 1]$.

Proof. The interpolation error for (multi-)linear finite elements with $d \in \{1, 2, 3\}$ satisfies [EG13, Example 1.111]

$$\|u - I_h u\|_{L^2(\Omega)} + h|u - I_h u|_{H^1(\Omega)} \leqslant C h^2 |u|_{H^2(\Omega)} \qquad \forall u \in H^2(\Omega). \tag{4.121}$$

Then, by (2.13a), (2.19), and (4.121), the interpolant $I_h u$ converges to u in $\|\cdot\|_a$ with order 1

$$
\begin{aligned}
\|u - I_h u\|_a^2 &= c_0 \|u - I_h u\|_{L^2(\Omega)}^2 + \tfrac{1}{2}\|u - I_h u\|_{L^2(|\mathbf{v}\cdot\mathbf{n}|;\partial\Omega)}^2 \\
&\leqslant c_0 \|u - I_h u\|_{L^2(\Omega)}^2 + \tfrac{1}{2}C_\gamma^2 \|u - I_h u\|_V^2 \\
&= (c_0 + \tfrac{1}{2}C_\gamma^2)\|u - I_h u\|_{L^2(\Omega)}^2 + \tfrac{1}{2}C_\gamma^2 \big\|\mathbf{v}\cdot\mathrm{grad}(u - I_h u)\big\|_{L^2(\Omega)}^2 \\
&\leqslant (c_0 + \tfrac{1}{2}C_\gamma^2)\|u - I_h u\|_{L^2(\Omega)}^2 \\
&\quad + \tfrac{1}{2}C_\gamma^2 \|\mathbf{v}\|_{L^\infty(\Omega)}^2 \big\|\mathrm{grad}(u - I_h u)\big\|_{L^2(\Omega)}^2 \\
&\leqslant C \|u - I_h u\|_{H^1(\Omega)}^2 \leqslant C h^2 |u|_{H^2(\Omega)}^2.
\end{aligned}
$$

Similar estimates yield

$$\left| a(u, \varphi_h) - a(I_h u, \varphi_h) \right|$$

$$= \left| \int_\Omega c(u - I_h u)\varphi_h + \int_\Omega \mathbf{v} \cdot \operatorname{grad}(u - I_h u)\varphi_h \right.$$

$$\left. + \int_{\partial\Omega} (\mathbf{v} \cdot \mathbf{n})_-(u - I_h u)\varphi_h \right|$$

$$\leqslant \left(\|c\|_{L^\infty(\Omega)} \|u - I_h u\|_{L^2(\Omega)} + \left\| \mathbf{v} \cdot \operatorname{grad}(u - I_h u) \right\|_{L^2(\Omega)} \right) \|\varphi_h\|_{L^2(\Omega)}$$

$$+ \int_{\partial\Omega} |\mathbf{v} \cdot \mathbf{n}| \, |u - I_h u| \, |\varphi_h|$$

$$\leqslant \max\left(\|c\|_{L^\infty(\Omega)}, 1 \right) \left(\|u - I_h u\|_{L^2(\Omega)} \right.$$

$$\left. + \left\| \mathbf{v} \cdot \operatorname{grad}(u - I_h u) \right\|_{L^2(\Omega)} \right) \|\varphi_h\|_{L^2(\Omega)}$$

$$+ \left(\int_{\partial\Omega} |\mathbf{v} \cdot \mathbf{n}| (u - I_h u)^2 \right)^{\frac{1}{2}} \left(\int_{\partial\Omega} |\mathbf{v} \cdot \mathbf{n}| \varphi_h^2 \right)^{\frac{1}{2}}$$

$$\leqslant C \|u - I_h u\|_V \|\varphi_h\|_{L^2(\Omega)} + \|u - I_h u\|_{L^2(|\mathbf{v}\cdot\mathbf{n}|;\partial\Omega)} \|\varphi_h\|_{L^2(|\mathbf{v}\cdot\mathbf{n}|;\partial\Omega)}$$

$$\leqslant C \|u - I_h u\|_V \|\varphi_h\|_{L^2(\Omega)} + \sqrt{2} C_\gamma \|u - I_h u\|_V \|\varphi_h\|_a$$

$$\leqslant C h |u|_{H^2(\Omega)} \|\varphi_h\|_a \qquad \forall \varphi_h \in W_h$$

and show that

$$\sup_{\varphi_h \in W_h} \frac{a(u, \varphi_h) - a(I_h u, \varphi_h)}{\|\varphi_h\|_h} \leqslant C h |u|_{H^2(\Omega)}.$$

To complete the proof based on (4.119), we note that

$$|a_{ij}| = \left| a(\varphi_j, \varphi_i) \right|$$

$$= \left| \int_\Omega c\varphi_j \varphi_i + \int_\Omega \mathbf{v} \cdot \operatorname{grad}(\varphi_j)\varphi_i + \int_{\partial\Omega} (\mathbf{v} \cdot \mathbf{n})_- \varphi_j \varphi_i \right|$$

$$\leqslant \|c\|_{L^\infty(\varpi_i \cap \varpi_j)} \|\varphi_j\|_{L^2(\varpi_i \cap \varpi_j)} \|\varphi_i\|_{L^2(\varpi_i \cap \varpi_j)}$$

$$+ \|\mathbf{v}\|_{L^\infty} |\varphi_j|_{H^1(\varpi_i \cap \varpi_j)} \|\varphi_i\|_{L^2(\varpi_i \cap \varpi_j)}$$

$$+ \|\mathbf{v}\|_{L^\infty} \|\varphi_j\|_{L^2(\partial\varpi_i \cap \partial\varpi_j \cap \partial\Omega)} \|\varphi_i\|_{L^2(\partial\varpi_i \cap \partial\varpi_j \cap \partial\Omega)}$$

$$\leqslant C \max_{e \in \mathcal{E}_i \cap \mathcal{E}_j} h_{K^e}^{d-1} \leqslant C h^{d-1}.$$

Consequently, the definition of the artificial diffusion coefficients, estimate (4.90) (only valid for linear finite elements), and the interpolation error estimate (4.121) yield the following result for the remaining term of (4.119)

$$d_h(u_h; I_h u, I_h u)$$
$$= - \sum_{i,j,\, j<i} \big(1 - \alpha_{ij}(u_h)\big) d_{ij} \big(u(\mathbf{x}_j) - u(\mathbf{x}_i)\big) \big(u(\mathbf{x}_i) - u(\mathbf{x}_j)\big)$$
$$\leqslant \sum_{i,j,\, j<i} d_{ij} \big(u(\mathbf{x}_j) - u(\mathbf{x}_i)\big)^2 = \tfrac{1}{2} \sum_{i,j,\, j\neq i} d_{ij} \big(u(\mathbf{x}_j) - u(\mathbf{x}_i)\big)^2$$
$$\leqslant C \sum_{e=1}^{E} \sum_{i,j \in \mathcal{N}^e} h^{d-1} h^{2-d} |I_h u|_{H^1(K^e)}^2$$
$$\leqslant C h \sum_{e=1}^{E} |I_h u|_{H^1(K^e)}^2 \leqslant C h |I_h u|_{H^1(\Omega)}^2 \leqslant C h \|u\|_{H^2(\Omega)}^2$$

$$(4.122)$$

and the statement of the theorem follows by Lemma 4.71.   □

Remark 4.73. Theorem 4.72 holds for arbitrary correction factors that satisfy the symmetry condition. In particular, the result is true if all correction factors are set to zero. Hence, the low order method corresponding to this trivial choice exhibits the following convergence behavior: The low order solution u_h^{L} converges to the exact solution $u \in H^2(\Omega)$ of (2.28) in $\| \cdot \|_h$, and consequently in $\| \cdot \|_{L^2(\Omega)}$, with order $\tfrac{1}{2}$.

Furthermore, the result of Theorem 4.72 implies that

$$\|u_h^{\max} - u_h^{\mathrm{L}}\|_{L^\infty(\Omega)} \leqslant C\sqrt{h}, \qquad \|u_h^{\min} - u_h^{\mathrm{L}}\|_{L^\infty(\Omega)} \leqslant C\sqrt{h},$$

where $u_h^{\max}, u_h^{\min} \in V_h$ are the finite element interpolants as defined in Theorem 4.60.

Without loss of generality, we prove the inequality for $u_h^{\max}$ and first show that $a_{ij} + a_{ji} \geqslant 2c_0 m_{ij}$. In view of (2.14), we have

$$
\begin{aligned}
a_{ij} + a_{ji} &= 2\int_{\Omega} c\varphi_i\varphi_j + \int_{\Omega} \mathbf{v} \cdot \mathrm{grad}(\varphi_j)\varphi_i \\
&\quad + \int_{\Omega} \mathbf{v} \cdot \mathrm{grad}(\varphi_i)\varphi_j + 2\int_{\partial\Omega} (\mathbf{v} \cdot \mathbf{n})_-\varphi_i\varphi_j \\
&= 2\int_{\Omega} c\varphi_i\varphi_j + \int_{\Omega} \mathrm{div}(\mathbf{v}\varphi_i\varphi_j) \\
&\quad + \int_{\Omega} (\mathbf{v} \cdot \mathrm{grad}(\varphi_i) - \mathrm{div}(\mathbf{v}\varphi_i))\varphi_j + 2\int_{\partial\Omega} (\mathbf{v} \cdot \mathbf{n})_-\varphi_i\varphi_j \\
&= 2\int_{\Omega} c\varphi_i\varphi_j + \int_{\partial\Omega} (\mathbf{v} \cdot \mathbf{n})\varphi_i\varphi_j \\
&\quad - \int_{\Omega} \mathrm{div}(\mathbf{v})\varphi_i\varphi_j + 2\int_{\partial\Omega} (\mathbf{v} \cdot \mathbf{n})_-\varphi_i\varphi_j \\
&\geqslant 2c_0 \int_{\Omega} \varphi_i\varphi_j + \int_{\partial\Omega} |\mathbf{v} \cdot \mathbf{n}|\varphi_i\varphi_j \geqslant 2c_0 m_{ij} > 0 \\
&\qquad\qquad\qquad\qquad\qquad \forall i \in \{1, \ldots, N\}, j \in \mathcal{N}_i.
\end{aligned}
$$

Therefore, the artificial diffusion coefficient d_{ij} satisfies

$$
d_{ij} = \max(a_{ij}, 0, a_{ji}) \geqslant \frac{a_{ij} + a_{ji}}{2} \geqslant c_0 m_{ij} \qquad \forall i \in \{1, \ldots, N\}, j \in \mathcal{N}_i \setminus \{i\}
$$

and we deduce that

$$
\frac{m_i}{d_{ij}} \leqslant c_0^{-1} \frac{m_i}{m_{ij}} \leqslant C \qquad \forall i \in \{1, \ldots, N\}, j \in \mathcal{N}_i \setminus \{i\}
$$

for some $C > 0$ independent of h. Invoking (4.91) and (4.122), we arrive at

$$\|u_h^{\max} - u_h^{\mathrm{L}}\|_{L^2(\Omega)}^2 \leqslant C\sum_{i=1}^N m_i(u_i^{\max} - u_i^{\mathrm{L}})^2 = C\sum_{i=1}^N m_i\big(\max_{j\in\mathcal{N}_i}(u_j^{\mathrm{L}} - u_i^{\mathrm{L}})\big)^2$$

$$\leqslant C\sum_{i=1}^N m_i \max_{j\in\mathcal{N}_i}(u_i^{\mathrm{L}} - u_j^{\mathrm{L}})^2 \leqslant C\sum_{i=1}^N \sum_{j\in\mathcal{N}_i} m_i(u_i^{\mathrm{L}} - u_j^{\mathrm{L}})^2$$

$$\leqslant C\sum_{i=1}^N \sum_{j\in\mathcal{N}_i\setminus\{i\}} \bigg(m_i\big((u_i^{\mathrm{L}} - u(\mathbf{x}_i)) - (u_j^{\mathrm{L}} - u(\mathbf{x}_j))\big)^2$$
$$+ m_i\big(u(\mathbf{x}_i) - u(\mathbf{x}_j)\big)^2 \bigg)$$

$$\leqslant C\sum_{i,j,\,j<i} \bigg(d_{ij}\big((u_i^{\mathrm{L}} - u(\mathbf{x}_i)) - (u_j^{\mathrm{L}} - u(\mathbf{x}_j))\big)^2$$
$$+ d_{ij}\big(u(\mathbf{x}_i) - u(\mathbf{x}_j)\big)^2 \bigg)$$

$$= Cd_h(u_h^{\mathrm{L}}; u_h^{\mathrm{L}} - u, u_h^{\mathrm{L}} - u) + Cd_h(u_h^{\mathrm{L}}; I_h u, I_h u)$$

$$\leqslant C\|u_h^{\mathrm{L}} - u\|_h^2 + Cd_h(u_h^{\mathrm{L}}; I_h u, I_h u) \leqslant Ch$$

by Theorem 4.72, where all correction factors of the diffusive term d_h are set to zero.

Next, we define the *linearity preserving* property of an AFC method in a way which is slightly different from the definition in [Bar+18, Assumption (A3)].

Definition 4.74. The AFC method (4.98) is called *linearity preserving* if

$$\alpha_{ij}(u_i - u_j) = u_i - u_j$$
$$\forall i \in \{1,\ldots,N\},\, j \in \mathcal{N}_i \setminus \{i\} \text{ s.t. } u_h|_{\varpi_i\cup\varpi_j} \in \mathbb{P}_1(\varpi_i \cup \varpi_j), \quad (4.123)$$

where ϖ_i is the patch of elements defined at the beginning of Chapter 3.

In contrast to [Bar+18, Assumption (A3)], this criterion does not require that the correction factors for constant functions $u_h|_{\varpi_i\cup\varpi_j} \in \mathbb{P}_0(\varpi_i \cup \varpi_j)$ be equal to one. Hence, there is no contradiction between the principle of linearity preservation for constant functions and the practice of using zero correction factors at local extrema.

Definition 4.74 implies that globally linear functions u are recovered exactly by the AFC algorithm. In [Bar+18], the authors focused on the convergence behavior of linearity-preserving AFC algorithms. Sufficient conditions of linearity preservation, improved a priori error estimates for AFC discretizations of advection-diffusion equations, and numerical examples illustrating the benefits of this design criterion for limiter functions can be found, for example, in [Bar+17b; Bar+17a; Kuz+17]. There is strong numerical evidence that linearity preservation is essential for achieving optimal accuracy in the context of hyperbolic problems [Kuz+17; Kuz18a]. However, no rigorous proof of an improved convergence rate is currently available for the case of vanishing physical diffusion and further theoretical analysis aimed at obtaining such a proof is beyond the scope of this work.

4.5.1.4 Examples

In the above sections, theoretical analysis of generic AFC schemes was performed under certain assumptions regarding the properties of correction factors α_{ij}. The following criteria play an important role in the design of limiter functions:

- Continuity of $\alpha_{ij}(u_j - u_i)$ is required to ensure existence of a solution;

- Lipschitz continuity of $\alpha_{ij}(u_j - u_i)$ is a useful tool for proving uniqueness;

- symmetry ($\alpha_{ij} = \alpha_{ji}$) is required for conservation and well-posedness;

- DMPs hold if $\alpha_{ij} = 0$ whenever u_i is a local extremum and $a_{ij} > 0$.

Furthermore, the correction factors $\alpha_{ij} \in [0, 1]$ should be defined to be as large as possible since the high order target solution corresponds to $\alpha_{ij} \equiv 1$.

In this section, we put theory into practice by presenting practical definitions of correction factors satisfying the above requirements. The way in which a particular definition meets the above requirements may influence the accuracy of the resulting solution, the convergence behavior of iterative solvers, and the computational costs.

Before presenting the first limiting operator, we formulate convenient criteria for proving the Lipschitz continuity of limited antidiffusive fluxes in edge-based AFC schemes. The analysis leading to these criteria is partially based on the following lemma.

Lemma 4.75. *The absolute value* $|\cdot| : \mathbb{R} \to \mathbb{R}_0^+$ *of a scalar quantity as well as the maximum and minimum* $\max(\cdot, \cdot), \min(\cdot, \cdot) : \mathbb{R} \times \mathbb{R} \to \mathbb{R}$ *are Lipschitz continuous functions.*

Proof. In fact, the Lipschitz continuity of the absolute value follows from the reverse triangle inequality

$$\big||v| - |\bar{v}|\big| \leqslant |v - \bar{v}| \qquad \forall v, \bar{v} \in \mathbb{R}. \tag{4.124a}$$

Hence, the following estimates hold [Loh19, Eq. (3)]

$$\begin{aligned}
\big|\max(u,v) - \max(\bar{u},\bar{v})\big| &= \tfrac{1}{2}\big|(u+v) + |u-v| - (\bar{u}+\bar{v}) - |\bar{u}-\bar{v}|\big| \\
&\leqslant \tfrac{1}{2}|u+v-\bar{u}-\bar{v}| + \tfrac{1}{2}\big||u-v| - |\bar{u}-\bar{v}|\big| \\
&\leqslant |u-\bar{u}| + |v-\bar{v}| \qquad \forall u,\bar{u},v,\bar{v} \in \mathbb{R},
\end{aligned} \tag{4.124b}$$

$$\begin{aligned}
\big|\min(u,v) - \min(\bar{u},\bar{v})\big| &= \tfrac{1}{2}\big|(u+v) - |u-v| - (\bar{u}+\bar{v}) + |\bar{u}-\bar{v}|\big| \\
&\leqslant \tfrac{1}{2}|u+v-\bar{u}-\bar{v}| + \tfrac{1}{2}\big||u-v| - |\bar{u}-\bar{v}|\big| \\
&\leqslant |u-\bar{u}| + |v-\bar{v}| \qquad \forall u,\bar{u},v,\bar{v} \in \mathbb{R}
\end{aligned} \tag{4.124c}$$

and complete the proof of the lemma. $\qquad\square$

Lemma 4.76. *Let $i,j \in \{1,\dots,N\}$, $j \neq i$, be arbitrary and $\xi : \mathbb{R}^N \to \mathbb{R}$ a function such that*

$$|\xi(v)| \leqslant C \qquad\qquad \forall v \in \mathbb{R}^N, \tag{4.125a}$$
$$\big|\xi(v) - \xi(\bar{v})\big|\,|v_j - v_i| \leqslant C\|v - \bar{v}\| \qquad \forall v, \bar{v} \in V_+ \wedge \forall v, \bar{v} \in V_-, \tag{4.125b}$$

where $C > 0$, $V_+ := \{v \in \mathbb{R}^N \,|\, v_i < v_j\}$, and $V_- := \{v \in \mathbb{R}^N \,|\, v_i > v_j\}$. Then we have

$$\big|\xi(v)(v_j - v_i) - \xi(\bar{v})(\bar{v}_j - \bar{v}_i)\big| \leqslant C\|v - \bar{v}\| \qquad \forall v, \bar{v} \in \mathbb{R}^N. \tag{4.126}$$

Examples of $\xi(v)$ satisfying (4.125) include

E1. Function values $\xi \in [0,C]$ that can be written as

$$\xi(v) = \frac{A(v)}{|v_j - v_i| + B(v)} \qquad \forall v \in V_+ \cup V_-, \tag{4.127}$$

where $A, B : V_+ \cup V_- \to \mathbb{R}_0^+$ are Lipschitz continuous functions which are nonnegative on V_+ and V_-;

E2. The minimum of functions ξ_l, $l \in \{1,\dots,\zeta\}$, satisfying condition (4.125);

E3. The product of functions ξ_l, $l \in \{1, \ldots, \zeta\}$, satisfying condition (4.125);

E4. A polynomial in functions ξ_l, $l \in \{1, \ldots, \zeta\}$, satisfying condition (4.125).

The Lipschitz continuity of limited differences $\alpha_{ij}(v)(v_i - v_j)$ with correction factors α_{ij} which can be written in the form (4.127) was originally shown in [Bar+16, Lemma 3.5].

Proof. Let us first show that (4.126) holds for arbitrary $v, \bar{v} \in \mathbb{R}^N$ if the function $\xi(v)$ satisfies conditions (4.125). In the case $(v_j - v_i)(\bar{v}_j - \bar{v}_i) \leqslant 0$, this is true because

$$\left|\xi(v)(v_j - v_i) - \xi(\bar{v})(\bar{v}_j - \bar{v}_i)\right| \leqslant \left|\xi(v)(v_j - v_i)\right| + \left|\xi(\bar{v})(\bar{v}_j - \bar{v}_i)\right|$$
$$\leqslant C|v_j - v_i| + C|\bar{v}_j - \bar{v}_i| \leqslant C\left|(v_j - v_i) - (\bar{v}_j - \bar{v}_i)\right| \leqslant C\|v - \bar{v}\|.$$

For $(v_j - v_i)(\bar{v}_j - \bar{v}_i) > 0$, we have $v, \bar{v} \in V_+$ or $v, \bar{v} \in V_-$ and, hence,

$$\left|\xi(v)(v_j - v_i) - \xi(\bar{v})(\bar{v}_j - \bar{v}_i)\right|$$
$$\leqslant \left|\xi(v)\right|\left|(v_j - v_i) - (\bar{v}_j - \bar{v}_i)\right| + \left|\xi(v) - \xi(\bar{v})\right|\left|\bar{v}_j - \bar{v}_i\right|$$
$$\leqslant C\left(|v_j - \bar{v}_j| + |v_i - \bar{v}_i|\right) + C\|v - \bar{v}\| \leqslant C\|v - \bar{v}\|.$$

To prove (4.125) for the examples E1–E4, it suffices to verify (4.125b) because condition (4.125a) is satisfied trivially.

E1. Assume that ξ can be written in the form (4.127). Then, for all $v, \bar{v} \in V_{\pm}$, we have

$$\xi(v) - \xi(\bar{v}) - \frac{A(v) - A(\bar{v})}{|v_j - v_i| + B(v)}$$
$$= \frac{A(\bar{v})}{|v_j - v_i| + B(v)} - \frac{A(\bar{v})}{|\bar{v}_j - \bar{v}_i| + B(\bar{v})}$$
$$= A(\bar{v})\frac{\left(|\bar{v}_j - \bar{v}_i| + B(\bar{v})\right) - \left(|v_j - v_i| + B(v)\right)}{\left(|v_j - v_i| + B(v)\right)\left(|\bar{v}_j - \bar{v}_i| + B(\bar{v})\right)}$$
$$= \xi(\bar{v})\frac{\left(B(\bar{v}) - B(v)\right) + \left(|\bar{v}_j - \bar{v}_i| - |v_j - v_i|\right)}{|v_j - v_i| + B(v)}.$$

Therefore, (4.125b) follows from the fact that $B \geqslant 0$, A and B are Lipschitz continuous, and $\xi \in [0, C]$.

E2. It is sufficient to prove the statement for two functions ξ_1 and ξ_2. Then the more general result follows by induction. According to the Lipschitz continuity of the minimum (4.124c), condition (4.125b) holds due to

$$\left| \min\big(\xi_1(v), \xi_2(v)\big) - \min\big(\xi_1(\bar{v}), \xi_2(\bar{v})\big) \right| |v_j - v_i|$$
$$\leqslant \left(\left| \min\big(\xi_1(v), \xi_2(v)\big) - \min\big(\xi_1(\bar{v}), \xi_2(v)\big) \right| \right.$$
$$\left. + \left| \min\big(\xi_1(\bar{v}), \xi_2(v)\big) - \min\big(\xi_1(\bar{v}), \xi_2(\bar{v})\big) \right| \right) |v_j - v_i|$$
$$\leqslant \left| \xi_1(v) - \xi_1(\bar{v}) \right| |v_j - v_i| + \left| \xi_2(v) - \xi_2(\bar{v}) \right| |v_j - v_i|$$
$$\leqslant C \|v - \bar{v}\| \quad \forall v, \bar{v} \in V_\pm.$$

E3. The result for the product $\prod_l \xi_l$ can be shown in exactly the same way as for the minimum exploiting the boundedness $|\xi_l| \leqslant C$ instead of (4.124c).

E4. Condition (4.125b) for this function can be easily shown by employing the result for the product and the triangle inequality. □

These results can be combined in an appropriate manner to prove the Lipschitz continuity of more complicated functions and, particularly, of the limiters presented in this section.

The following algorithm for calculating the correction factors of the non-linear system (4.98) was proposed by Kuzmin [Kuz07] without presenting any theoretical results on the validity of DMPs or solvability of the discrete problem.

Algorithm 4.77 ([Kuz07]). *The correction factors for $f_{ij} := d_{ij}(u_i - u_j)$ can be determined as follows:*

1. *Compute the sums of positive and negative unlimited fluxes into node i*

$$P_i^+ := \sum_{\substack{j \neq i, \\ a_{ji} \leqslant a_{ij}}} \max(0, f_{ij}), \quad P_i^- := \sum_{\substack{j \neq i, \\ a_{ji} \leqslant a_{ij}}} \min(0, f_{ij}) \quad \forall i \in \{1, \ldots, N\};$$

$$(4.128)$$

2. *Compute the upper and lower bounds to be imposed on the sums $P_i^\pm$*

$$Q_i^+ := \sum_{j \neq i} \max(0, -f_{ij}), \quad Q_i^- := \sum_{j \neq i} \min(0, -f_{ij}) \quad \forall i \in \{1, \ldots, N\};$$

$$(4.129)$$

3. Calculate the nodal correction factors for undershoot/overshoot limiting

$$R_i^+ := \min\left(1, \frac{Q_i^+}{P_i^+}\right), \quad R_i^- := \min\left(1, \frac{Q_i^-}{P_i^-}\right) \quad \forall i \in \{1, \dots, N\}; \quad (4.130)$$

4. Check the sign of f_{ij} and apply the symmetry-preserving correction factor

$$\alpha_{ij} := \begin{cases} R_i^+ & : f_{ij} > 0 \wedge a_{ji} \leqslant a_{ij}, \\ R_j^- & : f_{ij} > 0 \wedge a_{ji} > a_{ij}, \\ 1 & : f_{ij} = 0, \\ R_i^- & : f_{ij} < 0 \wedge a_{ji} \leqslant a_{ij}, \\ R_j^+ & : f_{ij} < 0 \wedge a_{ji} > a_{ij} \end{cases} \quad \forall i, j \in \{1, \dots, N\}, j \neq i.$$

$$(4.131)$$

The residual of the AFC system that uses Algorithm 4.77 to calculate the correction factors is Lipschitz continuous by Lemma 4.76. Due to the dependence of α_{ij} on the sign of $f_{ij} = d_{ij}(u_i - u_j)$, the cases $u_i < u_j$ and $u_i > u_j$ need to be considered separately. A detailed proof of the Lipschitz continuity can be found in [Bar+16].

Furthermore, the AFC solution possesses the local and global DMP properties if

$$a_{ij} \leqslant 0 \quad \text{or} \quad a_{ji} \leqslant 0 \quad \forall i, j \in \{1, \dots, N\}, j \neq i. \quad (4.132)$$

For instance, at a local maximum, we have $Q_i^+ = 0$ by definition of the antidiffusive fluxes $f_{ij} = d_{ij}(u_i - u_j) \geqslant 0$. Then we obtain $R_i^+ = 0$ and find that

$$\alpha_{ij} f_{ij} = 0 \text{ if } a_{ji} \leqslant a_{ij} \quad \implies \quad \alpha_{ij} f_{ij} = 0 \text{ if } a_{ij} > 0$$
$$\forall j \in \{1, \dots, N\}, j \neq i.$$

This implies (4.112a) due to $d_{ij} > 0$.

Another (more technical) proof of the local DMP property was obtained in [Bar+16] under the assumption that $a_{ij} + a_{ji} \leqslant 0$ for all $i, j \in \{1, \dots, N\}$, $j \neq i$.

General discretizations of the advection-reaction equation (2.28) may violate criterion (4.132). Especially in the case of sufficiently large reactive rates c, we have $a_{ij}, a_{ji} > 0$, $j \in \mathcal{N}_i \setminus \{i\}$, and the validity of local DMPs is not guaranteed by Algorithm 4.77. However, an AFC scheme satisfying discrete maximum principles can also be designed for target discretizations violating requirement (4.132). To that end, the sums in (4.128) are taken over all $j \neq i$ and the formula for the correction factors α_{ij} is modified as follows:

$$\alpha_{ij} := \begin{cases} \min(R_i^+, R_j^-) & : f_{ij} > 0, \\ 1 & : f_{ij} = 0, \\ \min(R_i^-, R_j^+) & : f_{ij} < 0 \end{cases}$$

$$\forall i, j \in \{1, \ldots, N\}, \ j \neq i, \ \text{s.t.} \ a_{ij}, a_{ji} \geqslant 0. \quad (4.133)$$

Using the minimum of the sign-dependent nodal correction factors for nodes i and j, this definition of $\alpha_{ij} = \alpha_{ji}$ satisfies the DMP requirements for both nodes.

Algorithm 4.77 is a classical representative of edge-based limiters for AFC schemes. It can be modified and improved in many ways. For example, the following modification was proposed in [Kuz12b].

Algorithm 4.78 ([Kuz12b]). *The local bounds $Q_i^\pm$ for Algorithm 4.77 using (4.131) and (4.133) can be redefined as*

$$Q_i^+ := q_i(u_i^{\max} - u_i), \qquad Q_i^- := q_i(u_i^{\min} - u_i) \qquad \forall i \in \{1, \ldots, N\}, \quad (4.134)$$

where $q_i > 0$ are adjustable parameters. The local extrema $u_i^{\max}$ and $u_i^{\min}$ are defined by (4.79) using u instead of u^{L}.

Larger values of the parameters $q_i > 0$ imply less restrictive bounds for the sums of limited antidiffusive fluxes. The reduced amount of artificial diffusion has the positive effect of keeping the numerical solution closer to the high order target. However, the Lipschitz constants grow and lead to an ill-posed problem as q_i increases. In contrast to Algorithm 4.77, linearity preservation is guaranteed for sufficiently large values of q_i. Barrenechea, John, and Knobloch [Bar+17b, Theorem 6.1] have proved linearity preservation for Algorithm 4.78 with

$$q_i \geqslant \left(\sum_{j \neq i} d_{ij} \right) \frac{\max\limits_{\mathbf{x} \in \partial \varpi_i} \|\mathbf{x}_i - \mathbf{x}\|_2}{\min\limits_{\mathbf{x} \in \partial(\mathrm{conv}(\varpi_i))} \|\mathbf{x}_i - \mathbf{x}\|_2} \qquad \forall i \in \{1, \ldots, N\}, \quad (4.135)$$

where $\varpi_i \subseteq \Omega$ is the local patch of node i (as defined at the beginning of Chapter 3) and $\mathrm{conv}(\varpi_i)$ denotes its convex hull. The ratio on the right hand side of inequality (4.135) depends only on the triangulation $\mathcal{T}_h$ and not on the solution values. Therefore, it can be calculated a priori and does not need to be updated along with the AFC solution. It is worth mentioning that the computation of the smallest parameters q_i satisfying (4.135) involves significant effort and produces rather restrictive bounds. A more practical linearity-preserving lower bound for q_i was derived by Kuzmin [Kuz18a, Eq. (1.10)] using the estimate

$$q_i \geqslant \left(\sum_{j \neq i} d_{ij}\right) \frac{\max\limits_{K \in \mathcal{T}_h} h_K}{\min\limits_{K \in \mathcal{T}_h} \rho_K} \geqslant \left(\sum_{j \neq i} d_{ij}\right) \frac{\max\limits_{\mathbf{x} \in \partial \varpi_i} \|\mathbf{x}_i - \mathbf{x}\|_2}{\min\limits_{\mathbf{x} \in \partial(\mathrm{conv}(\varpi_i))} \|\mathbf{x}_i - \mathbf{x}\|_2}$$

$$\forall i \in \{1, \ldots, N\}.$$

Algorithms 4.77 and 4.78 differ in the way in which the auxiliary quantities $Q_i^{\pm}$ are determined. The replacement of u_j by a local extremum makes the bounds less restrictive and decreases the cost of calculating $Q_i^{\pm}$ in Algorithm 4.78.

Instead of limiting the sums of antidiffusive fluxes, each flux can be constrained independently in a manner which provides Lipschitz continuity. Moreover, the need for checking the sign of f_{ij} can be avoided by using the minimum of the correction factors for positive and negative fluxes. Introducing free parameters $p \in \mathbb{N}$, $q > 0$, and $\varepsilon \geqslant 0$, we consider the following algorithm, which we later extend to unsteady advection problems and tensor fields.

Algorithm 4.79. *For each pair of nodes i and $j \neq i$, calculate the auxiliary quantities*

$$\beta_{ij} := \begin{cases} 1 & : a_{ij} \leqslant 0, \\ 1 - \left(1 - \min\left(1, \frac{q(u_i^{\max} - u_i)}{|u_j - u_i| + \varepsilon}, \frac{q(u_i - u_i^{\min})}{|u_j - u_i| + \varepsilon}\right)\right)^p & : a_{ij} > 0 \end{cases} \qquad (4.136)$$

and define the correction factors α_{ij} as follows:

$$\alpha_{ij} := \min(\beta_{ij}, \beta_{ji}) \qquad \forall i, j \in \{1, \ldots, N\}, \, j \neq i. \qquad (4.137)$$

By definition of β_{ij}, we have $\alpha_{ij} = \beta_{ij} = 0$ if u_i is a local extremum and $a_{ij} > 0$. Similarly, we have $\alpha_{ij} = \beta_{ji} = 0$ if u_j is a local extremum and $a_{ji} > 0$. Since $\beta_{ji} \neq \beta_{ij}$ in general, the minimum of the auxiliary correction factors is taken in (4.137) to satisfy the symmetry condition $\alpha_{ij} = \alpha_{ji}$.

Remark 4.80. All limiting strategies presented so far produce Lipschitz continuous residuals of the AFC system, which guarantees the convergence of the defect correction scheme under suitable additional assumptions (cf.

Section 4.5.1.1). However, the constrained fluxes are not twice continuously differentiable and, hence, more advanced solvers like the Newton's method are not readily applicable to nonlinear problems of this kind. Therefore, the use of Anderson acceleration [BB17; Kuz12b] and more advanced quasi-Newton approaches to solving (4.98) have been proposed in the literature. For example, Möller [Möl08] approximated the Jacobian matrix by finite differences, while Badia and Bonilla [BB17] employed suitable regularizations of the absolute value to make the residual differentiable.

Another way to ensure that the function $\mathbb{R}^N \ni u \mapsto \alpha_{ij}(u)(u_j - u_i) \in \mathbb{R}$ is in $C^{p-1}(\mathbb{R}^N)$, $p \in \mathbb{N}$, is to use the correction factors

$$\alpha_{ij} := \beta_{ij}^+ \beta_{ij}^- \beta_{ji}^+ \beta_{ji}^- \qquad \forall i, j \in \{1, \dots, N\}, \, j \neq i,$$

where $\beta_{ij}^{\pm} \in [0,1]$ (and $\beta_{ji}^{\pm}$) are defined in terms of the parameters $q, \varepsilon > 0$ by

$$\beta_{ij}^+ := \begin{cases} 1 & : a_{ij} \leqslant 0, \\ 1 - \max\left(0, 1 - q \sum_{k \in \mathcal{N}_i \backslash \{i\}} \frac{\tilde{m}_{ik} \max(0, u_k - u_i)^p}{m_i |u_j - u_i|^p + \varepsilon}\right)^p & : a_{ij} > 0, \end{cases}$$

$$\beta_{ij}^- := \begin{cases} 1 & : a_{ij} \leqslant 0, \\ 1 - \max\left(0, 1 - q \sum_{k \in \mathcal{N}_i \backslash \{i\}} \frac{\tilde{m}_{ik} \max(0, u_i - u_k)^p}{m_i |u_j - u_i|^p + \varepsilon}\right)^p & : a_{ij} > 0. \end{cases}$$

The so defined constrained fluxes are twice continuously differentiable only if $p \geqslant 3$ and $\varepsilon > 0$. As in [BB17], the latter requirement causes the lack of strict linearity preservation no matter how q is chosen. In a practical implementation, the parameters q and ε may be adapted in an iterative way while solving the AFC system using Newton's method.

Element-based versions of the presented limiters can be configured to constrain antidiffusive element contributions (as defined in Section 4.4.1 for FCT algorithms) using the same nodal correction factors. A detailed description and analysis of resulting algorithms is beyond the scope of this work. For examples of element-based monolithic AFC schemes, the interested reader is referred, e.g., to [Kuz+17; Kuz18a].

4.5.2 Unsteady problem

In the previous section, the nonlinear AFC methodology was presented in the context of DMP satisfying finite element approximations to the steady advection-reaction equation. The main idea behind the AFC approach is the incorporation of diffusive and antidiffusive corrections into the residual of a nonlinear system. In this section, we present and analyze such algebraic correction tools for the semi-discrete form of the unsteady advection-reaction equation. If discretization in time is performed using a strong stability preserving (SSP) time integrator [Got+11], DMP properties of the fully discrete nonlinear problem can be shown, at least for sufficiently small time steps. The fully implicit backward Euler method is the only SSP scheme which yields bound-preserving solutions without any time step restriction. Other SSP time integrators guarantee the validity of DMPs under CFL-like conditions.

The use of the nonlinear AFC methodology for unsteady conservation laws goes back to the publications [KT04; KM05]. The AFC approaches developed therein are based on the semi-discrete LED criterion [Jam93; Cou+98; Lyr95], which leads to many useful multidimensional generalizations of total variation diminishing (TVD) schemes [Har84; Yee87] to edge-based finite element discretizations on unstructured meshes [Löh08; Lyr+94].

The semi-discrete low order scheme (4.63) was constructed using mass lumping and discrete upwinding for convection matrices in (3.18). This algebraic manipulation of the discretization at hand leads to an ODE system satisfying the LED condition. Furthermore, the use of an SSP time integrator guarantees the DMP property of the fully discrete AFC method, possibly under CFL-like conditions. However, the introduced artificial diffusion is too strong to achieve acceptable accuracy on coarse meshes. To reduce the error in smooth regions without local extrema, we proceed as in the steady case: Decomposing the antidiffusive correction terms into fluxes, multiplying them by adaptively chosen correction factors, and incorporating the result into the residual of the semi-discrete problem, we obtain

$$m_i d_t u_i + \sum_{j \neq i} \dot{\alpha}_{ij} m_{ij} (d_t u_j - d_t u_i) - \sum_{j=1}^{N} k_{ij} u_j - \sum_{j \neq i} (1 - \alpha_{ij}) d_{ij} (u_j - u_i) = g_i$$

$$\forall i \in \{1, \ldots, N\}, \quad (4.138)$$

where $\alpha_{ij}, \dot{\alpha}_{ij} \in [0, 1]$ denote symmetric correction factors associated with the spatial and temporal discrete diffusion operators $\mathcal{D}$ and $\mathcal{M} - \mathcal{M}_L$, respectively. By setting $\alpha_{ij} = \dot{\alpha}_{ij} = 1$ for all $i, j \in \{1, \ldots, N\}$, $j \neq i$, the semi-discrete target method (3.18) is recovered. Conversely, system (4.138) reverts to the low order method (4.63) if all correction factors vanish.

In general, the correction factors α_{ij} and $\dot{\alpha}_{ij}$ may depend nonlinearly on the solution u and its temporal derivative $d_t u$. Their definition should guarantee desirable properties like the Lipschitz continuity of residuals and the LED property of the solution. Due to the presence of time derivatives in the antidiffusive fluxes depending on m_{ij}, the semi-discrete problem (4.138) represents an unwieldy nonlinear system of differential-algebraic equations (DAEs) with a partially lumped solution-dependent mass matrix. To express our problem in the canonical ODE form, we replace $d_t u$ in the nonlinear correction term by a suitable approximation $\dot{u} = \dot{u}(u) \approx d_t u$. Then the general AFC system for the unsteady advection-reaction equation (2.30) simplifies to

$$m_i \frac{du_i}{dt} = g_i + z_i(u) \qquad \forall i \in \{1, \dots, N\}, \qquad (4.139a)$$

where the right hand side z is given by

$$z_i(u) := \sum_{j=1}^{N} k_{ij} u_j + \sum_{j \neq i} (1 - \alpha_{ij}) d_{ij}(u_j - u_i) - \sum_{j \neq i} \dot{\alpha}_{ij} m_{ij}(\dot{u}_j - \dot{u}_i)$$
$$\forall i \in \{1, \dots, N\} \quad (4.139b)$$

and does not depend on $d_t u$ any more. Appropriate approximations of the time derivative $d_t u$ can be defined for example, by invoking (3.18) or (4.63). The corresponding high and low order approximations are given by

$$\dot{u}^{\mathrm{H}} := \mathcal{M}^{-1}(g + \mathcal{K}u), \qquad \dot{u}^{\mathrm{L}} := \mathcal{M}_{\mathrm{L}}^{-1}(g + (\mathcal{K} + \mathcal{D})u). \qquad (4.140)$$

Note that the former definition of $\dot{u}$ still requires inversion of the mass matrix. However, since $\mathcal{M}$ is independent of the solution, it can be approximated conveniently by a truncated Neumann series [Gue+14; Kuz18a]. The use of the low order lumped approximation $\dot{u}^{\mathrm{L}}$ implies that the semi-discrete Galerkin system (3.18) can no longer be recovered by setting all correction factors equal to one. However, the target scheme corresponding to $\alpha_{ij} \equiv 1$ in this version is sufficiently accurate and more stable than the original Galerkin approximation. At the moment, the exact definition of $\dot{u}$ is of minor interest but we assume that $\dot{u}$ is a Lipschitz continuous function of u.

The following theorem formulates requirements that are sufficient to guarantee the LED property of the solution to (4.139). For the proof it is essential that the antidiffusive fluxes vanish at local extrema. As in Section 4.3.2, we hereafter assume that (4.62) holds because the exact solution satisfies maximum principles under this assumption.

Theorem 4.81. *The solution of* (4.139) *is LED if*

$$\alpha_{ij}(u_i - u_j) = 0 \qquad \forall i,j \in \{1,\dots,N\},\ j \neq i,\ s.t.\ u_i = \max_{l \in \mathcal{N}_i} u_l \wedge k_{ij} < 0,$$
$$(4.141a)$$

$$\alpha_{ij}(u_i - u_j) = 0 \qquad \forall i,j \in \{1,\dots,N\},\ j \neq i,\ s.t.\ u_i = \min_{l \in \mathcal{N}_i} u_l \wedge k_{ij} < 0,$$
$$(4.141b)$$

$$\dot{\alpha}_{ij}(\dot{u}_i - \dot{u}_j) = 0 \qquad \forall i,j \in \{1,\dots,N\},\ j \neq i,\ s.t.\ u_i = \max_{l \in \mathcal{N}_i} u_l, \qquad (4.141c)$$

$$\dot{\alpha}_{ij}(\dot{u}_i - \dot{u}_j) = 0 \qquad \forall i,j \in \{1,\dots,N\},\ j \neq i,\ s.t.\ u_i = \min_{l \in \mathcal{N}_i} u_l \qquad (4.141d)$$

and the functions

$$\mathbb{R}^N \ni v \ \mapsto\ \alpha_{ij}(v)(v_i - v_j) \in \mathbb{R}, \qquad \mathbb{R}^N \ni v \ \mapsto\ \dot{\alpha}_{ij}(v)(\dot{v}_i - \dot{v}_j) \in \mathbb{R}$$
$$(4.142)$$
are Lipschitz continuous for all $i,j \in \{1,\dots,N\}$, $j \neq i$.

Proof. The result can easily be shown combining the proof techniques used in Theorem 4.19 and Lemma 4.69. $\qquad\qquad\qquad\qquad\qquad\qquad\qquad\qquad \square$

The LED property of the solution to the semi-discrete problem (4.139) makes it possible to guarantee the validity of local and global DMPs for its fully discretized counterpart. If the employed time integrator is not fully implicit, we expect that a CFL-like time step restriction like (4.66) will apply. To illustrate this point, we discretize (4.139) in time using the θ-scheme and obtain the fully discrete method

$$m_i u_i^{n+1} = \underline{g}_i + m_i u_i^n + \Delta t \theta z_i^{n+1}(u^{n+1}) + \Delta t(1-\theta)z_i^n(u^n) \qquad \forall i \in \{1,\dots,N\},$$
$$(4.143)$$
where $\underline{g}$ is defined as in (4.65). Furthermore, we redefine the auxiliary quantity (cf. Section 4.2.3)

$$\underline{p}_i := -\Delta t \theta \left(\sum_{j=1}^N k_{ij}^{n+1} \right) - \Delta t(1-\theta)\left(\sum_{j=1}^N k_{ij}^n \right) \qquad \forall i \in \{1,\dots,N\}$$

in terms of $\mathcal{K}^{n+1}$ because the matrices $\mathcal{A}$ and $\mathcal{B}$ may no longer be readily available for the AFC system with antidiffusive terms depending on $\partial_t u$ or its approximation $\dot{u}$.

Using this notation, the result of Lemma 4.24 can be adapted to the fully discrete AFC system (4.143).

Lemma 4.82. *Let the correction factors α_{ij} and $\dot{\alpha}_{ij}$ of (4.143) satisfy (4.141). For $\theta \in [0,1)$, we additionally require the existence of constants $L_1, L_2 > 0$ such that*

$$\alpha_{ij}(u_i - u_j) \leqslant L_1(u_i^{\max} - u_i) \qquad \forall i,j \in \{1,\ldots,N\},\ j \neq i\ \text{if}\ k_{ij} < 0,$$
$$\text{(4.144a)}$$

$$\alpha_{ij}(u_j - u_i) \leqslant L_1(u_i - u_i^{\min}) \qquad \forall i,j \in \{1,\ldots,N\},\ j \neq i\ \text{if}\ k_{ij} < 0,$$
$$\text{(4.144b)}$$

$$\dot{\alpha}_{ij}(\dot{u}_i - \dot{u}_j) \leqslant L_2(u_i^{\max} - u_i) \qquad \forall i,j \in \{1,\ldots,N\},\ j \neq i, \qquad \text{(4.144c)}$$

$$\dot{\alpha}_{ij}(\dot{u}_j - \dot{u}_i) \leqslant L_2(u_i - u_i^{\min}) \qquad \forall i,j \in \{1,\ldots,N\},\ j \neq i. \qquad \text{(4.144d)}$$

For $\theta = 1$, the constants $L_1, L_2 > 0$ can be chosen arbitrarily. Assume that the CFL-like condition

$$\Delta t(1 - \theta)\big(L_2(m_i - m_{ii}) + (1 + L_1)|d_{ii}^n| - k_{ii}^n\big) \leqslant m_i \qquad \text{(4.145)}$$

is satisfied for every $i \in \{1,\ldots,N\}$. Then the statements of Lemma 4.24 without requirements (i)–(iv) hold for the solution u^{n+1} of (4.143).

Note that, strictly speaking, the validity of conditions (4.141), (4.144), and (4.145) for all $i \in I \setminus J$ (or all $i \in \{1,\ldots,N\} \setminus J$ in the case of (4.34)) is sufficient to guarantee that the corresponding generalized maximum principles hold.

Proof. Without loss of generality, we assume that $\alpha_{ij} = 0$ (resp. $\dot{\alpha}_{ij} = 0$) for $i,j \in \{1,\ldots,N\}$, $j \neq i$, if $u_i = u_j$ (resp. $\dot{u}_i = \dot{u}_j$).

We first consider the case $\theta \in [0,1)$ in which

$$\big(k_{ij}^n + (1 - \alpha_{ij}^n)d_{ij}^n\big)(u_j^n - u_i^n) \leqslant \big(k_{ij}^n + (1 + L_1)d_{ij}^n\big)(u_i^{\max,n} - u_i^n)$$
$$\forall i,j \in \{1,\ldots,N\},\ j \neq i. \quad \text{(4.146)}$$

Indeed, if $k_{ij}^n < 0$, we have

$$(k_{ij}^n + d_{ij}^n)(u_j^n - u_i^n) - d_{ij}^n \alpha_{ij}^n(u_j^n - u_i^n)$$
$$\leqslant (k_{ij}^n + d_{ij}^n)(u_i^{\max,n} - u_i^n) + d_{ij}^n L_1(u_i^{\max,n} - u_i^n)$$

by virtue of (4.144a). In the case of $k_{ij}^n \geqslant 0$, we obtain

$$k_{ij}^n(u_j^n - u_i^n) + (1 - \alpha_{ij}^n)d_{ij}^n(u_j^n - u_i^n) \leqslant k_{ij}^n(u_i^{\max,n} - u_i^n) + d_{ij}^n(u_i^{\max,n} - u_i^n).$$

According to (4.146) and the assumption that $\theta \in [0, 1)$, we deduce

$$z_i^n(u^n) - \Big(\sum_{j=1}^{N} k_{ij}^n\Big)u_i^n$$

$$= \sum_{j \neq i}(k_{ij}^n + (1 - \alpha_{ij}^n)d_{ij}^n)(u_j^n - u_i^n) + \sum_{j \neq i} m_{ij}\dot{\alpha}_{ij}^n(\dot{u}_i^n - \dot{u}_j^n)$$

$$\leqslant \Big(\sum_{j \neq i}(k_{ij}^n + (1 + L_1)d_{ij}^n) + \sum_{j \neq i} m_{ij}L_2\Big)(u_i^{\max,n} - u_i^n)$$

$$= \Big(L_2(m_i - m_{ii}) + (1 + L_1)|d_{ii}^n| + \sum_{j \neq i} k_{ij}^n\Big)(u_i^{\max,n} - u_i^n)$$

$$= \Big(L_2(m_i - m_{ii}) + (1 + L_1)|d_{ii}^n| - k_{ii}^n\Big)(u_i^{\max,n} - u_i^n)$$

$$+ \Big(\sum_{j=1}^{N} k_{ij}^n\Big)(u_i^{\max,n} - u_i^n) \qquad \forall i \in \{1, \dots, N\}$$

by (4.144c) and the definition of $\mathcal{D}^n$. If (4.145) holds as well, this leads to

$$m_i u_i^n + \Delta t(1 - \theta)z_i^n(u^n)$$

$$\leqslant m_i u_i^n + (u_i^{\max,n} - u_i^n)m_i + \Delta t(1 - \theta)\Big(\sum_{j=1}^{N} k_{ij}^n\Big)u_i^{\max,n} \tag{4.147}$$

$$= \Big(m_i + \Delta t(1 - \theta)(\sum_{j=1}^{N} k_{ij}^n)\Big)u_i^{\max,n} \qquad \forall i \in \{1, \dots, N\}$$

regardless of the choice of $\theta \in [0, 1]$. Furthermore, we have

$$m_i + \Delta t(1 - \theta)\sum_{j=1}^{N} k_{ij}^n = m_i > 0 \qquad \forall i \in \{1, \dots, N\}, \theta = 1 \tag{4.148a}$$

and

$$m_i + \Delta t(1 - \theta)\sum_{j=1}^{N} k_{ij}^n$$

$$= m_i + \Delta t(1 - \theta)(k_{ii}^n + d_{ii}^n) + \Delta t(1 - \theta)\sum_{j \neq i}(k_{ij}^n + d_{ij}^n)$$

$$\geqslant \Delta t(1 - \theta)\big(L_2(m_i - m_{ii}) + L_1|d_{ii}^n|\big) > 0$$

$$\forall i \in \{1, \dots, N\}, \theta \in [0, 1) \tag{4.148b}$$

according to (4.145) and in view of the fact that $k_{ij}^n + d_{ij}^n \geqslant 0$ for all $j \neq i$.

To prove (4.32a), let us suppose that there exists an index $i \in I \setminus J$ such that

$$u_i^{n+1} = \max_{j \in I \setminus J} u_j^{n+1} > c := \max\left(\max_{i \in \bar{I}} u_i^n, \max_{i \in (J \cap I) \cup \partial I} u_i^{n+1}\right). \qquad (4.149)$$

If there is no such index i, then condition (4.32a) is satisfied and we are done. According to (4.149), we have $u_i^{n+1} = u_i^{\max,n+1}$ and

$$\hat{k}_{ij}^{n+1} := k_{ij}^{n+1} + (1 - \alpha_{ij}^{n+1})d_{ij}^{n+1} \geqslant 0, \quad \dot{\alpha}_{ij}^{n+1} = 0 \qquad \forall j \in \mathcal{N}_i \setminus \{i\}.$$

Consequently, the vector of unknowns u^{n+1} satisfies

$$\sum_{j=1}^{N} \hat{a}_{ij} u_j^{n+1} := \left(m_i - \Delta t\theta\left(\sum_{j=1}^{N} k_{ij}^{n+1}\right) + \Delta t\theta \sum_{j \neq i} \hat{k}_{ij}^{n+1}\right) u_i^{n+1}$$
$$- \Delta t\theta \sum_{j \neq i} \hat{k}_{ij}^{n+1} u_j^{n+1}$$
$$= m_i u_i^{n+1} - \Delta t\theta \sum_{j=1}^{N} k_{ij}^{n+1} u_j^{n+1}$$
$$- \Delta t\theta \sum_{j \neq i} (1 - \alpha_{ij}^{n+1})d_{ij}^{n+1}(u_j^{n+1} - u_i^{n+1})$$
$$= m_i u_i^{n+1} - \Delta t\theta z_i^{n+1}(u^{n+1}) = \underline{g}_i + m_i u_i^n + \Delta t(1 - \theta)z_i^n(u^n)$$

by (4.143), where $\hat{a}_{ij} \leqslant 0$ if $j \neq i$. Recalling (4.147) and the fact that $i \in I \setminus J$, we find that

$$\left(m_i - \Delta t\theta \sum_{j=1}^{N} k_{ij}^{n+1}\right) u_i^{n+1} = \left(\sum_{j=1}^{N} \hat{a}_{ij}\right) u_i^{n+1}$$
$$\leqslant \sum_{j=1}^{N} \hat{a}_{ij} u_j^{n+1} = \underline{g}_i + m_i u_i^n + \Delta t(1 - \theta)z_i^n(u^n)$$
$$\leqslant \left(m_i + \Delta t(1 - \theta)\left(\sum_{j=1}^{N} k_{ij}^n\right)\right) u_i^{\max,n} < \left(m_i + \Delta t(1 - \theta)\sum_{j=1}^{N} k_{ij}^n\right) u_i^{n+1}$$

by virtue of (4.148) and (4.149). Therefore, we have $\underline{p}_i u_i^{n+1} < 0$, which cannot be the case since $\underline{p}_i = 0$ for $i \in I \setminus J$. This contradiction completes the proof of (4.32a).

In the proof of (4.32c), we redefine $c \geqslant 0$ as the right hand side of (4.32c). The same argumentation leads to $p_i u_i^{n+1} < 0$ again. In this case, the statement follows from a contradiction to $p_i \geqslant 0$ and $u_i^{n+1} > c \geqslant 0$.

To show (4.34a), we note that (4.32a) with $I = \{1, \ldots, N\}$ implies the existence of an index $i \in J$ such that

$$u_i^{n+1} = \max_{j \in \{1, \ldots, N\}} u_j^{n+1}.$$

By (4.33) and the assumption that $\underline{p} \geqslant 0$, we have $\underline{p}_i > 0$. Without loss of generality, let us suppose that

$$u_i^{n+1} = \max_{j \in \{1, \ldots, N\}} u_j^{n+1} > \max_{j \in \{1, \ldots, N\}} u_j^n.$$

Otherwise, the result follows immediately. The same arguments as above yield

$$\Big(m_i - \Delta t \theta \sum_{j=1}^N k_{ij}^{n+1}\Big) u_i^{n+1} \leqslant \underline{g}_i + \Big(m_i + \Delta t (1-\theta) \sum_{j=1}^N k_{ij}^n\Big) u_i^{n+1}$$

and the result follows by rearranging.

The lower bounds of Lemma 4.24 can be proved similarly.   □

If all correction factors are set to zero, problem (4.143) is equivalent to the low order system (4.65) and condition (4.144) holds for $L_1, L_2 = 0$. Therefore, the CFL-like condition (4.145) coincides with that of the low order scheme (4.66) and Lemma 4.82 yields the same results as Lemma 4.24 in the context of (4.65).

Lemma 4.82 makes it possible to prove discrete maximum principles for any algebraic problem that can be written in the form (4.143). For the unsteady advection equation satisfying assumption (4.62), the DMP property of the solution to (4.143) can be shown as for its low order counterpart in Section 4.3.2 if a more restrictive time step condition is satisfied.

Any algorithm presented in Section 4.5.1.4 can be used to define the correction factors α_{ij} and $\dot{\alpha}_{ij}$ for problem (4.143) in a way which guarantees the validity of conditions (4.141) and (4.144). For example, the unsteady counterpart of Algorithm 4.79 operates as follows.

Algorithm 4.83. *For each pair of nodes i and $j \neq i$, calculate the auxiliary quantities*

$$\beta_{ij} := \begin{cases} 1 & : k_{ij} \geqslant 0, \\ 1 - \left(1 - \min\left(1, \frac{q_1(u_i^{\max} - u_i)}{|u_j - u_i| + \varepsilon}, \frac{q_1(u_i - u_i^{\min})}{|u_j - u_i| + \varepsilon}\right)\right)^{p_1} & : k_{ij} < 0, \end{cases} \quad (4.150a)$$

$$\dot{\beta}_{ij} := 1 - \left(1 - \min\left(1, \frac{q_2(u_i^{\max} - u_i)}{|\dot{u}_j - \dot{u}_i| + \varepsilon}, \frac{q_2(u_i - u_i^{\min})}{|\dot{u}_j - \dot{u}_i| + \varepsilon}\right)\right)^{p_2} \quad (4.150b)$$

and limit the antidiffusive fluxes using

$$\alpha_{ij} := \min(\beta_{ij}, \beta_{ji}), \quad \dot{\alpha}_{ij} := \min(\dot{\beta}_{ij}, \dot{\beta}_{ji}) \qquad \forall i,j \in \{1, \dots, N\}, \, j \neq i. \quad (4.151)$$

The definition of β_{ij} and $\dot{\beta}_{ij}$ is based on DMP constraints and continuity requirements for node i. The correction factors $\alpha_{ij} = \alpha_{ji}$ for antidiffusive fluxes associated with the matrix $\mathcal{K} = -\mathcal{A}$ are defined as in Algorithm 4.79. The correction factors $\dot{\alpha}_{ij} = \dot{\alpha}_{ji}$ for fluxes depending on the approximate time derivatives $\dot{u}$ are defined so as to guarantee the validity of conditions (4.144c) and (4.144d).

Remark 4.84. The correction factors α_{ij} and $\dot{\alpha}_{ij}$ defined by Algorithm 4.83 satisfy conditions (4.141) and (4.144) with $L_1 := p_1 q_1$ and $L_2 := p_2 q_2$.

The crucial part of the argumentation leading to this statement is the derivation of the constants L_1 and L_2 because (4.141) holds by definition. For the sake of simplicity, we only focus on condition (4.144c) and the definition of L_2. Using *Bernoulli's inequality*

$$(1 + x)^p \geqslant 1 + px \qquad \forall x \geqslant -1, \, p \in \mathbb{N} \cup \{0\} \quad (4.152)$$

for estimation of $\dot{\beta}_{ij}$, we find that

$$\dot{\alpha}_{ij}(\dot{u}_i - \dot{u}_j)$$
$$\leqslant \dot{\beta}_{ij}|\dot{u}_j - \dot{u}_i|$$
$$= |\dot{u}_j - \dot{u}_i| - \left(1 - \min\left(1, \frac{q_2(u_i^{\max} - u_i)}{|\dot{u}_j - \dot{u}_i| + \varepsilon}, \frac{q_2(u_i - u_i^{\min})}{|\dot{u}_j - \dot{u}_i| + \varepsilon}\right)\right)^{p_2}|\dot{u}_j - \dot{u}_i|$$
$$\leqslant |\dot{u}_j - \dot{u}_i| - \left(1 - p_2\min\left(1, \frac{q_2(u_i^{\max} - u_i)}{|\dot{u}_j - \dot{u}_i| + \varepsilon}, \frac{q_2(u_i - u_i^{\min})}{|\dot{u}_j - \dot{u}_i| + \varepsilon}\right)\right)|\dot{u}_j - \dot{u}_i|$$
$$\leqslant p_2 \frac{q_2(u_i^{\max} - u_i)}{|\dot{u}_j - \dot{u}_i| + \varepsilon}|\dot{u}_j - \dot{u}_i| \leqslant p_2 q_2(u_i^{\max} - u_i).$$

This yields $L_2 := p_2 q_2$.

In practical implementations, we define the limiting parameters $\dot{\beta}_{ij}$ in a dimensionless manner using $q_2 = \Delta t^{-1}$ and $p_2 = 1$. Then $L_1 \geqslant 0$ can be determined such that the CFL-like condition (4.145) is satisfied since $\Delta t(1-\theta)L_2 = 1 - \theta \leqslant 1$ for $\theta \in [0,1]$ regardless of Δt.

4.6 Numerical examples

In what follows, we perform numerical investigations of the above methods using standard steady and unsteady two dimensional benchmarks. In our presentation of the results, we call $u_h \in V_h$ the *(monolithic) AFC solution* if the degrees of freedom solve (4.98) or (4.143). The nodal function values of the *FCT solution* $u_h \in V_h$ are computed as described in Section 4.4. Furthermore, we are just interested in numerical studies of the localized edge-based approaches involving Algorithms 4.58, 4.79, and 4.83.

4.6.1 Steady problem

This section deals with the bound-preserving treatment of the circular convection benchmark as described in Section 3.4. For simplicity, we restrict our attention to the limit of vanishing reactivity ($c = 0$ and $f = 0$), in which the coercivity condition (2.14) is violated and the Galerkin discretization struggles with the possible lack of convergence. Note that in this case some elements of the theory behind the monolithic AFC approach are not applicable either.

The solution to (4.98) is computed using a pseudo time stepping approach as described in [Loh19, Section 8], where the pseudo time increment Δt is chosen sufficiently small to obtain convergence. In a nutshell, the solution $u^{s+1} \in \mathbb{R}^N$ of the $s+1$-st cycle satisfies

$$\left(m_i + \Delta t \sum_{j=1}^N (a_{ij} - d_{ij})\right) u_i^{s+1} = m_i u_i^s + \Delta t g_i - \Delta t \sum_{j \neq i} \alpha_{ij} d_{ij}(u_j^s - u_i^s)$$

$$\forall i \in \{1, \dots, N\}.$$

We first consider the circular convection benchmark with vanishing reactivity parameter and exact solution given by (3.24). In Table 4.1, the $\|\cdot\|_{L^2(\Omega)}$-errors and experimental orders of convergence are presented for the low order approximation and the AFC scheme with parameters $p = q = 2$ (cf. Algorithm 4.79). While the Galerkin method is very unstable on $\mathcal{T}_h^1$ and the corresponding solution exhibits spurious oscillations (cf. Section 3.4), the low order solution enjoys the DMP properties and converges to the exact solution. However, the errors are quite large due to the tremendous amount of artificial diffusion. The addition of limited antidiffusive fluxes in the nonlinear AFC problem improves the accuracy of the solution and results in higher EOC. However, no a priori error estimates for AFC discretizations of hyperbolic problems are currently available to confirm that optimal convergence behavior can be achieved for sufficiently smooth data. While the accuracy of the Galerkin solution to problem (3.22) depends on the triangulation at hand, the impact of the mesh seems to be negligible for the bound-preserving approximations under investigation.

The finite element AFC solution remains more accurate than its low order counterpart if the smooth exact solution is replaced by the composition (3.25) of smooth and discontinuous functions (cf. Fig. 4.6b). The outflow profile of u_h^L is very diffusive and the discontinuities are completely smeared out. In the case of the AFC solution with $p = q = 1$, the amount of artificial diffusion is reduced significantly enough for the profile of the step function to become recognizable. However, both nonzero parts of the exact solution are still blurred and significant peak clipping effects are observed at the smooth summit. The overall accuracy of the AFC solution improves as the parameters p and q are increased.

As mentioned in Section 4.4, the pseudo time stepping approach makes it possible to solve stationary problems using predictor-corrector algorithms as well. Although monolithic AFC schemes are better suited for this purpose, we also present the FCT results for the steady circular convection benchmark. In this numerical experiment, the transient counterpart of the steady advection equation is solved using the methodology of Section 4.4.2 and $\theta = 0$ until the difference between the FCT solutions at two consecutive time steps becomes sufficiently small. The maximal pseudo time increment $\Delta t = 5.21 \cdot 10^{-3}$ is chosen as large as possible without violating the CFL-like condition (4.66) for any index $i \in \{1, \ldots, N\}$. Since the low order method was shown to satisfy discrete maximum principles under this condition and the FCT correction is bound-preserving, the use of slack stopping criteria does not result in overshoots or undershoots. While the peak preservation property of the FCT algorithm improves as Δt decreases, the resolution of the step function remains virtually unchanged (cf. Fig. 4.6a). The lack of improvement in this region can be attributed to the fact that the overall accuracy of FCT-constrained approximations depends both on the definition of the antidiffusive fluxes and on the way in which these fluxes are limited. The use of $\dot{u}^L$ in the fluxes of the FCT algorithm (cf. (4.97)) stabilizes the target

Table 4.1: Circular convection (smooth test): Convergence of $\|\cdot\|_{L^2(\alpha)}$-errors for different numerical solutions on uniform and distorted meshes $\mathcal{T}_h^1$ and $\tilde{\mathcal{T}}_h^1$ (cf. Section 3.4) with reactivity parameter $c = 0$.

	mesh level	Galerkin		low order		AFC with $p = q = 2$	
		error	EOC	error	EOC	error	EOC
$\mathcal{T}_h^1$	5	$1.96 \cdot 10^{-3}$		$1.35 \cdot 10^{-1}$		$2.71 \cdot 10^{-2}$	
	6	$1.05 \cdot 10^{0}$	-9.06	$9.33 \cdot 10^{-2}$	0.54	$8.69 \cdot 10^{-3}$	1.64
	7	$2.30 \cdot 10^{-1}$	2.19	$5.89 \cdot 10^{-2}$	0.66	$2.56 \cdot 10^{-3}$	1.76
	8	$1.87 \cdot 10^{0}$	-3.02	$3.43 \cdot 10^{-2}$	0.78	$6.95 \cdot 10^{-4}$	1.88
$\tilde{\mathcal{T}}_h^1$	5	$1.56 \cdot 10^{-2}$		$1.37 \cdot 10^{-1}$		$2.91 \cdot 10^{-2}$	
	6	$5.21 \cdot 10^{-3}$	1.58	$9.46 \cdot 10^{-2}$	0.53	$9.53 \cdot 10^{-3}$	1.61
	7	$2.07 \cdot 10^{-3}$	1.33	$5.98 \cdot 10^{-2}$	0.66	$2.96 \cdot 10^{-3}$	1.69
	8	$7.48 \cdot 10^{-4}$	1.47	$3.48 \cdot 10^{-2}$	0.78	$8.65 \cdot 10^{-4}$	1.77

scheme by introducing numerical diffusion. Therefore, the gradients of the finite element approximation to the step function cannot become steeper than those of the stabilized target no matter how the correction factors are defined. In fact, very similar results are produced by the monolithic AFC scheme using the pseudo time stepping approach and the time derivative approximation $\dot{u}^L$ as defined in (4.140).

4.6.2 Unsteady problems

4.6.2.1 Transient circular convection

Let us now consider the transient counterpart of the circular convection benchmark. In contrast to the previous example, in which we marched the solution of an unsteady advection-reaction equation to the steady state, we abort the simulation at the final time $T = \frac{\pi}{3}$. Initializing the solution by $u_0 \equiv 0$ and considering a time independent inflow boundary condition, we find that the exact solution is given by

$$u(\mathbf{x}, t) = \begin{cases} u_{\text{in}}\big(\|\mathbf{x}\|_2, 0\big) & : \|\mathbf{x}\|_2 \leqslant 1 \wedge \frac{x_1}{x_2} \leqslant \tan(t), \\ u_{\text{in}}\big(1, \sqrt{\|\mathbf{x}\|_2^2 - 1}\big) & : \|\mathbf{x}\|_2 > 1 \wedge \frac{x_1}{x_2} \leqslant \tan(t), \\ 0 & : \text{otherwise} \end{cases}$$

$$\forall \mathbf{x} \in \Omega, \, t \in [0, T]. \quad (4.153)$$

We impose the inflow boundary function $u_{\text{in}} = u^\infty|_{\Gamma_{\text{in}}}$, where u^∞ is defined by (3.25). To avoid iterative solution of the nonlinear AFC problem (4.143) at each time step and illustrate stability issues associated with the use of the Galerkin approximation as the high order target, we discretize in time using the forward Euler method ($\theta = 0$). In this example, the parameters of the AFC limiter are given by $p_1 = q_1 = 2$ (and $p_2 = 1$, $q_2 = \Delta t^{-1}$ as defined in Remark 4.84) while the time increment Δt is chosen so that the CFL-like condition (4.145) holds for all $i \in \{1, \ldots, N\}$. This choice of Δt suffices to satisfy the CFL-like condition (4.66), too. Three possible definitions of the approximate time derivative $\dot{u}$ in (4.143) are considered in this study: $\dot{u}^L$ of (4.140), $\dot{u} = 0$ (denoted by $\dot{\alpha}_{ij} = 0$), and $\dot{u}^*$ defined by (4.138) with $\dot{\alpha}_{ij} = 0$ (cf. [Kuz18a]), i.e.,

$$\dot{u}_i^* = m_i^{-1}\Big(g_i + \sum_{j=1}^{N} k_{ij} u_j + \sum_{j \neq i}(1 - \alpha_{ij}) d_{ij}(u_j - u_i)\Big) \qquad \forall i \in \{1, \ldots, N\}.$$

Note that $\dot{u}^*$ vanishes for converged steady state solutions, whereas $\dot{u}^L$ is generally nonvanishing even in the steady state limit.

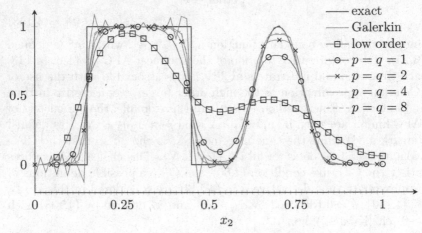

(a) FCT solutions obtained with different pseudo time increments Δt.

(b) Monolithic AFC solutions obtained with different parameters p and q.

Figure 4.6: Circular convection: Cutline $x_1 = 0$ profiles (cf. Fig. 3.3) of different numerical solutions on level 6 of $\mathcal{T}_h^3$.

The standard Galerkin method for the transient advection equation discretized in time using the forward Euler method is unconditionally unstable. The approximation obtained on level 6 of $\mathcal{T}_h^3$ is polluted by spurious oscillations and violates global bounds of the prescribed boundary condition (cf. Fig. 4.7a). Stabilization of this 'high order' target scheme as described in Section 4.3.2 leads to a low order solution possessing DMP properties if condition (4.66) is satisfied (cf. Fig. 4.7b). However, even though the correct amount of mass enters the domain, its distribution is highly smeared and has a larger support than the exact solution. The FCT correction of the low order predictor reduces the amount of artificial diffusion considerably. In this example, we constrain the linearized antidiffusive fluxes using the localized edge-based limiting procedure of Algorithm 4.58. The resulting FCT approximation satisfies local discrete maximum principles w.r.t. extended stencils in Definition 4.21 (cf. Fig. 4.7c). Both nonzero parts of the solution are clearly separated and their contour lines are very smooth and symmetric. Comparable results can be obtained with the monolithic AFC approach if $d_t u$ is approximated by $\dot{u}^L$ (cf. Fig. 4.7d). If the approximation $d_t u \approx \dot{u}^*$ is used, ripples occur in the contour plots of the AFC solution and the symmetry of the exact solution is not reproduced (cf. Fig. 4.7e). The spurious distortions of contour lines are caused by numerical dispersion effects which become more pronounced if all antidiffusive fluxes associated with $\dot{u}$ are canceled by setting $\dot{\alpha}_{ij} = 0$ (cf. Fig. 4.7f). Additionally, the poor phase accuracy of the lumped mass target results in a lack of monotonicity along the trajectory of the smooth peak. Note that this behavior does not violate DMPs. Global bounds are still satisfied and the local extremum forms in the evolution process due to dispersive crosswind fluxes which change the height of the traveling peak without violating local DMPs.

4.6.2.2 Solid body rotation

Another test that is commonly used to evaluate the behavior of bound-preserving methods is the *solid body rotation* benchmark which was originally introduced by Zalesak [Zal79] and extended by LeVeque [LeV96]. Three solid bodies are rotated around the center of the domain $\Omega = (0,1)^2$ using the stationary velocity field $\mathbf{v} = (\frac{1}{2} - x_2, x_1 - \frac{1}{2})^\top$ (cf. Fig. 4.8). Initially, the three circular objects with radius $r = 0.15$ are centered at $\mathbf{x}^{(1)} = (0.25, 0.5)^\top$,

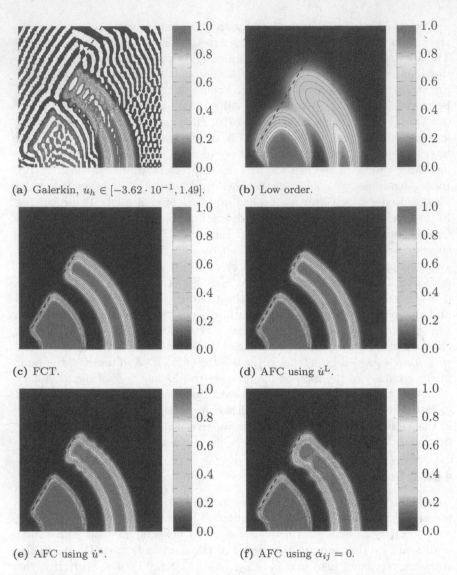

(a) Galerkin, $u_h \in [-3.62 \cdot 10^{-1}, 1.49]$.   (b) Low order.

(c) FCT.   (d) AFC using $\dot{u}^L$.

(e) AFC using $\dot{u}^*$.   (f) AFC using $\dot{\alpha}_{ij} = 0$.

Figure 4.7: Unsteady circular convection: Solutions obtained using different numerical methods on level 6 of $\mathcal{T}_h^3$ with time increment $\Delta t = 6.39 \cdot 10^{-4}$. Overshoots and undershoots are plotted in white.

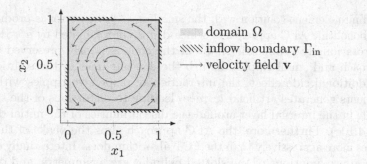

Figure 4.8: Solid body rotation: Geometry of domain Ω, velocity field $\mathbf{v} = (\frac{1}{2} - x_2, x_1 - \frac{1}{2})^\top$, and inflow boundary Γ_{in}.

$\mathbf{x}^{(2)} = (0.5, 0.25)^\top$, and $\mathbf{x}^{(3)} = (0.5, 0.75)^\top$. More specifically, the initial condition u_0 is given by

$$u_0(\mathbf{x}) := \begin{cases} u^{(1)}\big((\mathbf{x} - \mathbf{x}^{(1)})r^{-1}\big) & : \|\mathbf{x} - \mathbf{x}^{(1)}\|_2 \leqslant r, \\ u^{(2)}\big((\mathbf{x} - \mathbf{x}^{(2)})r^{-1}\big) & : \|\mathbf{x} - \mathbf{x}^{(2)}\|_2 \leqslant r, \\ u^{(3)}\big((\mathbf{x} - \mathbf{x}^{(3)})r^{-1}\big) & : \|\mathbf{x} - \mathbf{x}^{(3)}\|_2 \leqslant r, \\ 0 & : \text{otherwise} \end{cases} \quad \forall \mathbf{x} \in \Omega, \quad (4.154)$$

where the functions

$$u^{(1)}(\mathbf{x}) = \tfrac{1}{4}\big(1 + \cos(\pi \|\mathbf{x}\|_2)\big) \qquad \forall \mathbf{x} \in \mathbb{R}^2 \text{ s.t. } \|\mathbf{x}\|_2 \leqslant 1,$$

$$u^{(2)}(\mathbf{x}) = 1 - \|\mathbf{x}\|_2 \qquad \forall \mathbf{x} \in \mathbb{R}^2 \text{ s.t. } \|\mathbf{x}\|_2 \leqslant 1,$$

$$u^{(3)}(\mathbf{x}) = \begin{cases} 1 & : |x_1| \geqslant \tfrac{1}{6} \vee x_2 \geqslant \tfrac{2}{3}, \\ 0 & : \text{otherwise} \end{cases} \qquad \forall \mathbf{x} \in \mathbb{R}^2 \text{ s.t. } \|\mathbf{x}\|_2 \leqslant 1$$

represent a 'smooth hump', a 'sharp cone', and a 'slotted cylinder'. At the final time $T = 2\pi$, the initial profile completes one full rotation cycle and coincides with the exact solution if the trivial boundary condition $u_{\mathrm{in}} = 0$ is imposed on the inflow boundary Γ_{in}. As in the previous section, we compare different numerical solutions calculated using $\theta = 0$, the monolithic AFC parameters $p_1 = q_1 = 2$, and Δt such that (4.145) holds for every $i \in \{1, \ldots, N\}$. A bound-preserving initial condition $u_{0,h}$ is obtained using the lumped L^2-projection.

Figures 4.9 and 4.10 show the final solutions delivered by the methods under investigation on levels 6 and 7 of $\mathcal{T}_h^3$. The corresponding $\|\cdot\|_{L^2(\Omega)}$-errors and the function values at global extrema are summarized in Table 4.2.

The finite element solution with the smallest $\| \cdot \|_{L^2(\Omega)}$-error is produced by the monolithic AFC approach in which $d_t u$ is approximated by $\dot{u}^*$. However, the rotationally symmetric shape of the cone is poorly preserved by this approach and small ripples disturb the contour lines of the solution. As an additional side effect, the interaction of dispersive ripples with steep gradients generates artificial *terraces* leading to distortions of the solution profile in the internal layer around the discontinuities of the initial data (cf. Fig. 4.10e). Furthermore, the AFC approach clips the peaks of the three bodies more aggressively than the FCT algorithm does. Interestingly enough, both approximations of the slotted cylinder are asymmetric and differ in the location of the summit. The global maximum of the FCT solution is attained in the left half which is resolved much better then the right half. The opposite is the case for the monolithic AFC approach using $\dot{u}^*$ (Fig. 4.9c vs. Fig. 4.9e). As in the case of the unsteady circular convection benchmark, the monolithic AFC solution calculated using $\dot{u}^L$ instead of $\dot{u}^*$ behaves similarly to the FCT solution. Especially on level 7, there is hardly any difference between Fig. 4.10c and Fig. 4.10d even though the $\| \cdot \|_{L^2(\Omega)}$-error is greater for the AFC solution. The most distorted solutions are again produced by the AFC approach in which antidiffusive fluxes depending on the time derivatives are neglected completely (denoted by $\dot{\alpha}_{ij} = 0$). In this case, even the shape of the smooth jump is distorted by spurious dispersive waves and the right half of the slotted cylinder is strongly smeared. Additionally, this solution exhibits the largest error and the smallest maximum among the 'high order' solutions under investigation.

So far, we have used fixed Δt and constant values of parameters like p_1 and q_1 in our comparative study of different schemes. Next, we investigate the dependence of FCT and AFC results on the time increment, as well as the influence of the parameters p_1, q_1 on the accuracy of the latter approach. The AFC version using $\dot{u}^L$ produces the most accurate results in terms of the $\| \cdot \|_{L^2(\Omega)}$-error when Δt is chosen as large as possible subject to the corresponding CFL-like condition (cf. Table 4.3). As Δt decreases, the accuracy of the AFC solution deteriorates. This adverse effect can be partially compensated by adapting the parameters p_1 and q_1 in the range of values that satisfy the CFL-like condition (4.145). In this study, the overall most accurate AFC solution is obtained using $p_1 = 1$ and $q_1 = 4$. In particular, the $\| \cdot \|_{L^2(\Omega)}$-error grows whenever p_1 is increased while keeping the product $L_1 = p_1 q_1$ constant. This behavior is caused by the fact that the correction factors α_{ij} shrink and, hence, the solution stays closer to its low order counterpart if greater emphasis is placed on p_1 instead of q_1.

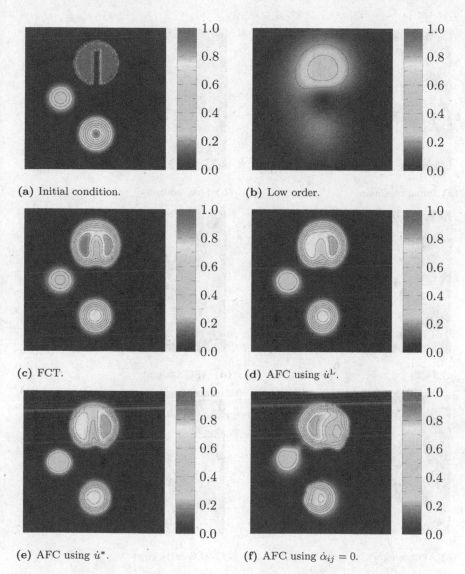

Figure 4.9: Solid body rotation: Projected initial condition and solutions produced by different numerical methods at $t = 2\pi$ on level 6 of $\mathcal{T}_h^3$ with time increment $\Delta t = 1.29 \cdot 10^{-3}$.

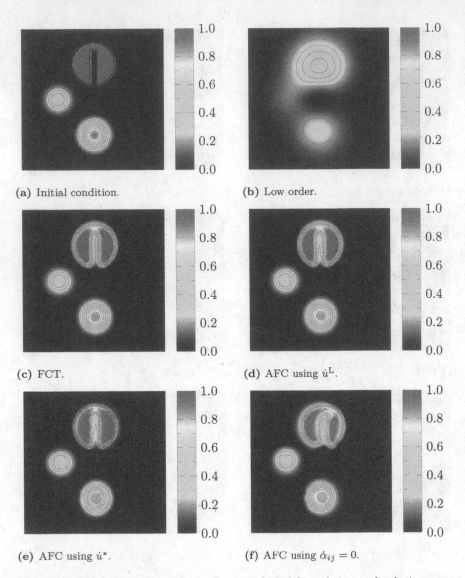

(a) Initial condition.

(b) Low order.

(c) FCT.

(d) AFC using $\dot{u}^{\mathrm{L}}$.

(e) AFC using $\dot{u}^*$.

(f) AFC using $\dot{\alpha}_{ij} = 0$.

Figure 4.10: Solid body rotation: Projected initial condition and solutions produced by different numerical methods at $t = 2\pi$ on level 7 of $\mathcal{T}_h^3$ with time increment $\Delta t = 6.39 \cdot 10^{-4}$.

Table 4.2: Solid body rotation: Convergence history for projected initial condition and different numerical methods on $\mathcal{T}_h^3$.

mesh level	Δt	method	$\| \cdot \|_{L^2(\Omega)}$-error	$\min_i u_i$	$\max_i u_i$
6	$1.29 \cdot 10^{-3}$	initial condition	$5.42 \cdot 10^{-2}$	0.00	$1.00 \cdot 10^0$
		low order	$2.10 \cdot 10^{-1}$	$1.17 \cdot 10^{-15}$	$3.67 \cdot 10^{-1}$
		FCT	$1.17 \cdot 10^{-1}$	0.00	$8.93 \cdot 10^{-1}$
		AFC using $\dot{u}^{\mathrm{L}}$	$1.24 \cdot 10^{-1}$	0.00	$8.14 \cdot 10^{-1}$
		AFC using $\dot{u}^*$	$1.16 \cdot 10^{-1}$	0.00	$8.17 \cdot 10^{-1}$
		AFC using $\dot{\alpha}_{ij} = 0$	$1.43 \cdot 10^{-1}$	0.00	$8.11 \cdot 10^{-1}$
7	$6.39 \cdot 10^{-4}$	initial condition	$4.00 \cdot 10^{-2}$	0.00	$1.00 \cdot 10^0$
		low order	$1.87 \cdot 10^{-1}$	0.00	$5.45 \cdot 10^{-1}$
		FCT	$8.04 \cdot 10^{-2}$	0.00	$1.00 \cdot 10^0$
		AFC using $\dot{u}^{\mathrm{L}}$	$8.51 \cdot 10^{-2}$	0.00	$9.97 \cdot 10^{-1}$
		AFC using $\dot{u}^*$	$7.69 \cdot 10^{-2}$	0.00	$9.88 \cdot 10^{-1}$
		AFC using $\dot{\alpha}_{ij} = 0$	$1.07 \cdot 10^{-1}$	0.00	$9.58 \cdot 10^{-1}$

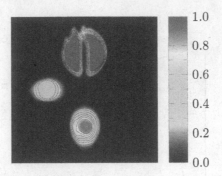

Figure 4.11: Solid body rotation: FCT solution at $t = 2\pi$ obtained on level 7 of
$\mathcal{T}_h^3$ using largest time increment $\Delta t = 3.92 \cdot 10^{-3}$ satisfying CFL-
like condition (4.66).

In contrast to the AFC approach described in Section 4.5.2, FCT approximations are guaranteed to be bound-preserving if the CFL-like condition (4.66) of the low order method is satisfied. This requirement is less
restrictive than its 'monolithic' counterpart and allows the use of larger
time increments. However, the FCT solution corresponding to the largest
time increment $\Delta t = 3.92 \cdot 10^{-3}$ satisfying (4.66) is distorted by strong
numerical dispersion within the range of values satisfying local and global
DMPs (cf. Fig. 4.11). When the time step size Δt is refined, these artifacts
disappear and the solution improves. However, for $\Delta t < 1.31 \cdot 10^{-3}$ the
FCT approximation behaves similarly to the AFC solution, i.e., its accuracy
deteriorates as Δt decreases.

Regardless of the time increment, FCT approximations to the solid body
rotation problem are more accurate than the corresponding AFC results
for any constellation of p_1 and q_1 considered in this study. The better
performance of FCT lies in the nature of the underlying limiting strategy
which is based on an *a posteriori* control of antidiffusive fluxes. It produces
the largest correction factors that guarantee the validity of local bounds
depending on the low order predictor and local DMPs hold only w.r.t. extended stencils. In contrast to this, the correction factors of the monolithic
AFC approach are determined *a priori* and no attempt is generally made to
maximize them subject to CFL-like conditions.

Table 4.3: Solid body rotation: Numerical approximations at $t = 2\pi$ obtained on level 7 of $\mathcal{T}_h^3$ with different time increments Δt.

Δt	method	$\|\cdot\|_{L^2(\Omega)}$-error	$\min_i u_i$	$\max_i u_i$
$3.92 \cdot 10^{-3}$	FCT	$7.71 \cdot 10^{-2}$	0.00	1.00
$1.31 \cdot 10^{-3}$	FCT	$7.49 \cdot 10^{-2}$	0.00	$1.00 \cdot 10^{0}$
	AFC using $\dot{u}^L$, $p_1 = 1, q_1 = 1$	$1.18 \cdot 10^{-1}$	0.00	$8.80 \cdot 10^{-1}$
$6.39 \cdot 10^{-4}$	FCT	$8.04 \cdot 10^{-2}$	0.00	$1.00 \cdot 10^{0}$
	AFC using $\dot{u}^L$, $p_1 = 1, q_1 = 1$	$1.20 \cdot 10^{-1}$	0.00	$8.76 \cdot 10^{-1}$
	AFC using $\dot{u}^L$, $p_1 = 4, q_1 = 1$	$8.77 \cdot 10^{-2}$	0.00	$9.92 \cdot 10^{-1}$
	AFC using $\dot{u}^L$, $p_1 = 2, q_1 = 2$	$8.51 \cdot 10^{-2}$	0.00	$9.97 \cdot 10^{-1}$
	AFC using $\dot{u}^L$, $p_1 = 1, q_1 = 4$	$8.31 \cdot 10^{-2}$	0.00	$9.99 \cdot 10^{-1}$
$1.95 \cdot 10^{-4}$	FCT	$8.47 \cdot 10^{-2}$	0.00	$1.00 \cdot 10^{0}$
	AFC using $\dot{u}^L$, $p_1 = 1, q_1 = 1$	$1.21 \cdot 10^{-1}$	0.00	$8.79 \cdot 10^{-1}$
	AFC using $\dot{u}^L$, $p_1 = 4, q_1 = 1$	$9.12 \cdot 10^{-2}$	0.00	$9.92 \cdot 10^{-1}$
	AFC using $\dot{u}^L$, $p_1 = 2, q_1 = 2$	$8.89 \cdot 10^{-2}$	0.00	$9.97 \cdot 10^{-1}$
	AFC using $\dot{u}^L$, $p_1 = 1, q_1 = 4$	$8.72 \cdot 10^{-2}$	0.00	$9.99 \cdot 10^{-1}$
	AFC using $\dot{u}^L$, $p_1 = 16, q_1 = 1$	$8.54 \cdot 10^{-2}$	0.00	$1.00 \cdot 10^{0}$
	AFC using $\dot{u}^L$, $p_1 = 4, q_1 = 4$	$8.51 \cdot 10^{-2}$	0.00	$1.00 \cdot 10^{0}$
	AFC using $\dot{u}^L$, $p_1 = 1, q_1 = 16$	$8.47 \cdot 10^{-2}$	0.00	$1.00 \cdot 10^{0}$

5 Limiting for tensors

So far, we presented bound-preserving algorithms for solving advection-reaction equations in the case of a scalar quantity of interest. The underlying design principles are motivated by physical properties of the continuous problem, which should be preserved by its (spatially) discretized counterpart. The methodology presented in the previous chapter fulfills this requirement using algebraic manipulations of a high order target method to enforce provable discrete maximum principles while preserving consistency and discrete conservation properties.

This chapter extends discrete upwinding and the algebraic flux correction framework to advection-reaction equations for symmetric tensor quantities. In Section 5.1, we discuss the proper formulation of tensorial maximum principles and explain the reasons for limiting the eigenvalues rather than tensor entries or principal invariants in this work. A brief introduction to basic properties of symmetric tensors is given in Section 5.2. As we show in Section 5.3, an extension of the low order method of Section 4.3 preserves the (local) eigenvalue range of an advected tensor field. In Sections 5.4 and 5.5, we propose and analyze tensorial extensions of the algebraic flux correction techniques presented in Sections 4.4 and 4.5, respectively. Finally, the properties of resulting algorithms are discussed and illustrated by numerical examples in Section 5.6.

The key results of this chapter were originally presented in the author's publications [Loh17b; Loh19] on bound-preserving eigenvalue range limiters for symmetric tensor quantities. In addition to summarizing these results, we improve and extend them making use of the theory developed in Chapter 4.

5.1 Introduction

This section gives an introduction to tensor limiting and summarizes some basic results, which will be used to prove certain theoretical properties of generalized algebraic flux correction schemes.

© Springer Fachmedien Wiesbaden GmbH, part of Springer Nature 2019
C. Lohmann, *Physics-Compatible Finite Element Methods for Scalar and Tensorial Advection Problems*, https://doi.org/10.1007/978-3-658-27737-6_5

5.1.1 Background

In numerous research areas, tensor variables occur as quantities of interest that may need to be constrained to achieve physically reasonable results. This need frequently arises, for example, in applications to image analysis (structure tensors), diffusion tensor magnetic resonance imaging (DT-MRI; diffusion tensors), fluid and solid dynamics (Cauchy stress tensors), civil engineering and solid mechanics (inertia, diffusion and permittivity tensors), and fiber suspensions (orientation tensors) [Bur+07; Mai+13; Klí+17; AT87; AT93].

Due to the great variety of applications, tensors can describe many different physical properties. Therefore, it is essential to clarify which characteristics of the tensor at hand are of primary interest and should be treated carefully in numerical approximations. For instance, the second principal invariant of deviatoric stress tensors is proportional to the elastic energy density. A tailor-made slope limiter was developed in Klíma et al. [Klí+17] to impose local bounds on this particular quantity of interest. In contrast to this, the second invariant of structure tensors seems to be of minor importance. Since their eigenvalues indicate the existence of corners and edges in images [HS88], an eigenvalue-based limiting strategy may be appropriate.

In addition to customized tensor limiters, as presented in [Sam+13; Klí+17], several general-purpose extensions of limiting approaches for scalar quantities can be found in literature. For instance, the limiting strategies proposed by Luttwak and Falcovitz [LF10; LF11] and Luttwak [Lut16] constrain tensors and vectors to be in the convex hull of physically admissible low order samples. Another promising approach to the development of bound-preserving numerical schemes for general tensor fields is based on the concept of Löwner ordering for symmetric matrices. Following this approach, Burgeth et al. [Bur+07] generalized the Osher-Sethian scheme and embedded it into a flux-corrected transport (FCT) algorithm [Bur+09] for matrix fields in the context of image processing.

5.1.2 Properties of interest

In contrast to the approaches summarized above, the main focus of this chapter is on limiting techniques that preserve the eigenvalue range. The corresponding generalization of the maximum principle for scalar fields imposes a lower bound on the minimal eigenvalue of the tensor quantity and an upper bound on the maximal eigenvalue. In two dimensions, both eigenvalues are effectively controlled by this approach, whereas the intermediate eigenvalue of a three dimensional tensor field may require additional stabilization to prevent spurious oscillations within the range determined by the local extrema of the maximal and minimal eigenvalues. In particular, the definiteness of the solution is preserved in applications to time dependent problems. The use of eigenvalue range limiters in algebraic flux correction schemes leads to very stable numerical algorithms for problems in which eigenvalues and invariants must have a certain sign for physical reasons. At the same time, the eigenvalue range preservation criterion may be less restrictive than other frame-invariant maximum principles, as we will see below.

In Chapter 6, we simulate fiber suspension flows using an Eulerian method, in which the Navier-Stokes equations are coupled with an evolution equation for a so called orientation tensor. The development of tensor limiters to be presented below was largely motivated by this application in which the advected tensor must remain positive semidefinite and have a unit trace.

To begin with, let us introduce the PDE that will serve as a model problem in this chapter: The non-conservative form of the steady advection-reaction equation for a symmetric tensor quantity $\mathbf{U} : \Omega \to \mathbb{S}_d$ reads

$$\mathbf{v} \cdot \operatorname{grad}(\mathbf{U}) + c\mathbf{U} = \mathbf{F} \qquad \text{in } \Omega, \qquad (5.1a)$$

$$\mathbf{U} = \mathbf{U}_{\text{in}} \qquad \text{on } \Gamma_{\text{in}}, \qquad (5.1b)$$

where $\mathbf{F} : \Omega \to \mathbb{S}_d$ and $\mathbf{U}_{\text{in}} : \Gamma_{\text{in}} \to \mathbb{S}_d$ are the tensorial source term and inflow boundary data, respectively. Following the convention introduced in Section 1.3, tensor quantities are denoted by capital and boldface letters. Furthermore, the notation $\mathbf{v} \cdot \operatorname{grad}(\mathbf{U})$ should be interpreted componentwise, i.e.,

$$\big(\mathbf{v} \cdot \operatorname{grad}(\mathbf{U})\big)_{k\ell} := \mathbf{v} \cdot \operatorname{grad}(u_{k\ell}) \qquad \forall k, \ell \in \{1, \dots, d\}.$$

The transient counterpart of (5.1) is given by

$$\frac{\partial \mathbf{U}}{\partial t} + \mathbf{v} \cdot \operatorname{grad}(\mathbf{U}) + c\mathbf{U} = \mathbf{F} \qquad \text{in } \Omega \times (0, T), \qquad (5.2a)$$

$$\mathbf{U} = \mathbf{U}_{\text{in}} \qquad \text{on } \Gamma_{\text{in}} \times (0, T), \qquad (5.2b)$$

$$\mathbf{U}(\cdot, 0) = \mathbf{U}_0 \qquad \text{on } \Omega, \qquad (5.2c)$$

where $\mathbf{U}_0 : \Omega \to \mathbb{S}_d$ is the initial data while $\mathbf{U}_{\text{in}} : \Gamma_{\text{in}} \times (0, T) \to \mathbb{S}_d$ and $\mathbf{F} : \Omega \times (0, T) \to \mathbb{S}_d$ may now depend on the time t.

Problems (5.1) and (5.2) are characterized by the fact that each tensor entry evolves independently of the others and satisfies a scalar advection-reaction equation. Consequently, each component of the exact tensor solution enjoys the maximum principle properties of Section 2.1.3. By summing up the equations for the diagonal entries, we obtain a scalar (initial-)boundary value problem for the trace and can formulate the corresponding maximum principles. Moreover, an evolution equation can be derived for any scalar differentiable function $\psi : \mathbb{S}_d \to \mathbb{R}$ of a symmetric tensor quantity. If $\mathbf{U}$ is the solution to (5.2), then we have

$$
\begin{aligned}
\partial_t \psi(\mathbf{U}) &+ \mathbf{v} \cdot \mathrm{grad}\big(\psi(\mathbf{U})\big) + c\psi(\mathbf{U}) \\
&= \big(\nabla_{\mathbf{U}} \psi(\mathbf{U})\big) : \partial_t \mathbf{U} + \big(\nabla_{\mathbf{U}} \psi(\mathbf{U})\big) : \big(\mathbf{v} \cdot \mathrm{grad}(\mathbf{U})\big) + c\psi(\mathbf{U}) \\
&= \big(\nabla_{\mathbf{U}} \psi(\mathbf{U})\big) : \Big(\partial_t \mathbf{U} + \big(\mathbf{v} \cdot \mathrm{grad}(\mathbf{U})\big)\Big) + c\psi(\mathbf{U}) \\
&= \big(\nabla_{\mathbf{U}} \psi(\mathbf{U})\big) : (\mathbf{F} - c\mathbf{U}) + c\psi(\mathbf{U}).
\end{aligned}
$$

In the case when $\mathbf{F}$ and c vanish, the unsteady scalar advection-reaction equation is satisfied for $\psi(\mathbf{U})$, too, and maximum principles hold again. Obviously, the result is the same in the case of the stationary PDE. On the continuous level, this property implies maximum principles for additional tensor quantities like the principal invariants. Even simple/distinct eigenvalues are differentiable functions of the tensor components [Kat95, Theorem 5.16] and the corresponding maximum principles apply.

However, enforcing all applicable maximum principles at the discrete level may give rise to unrealistic constraints or overly diffusive numerical approximations. The following example illustrates the behavior of limiting techniques for principal invariants and eigenvalues. We use it to explain why constraining the former set of variables can be problematic.

Example 5.1 ([Loh17b, Section 2]). Instead of making the usual assumption that the dimension of the tensorial solution coincides with the number of possible transport directions, we consider a one dimensional Riemann problem for two dimensional tensor quantities. The exact solution $\mathbf{U} : \mathbb{R} \to \mathbb{S}_2$ is piecewise constant with a discontinuity at $x = 0$ separating the two constant states

$$
\mathbf{U}_L = \begin{pmatrix} 1 & 0 \\ 0 & 2 \end{pmatrix}, \qquad \mathbf{U}_R = \begin{pmatrix} 2 & 0 \\ 0 & 1 \end{pmatrix}.
$$

We seek a numerical approximation $\mathbf{U}_h$ of the exact solution $\mathbf{U}$ such that scalar quantities of interest $\psi(\mathbf{U}_h)$ are bounded by the extrema of $\psi(\mathbf{U})$. For example, imposing global maximum principles on the range of eigenvalues leads to the restriction

$$
1 \leqslant u_{h,1}(x) \leqslant u_{h,2}(x) \leqslant 2 \qquad \forall x \in \mathbb{R},
$$

where $u_{h,1}(x) := \lambda_1\big(\mathbf{U}_h(x)\big)$ and $u_{h,2}(x) := \lambda_2\big(\mathbf{U}_h(x)\big)$ are the minimal and maximal eigenvalues of the approximation $\mathbf{U}_h$ following the notation of Section 1.3. Restricting ourselves to frame invariant approximations which must remain diagonal at any time, we obtain

$$u_{h,12}(x) = u_{h,21}(x) = 0, \quad 1 \leqslant u_{h,11}(x), u_{h,22}(x) \leqslant 2 \qquad \forall x \in \mathbb{R}. \quad (5.3)$$

Condition (5.3) guarantees the validity of global DMPs for the range of eigenvalues and seems to be not very restrictive. In particular, smooth blendings $\mathbf{U}_h(x)$ given by convex combinations of $\mathbf{U}_L$ and $\mathbf{U}_R$ are admissible. Overshoots and undershoots are only possible in the restricted range of eigenvalues guaranteeing a stable solution from this point of view.

On the other hand, maximum principles for the principal invariants I_1 and I_2 require

$$I_1\big(\mathbf{U}_h(x)\big) = u_{h,11}(x) + u_{h,22}(x) = 3 \qquad\qquad \forall x \in \mathbb{R}, \qquad (5.4a)$$
$$I_2\big(\mathbf{U}_h(x)\big) = u_{h,11}(x)u_{h,22}(x) - u_{h,12}(x)^2 = 2 \qquad \forall x \in \mathbb{R} \qquad (5.4b)$$

because the invariants of both Riemann states coincide. The only tensors possessing these properties are given by rotations of $\mathbf{U}_L$ and $\mathbf{U}_R$. More precisely, they are given by

$$\mathbf{U}_\theta = \begin{pmatrix} 1 + \sin^2\theta & -\sin\theta\cos\theta \\ -\sin\theta\cos\theta & 1 + \cos^2\theta \end{pmatrix} \qquad \forall \theta \in [0, \pi)$$

and coincide with $\mathbf{U}_L$ and $\mathbf{U}_R$ for $\theta = 0$ and $\theta = \frac{\pi}{2}$, respectively. Unfortunately, $\mathbf{U}_\theta$ is not diagonal and, hence, frame dependent for $\theta \notin \{0, \frac{\pi}{2}\}$. Thus, a smooth and continuous blending between $\mathbf{U}_L$ and $\mathbf{U}_R$ is impossible without violating (5.4).

The above example demonstrates why the use of limiters for principal invariants may have an adverse effect on the accuracy of a continuous finite element approximation: Frame invariant solutions possessing the corresponding DMP properties may not exist and the set of admissible solutions must be extended by waiving the objectivity requirement. On the other hand, scalar limiters for tensor entries are not frame invariant and may fail to preserve the sign of extremal eigenvalues. For that reason, they are not to be recommended for numerical treatment of definite tensors. In view of these considerations, we henceforth focus on limiting the eigenvalue range rather than principal invariants or tensor entries. However, criteria involving principal invariants of auxiliary tensors turn out to be a useful tool for the design of eigenvalue range preserving limiters which we present in Section 5.4.

5.2 Preliminaries

Before embarking on the design of generalized limiting techniques, we summarize some fundamental properties of (symmetric) tensor quantities. Most of them are used in below derivations and facilitate the compact presentation of numerical algorithms, as well as theoretical analysis.

The eigenvalues and invariants of two dimensional symmetric tensors $\mathbf{V} \in \mathbb{S}_2$ satisfy

$$v_1 = \frac{1}{2}\left(I_1 - \sqrt{I_1^2 - 4I_2}\right), \qquad v_2 = \frac{1}{2}\left(I_1 + \sqrt{I_1^2 - 4I_2}\right), \tag{5.5a}$$

$$\begin{aligned} I_1(\mathbf{V}) &= v_1 + v_2 = \operatorname{tr}(\mathbf{V}) = v_{11} + v_{22}, \\ I_2(\mathbf{V}) &= v_1 v_2 = \det(\mathbf{V}) = v_{11}v_{22} - v_{12}^2. \end{aligned} \tag{5.5b}$$

In the three dimensional case, the principal invariants I_1, I_2, and I_3 of $\mathbf{V} \in \mathbb{S}_3$ are given by

$$\begin{aligned} I_1(\mathbf{V}) &= \operatorname{tr}(\mathbf{V}) = v_1 + v_2 + v_3 = v_{11} + v_{22} + v_{33}, \\ I_2(\mathbf{V}) &= \frac{1}{2}\left((\operatorname{tr}\mathbf{V})^2 - \operatorname{tr}(\mathbf{V}^2)\right) = v_1 v_2 + v_1 v_3 + v_2 v_3 \\ &= v_{11}v_{22} + v_{11}v_{33} + v_{22}v_{33} - v_{12}^2 - v_{13}^2 - v_{23}^2, \\ I_3(\mathbf{V}) &= \det(\mathbf{V}) = v_1 v_2 v_3 \\ &= v_{11}v_{22}v_{33} + 2v_{12}v_{13}v_{23} - v_{22}v_{13}^2 - v_{33}v_{12}^2 - v_{11}v_{23}^2, \end{aligned} \tag{5.5c}$$

while the eigenvalues of $\mathbf{V}$ can be determined using *Cardano's formula* [Has+01; Smi61]

$$\begin{aligned} v_1 &= \tfrac{I_1}{3} - 2\sqrt{v}\cos(\tfrac{\pi}{3} - \phi), & v &= \left(\tfrac{I_1}{3}\right)^2 - \tfrac{I_2}{3}, \\ v_2 &= \tfrac{I_1}{3} - 2\sqrt{v}\cos(\tfrac{\pi}{3} + \phi), & s &= \left(\tfrac{I_1}{3}\right)^3 - \tfrac{I_1 I_2}{6} + \tfrac{I_3}{2}, \\ v_3 &= \tfrac{I_1}{3} + 2\sqrt{v}\cos(\phi), & \phi &= \tfrac{1}{3}\arccos\left(s\sqrt{v^{-3}}\right). \end{aligned} \tag{5.5d}$$

Even if explicit formulas for the eigenvalue calculation exist, they can be practically ill-posed and should not be used whenever the absolute tensor entries differ by orders of magnitude. In this case, iterative approaches like the Jacobi method produce more accurate results and additional computational costs are fully justified. Several iterative and direct algorithms are comprehensively discussed in [Kop08].

As already mentioned, eigenvalues are differentiable functions of the tensor entries as long as they are distinct [Kat95, Theorem 5.16]. If a tensor quantity possesses multiple eigenvalues, they are at least Lipschitz continuous.

Theorem 5.2 (Wielandt-Hoffman theorem [Wil88, pp. 104–109]). *Let* $\mathbf{V}, \bar{\mathbf{V}} \in \mathbb{S}_d$ *be symmetric tensors having the eigenvalues* v_k *and* $\bar{v}_k$, $k \in \{1, \dots, d\}$, *respectively, arranged in non-decreasing order. Then the following estimate holds*

$$|v_k - \bar{v}_k|^2 \leqslant \sum_{\ell=1}^{d} |v_\ell - \bar{v}_\ell|^2 \leqslant \|\mathbf{V} - \bar{\mathbf{V}}\|_\mathrm{F}^2 \qquad \forall k \in \{1, \dots, d\}. \tag{5.6}$$

Proof. The proof can be found, e.g., in [Wil88, p. 104–109; HW53]. $\square$

The *Frobenius norm* $\| \cdot \|_\mathrm{F} : \mathbb{R}^{d \times d} \to \mathbb{R}_0^+$ that appears in (5.6) is defined by

$$\|\mathbf{V}\|_\mathrm{F}^2 := \sum_{k,\ell=1}^{d} v_{k\ell}^2 \qquad \forall \mathbf{V} \in \mathbb{R}^{d \times d}$$

and associated with the Frobenius inner product $(\cdot, \cdot)_\mathrm{F}$

$$(\mathbf{U}, \mathbf{V})_\mathrm{F} := \mathbf{U} : \mathbf{V} := \sum_{k,\ell=1}^{d} u_{k\ell} v_{k\ell} = \mathrm{tr}(\mathbf{U}^\top \mathbf{V}) = \mathrm{tr}(\mathbf{V}^\top \mathbf{U}) \quad \forall \mathbf{U}, \mathbf{V} \in \mathbb{R}^{d \times d}. \tag{5.7}$$

The invariance of the trace $\mathrm{tr}(\cdot)$ under cyclic permutations

$$\mathrm{tr}(\mathbf{U}\mathbf{V}) = \sum_{k,\ell=1}^{d} u_{k\ell} v_{\ell k} = \mathrm{tr}(\mathbf{V}\mathbf{U}) \qquad \forall \mathbf{U}, \mathbf{V} \in \mathbb{R}^{d \times d} \tag{5.8}$$

implies the frame invariance of the trace which is given by

$$\mathrm{tr}(\mathbf{V}) = \mathrm{tr}(\mathbf{Q}\tilde{\mathbf{V}}\mathbf{Q}^\top) = \mathrm{tr}(\tilde{\mathbf{V}}\mathbf{Q}^\top\mathbf{Q}) = \mathrm{tr}(\tilde{\mathbf{V}}) = \sum_{k=1}^{d} v_k \qquad \forall \mathbf{V} \in \mathbb{S}_d$$

for the spectral decomposition $\mathbf{V} = \mathbf{Q}\tilde{\mathbf{V}}\mathbf{Q}^\top$ (cf. Section 1.3). Furthermore, we obtain

$$\|\mathbf{V}\|_\mathrm{F}^2 = \mathrm{tr}(\mathbf{V}^2) = \mathrm{tr}(\mathbf{Q}\tilde{\mathbf{V}}^2\mathbf{Q}^\top) = \mathrm{tr}(\tilde{\mathbf{V}}^2\mathbf{Q}^\top\mathbf{Q}) = \mathrm{tr}(\tilde{\mathbf{V}}^2) = \sum_{k=1}^{d} v_k^2$$

$$\forall \mathbf{V} \in \mathbb{S}_d. \tag{5.9}$$

Hence, the Frobenius norm is frame invariant and can be expressed in terms of the eigenvalues.

Another commonly used tensor norm is the *spectral norm* $\|\cdot\|_2 : \mathbb{R}^{d \times d} \to \mathbb{R}_0^+$ induced by the Euclidean vector norm

$$\|\mathbf{V}\|_2 := \sup_{\mathbf{x} \in \mathbb{R}^d \setminus \{\mathbf{0}\}} \frac{\|\mathbf{V}\mathbf{x}\|_2}{\|\mathbf{x}\|_2} = \sqrt{\lambda_d(\mathbf{V}^\top \mathbf{V})} \qquad \forall \mathbf{V} \in \mathbb{R}^{d \times d}.$$

By definition, the inequality

$$\|\mathbf{V}\mathbf{x}\|_2 \leqslant \|\mathbf{V}\|_2 \|\mathbf{x}\|_2 \qquad \forall \mathbf{V} \in \mathbb{R}^{d \times d}, \mathbf{x} \in \mathbb{R}^d$$

holds and implies the submultiplicativity due to

$$\|\mathbf{U}\mathbf{V}\|_2 = \sup_{\mathbf{x} \in \mathbb{R}^d \setminus \{\mathbf{0}\}} \frac{\|\mathbf{U}\mathbf{V}\mathbf{x}\|_2}{\|\mathbf{x}\|_2} \leqslant \sup_{\mathbf{x} \in \mathbb{R}^d \setminus \{\mathbf{0}\}} \|\mathbf{U}\|_2 \frac{\|\mathbf{V}\mathbf{x}\|_2}{\|\mathbf{x}\|_2} = \|\mathbf{U}\|_2 \|\mathbf{V}\|_2$$
$$\forall \mathbf{U}, \mathbf{V} \in \mathbb{R}^{d \times d}.$$

For symmetric tensors, the spectral norm coincides with the largest absolute eigenvalue

$$\|\mathbf{V}\|_2 = \max(|v_1|, |v_d|) \qquad \forall \mathbf{V} \in \mathbb{S}_d$$

and, hence, is frame invariant, too.

It is well known that all norms are equivalent in finite dimensional vector spaces. In particular, this is the case for the Frobenius and spectral norm

$$\|\mathbf{V}\|_2^2 = \lambda_d(\mathbf{V}^\top \mathbf{V}) \leqslant \sum_{k=1}^{d} \lambda_k(\mathbf{V}^\top \mathbf{V}) = \operatorname{tr}(\mathbf{V}^\top \mathbf{V}) = \|\mathbf{V}\|_F^2 \leqslant d \|\mathbf{V}\|_2^2$$
$$\implies \qquad \|\mathbf{V}\|_2 \leqslant \|\mathbf{V}\|_F \leqslant \sqrt{d} \|\mathbf{V}\|_2 \qquad \forall \mathbf{V} \in \mathbb{R}^{d \times d}.$$

Furthermore, the following inequalities hold [Loh19, Eq. (2)]

$$\|\mathbf{U}\mathbf{V}\|_F^2 = \sum_{k=1}^{d} \|\mathbf{U}\mathbf{v}_k\|_2^2 \leqslant \|\mathbf{U}\|_2^2 \sum_{k=1}^{d} \|\mathbf{v}_k\|_2^2 = \|\mathbf{U}\|_2^2 \|\mathbf{V}\|_F^2 \qquad \forall \mathbf{U}, \mathbf{V} \in \mathbb{R}^{d \times d},$$
$$(5.10a)$$

$$\|\mathbf{V}\mathbf{U}\|_F^2 = \|\mathbf{U}^\top \mathbf{V}^\top\|_F^2 \leqslant \|\mathbf{U}^\top\|_2^2 \|\mathbf{V}^\top\|_F^2 = \|\mathbf{U}\|_2^2 \|\mathbf{V}\|_F^2 \qquad \forall \mathbf{U}, \mathbf{V} \in \mathbb{R}^{d \times d},$$
$$(5.10b)$$

where we used the fact that $\|\mathbf{U}^\top\|_2 = \|\mathbf{U}\|_2$ for all $\mathbf{U} \in \mathbb{R}^{d \times d}$ and the notation that $\mathbf{v}_k \in \mathbb{R}^d$ is the k-th column vector of $\mathbf{V} \in \mathbb{R}^{d \times d}$.

The sum of two tensors $\mathbf{U}, \mathbf{V} \in \mathbb{S}_d$ satisfies

$$u_1 + v_1 \leqslant \mathbf{x}^\top (\mathbf{U} + \mathbf{V})\mathbf{x} \leqslant u_d + v_d \quad \forall \mathbf{x} \in \mathbb{R}^d \text{ s.t. } \|\mathbf{x}\|_2 = 1, \qquad (5.11\text{a})$$

$$u_1 + v_1 \leqslant \lambda_k(\mathbf{U} + \mathbf{V}) \quad \leqslant u_d + v_d \quad \forall k \in \{1, \dots, d\}. \qquad (5.11\text{b})$$

For example, the first inequality of (5.11a) follows from the fact that

$$\mathbf{x}^\top (\mathbf{U} + \mathbf{V} - (u_1 + v_1)\mathbf{I})\mathbf{x} = \mathbf{x}^\top (\mathbf{U} - u_1\mathbf{I})\mathbf{x} + \mathbf{x}^\top (\mathbf{V} - v_1\mathbf{I})\mathbf{x} \geqslant 0$$
$$\forall \mathbf{x} \in \mathbb{R}^d \text{ s.t. } \|\mathbf{x}\|_2 = 1$$

by virtue of the positive semidefiniteness of $\mathbf{U} - u_1\mathbf{I}$ and $\mathbf{V} - v_1\mathbf{I}$.

In what follows, we frequently exploit the concept of *Löwner ordering* [Löw34], which represents a partial relation between two symmetric tensors: A tensor quantity $\mathbf{U} \in \mathbb{S}_d$ is said to be 'smaller than or equal to' $\mathbf{V} \in \mathbb{S}_d$ if and only if all eigenvalues of $\mathbf{V} - \mathbf{U}$ are nonnegative, denoted by $\mathbf{V} - \mathbf{U} \succcurlyeq \mathbf{0}$ or $\mathbf{V} \succcurlyeq \mathbf{U}$. If all eigenvalues of $\mathbf{V} - \mathbf{U}$ are greater than zero, we write $\mathbf{V} - \mathbf{U} \succ \mathbf{0}$ or $\mathbf{V} \succ \mathbf{U}$. The relations '$\preccurlyeq$' and '$\prec$' are defined similarly.

Obviously, these relations are transitive

$$\mathbf{V}_1 \succcurlyeq \mathbf{V}_3 \quad \forall \mathbf{V}_1, \mathbf{V}_2, \mathbf{V}_3 \in \mathbb{S}_d \text{ s.t. } \mathbf{V}_1 \succcurlyeq \mathbf{V}_2, \mathbf{V}_2 \succcurlyeq \mathbf{V}_3,$$

and satisfy

$$\mathbf{U} + \beta\mathbf{V} = (1 - \beta)\mathbf{U} + \beta(\mathbf{U} + \mathbf{V}) \succcurlyeq \mathbf{0}$$
$$\forall \beta \in [0, 1], \mathbf{U}, \mathbf{V} \in \mathbb{S}_d \text{ s.t. } \mathbf{U} \succcurlyeq \mathbf{0}, \mathbf{U} + \mathbf{V} \succcurlyeq \mathbf{0}. \qquad (5.12)$$

In the spirit of Löwner ordering, the absolute value of a symmetric tensor can be defined by

$$|\mathbf{V}| := \mathbf{Q}|\tilde{\mathbf{V}}|\mathbf{Q}^\top \quad \forall \mathbf{V} = \mathbf{Q}\tilde{\mathbf{V}}\mathbf{Q}^\top \in \mathbb{S}_d,$$

where the entries of the diagonal tensors $|\tilde{\mathbf{V}}| = \mathrm{diag}(|v_1|, \dots, |v_d|) \succcurlyeq \mathbf{0}$ and $\tilde{\mathbf{V}} = \mathrm{diag}(v_1, \dots, v_d)$ have the same absolute values. Using this notation, the minimum and maximum of two tensors may be defined by [Loh19, Eq. (4)]

$$\max(\mathbf{U}, \mathbf{V}) := \tfrac{1}{2}(\mathbf{U} + \mathbf{V}) + \tfrac{1}{2}|\mathbf{U} - \mathbf{V}| \quad \forall \mathbf{U}, \mathbf{V} \in \mathbb{S}_d,$$
$$\min(\mathbf{U}, \mathbf{V}) := \tfrac{1}{2}(\mathbf{U} + \mathbf{V}) - \tfrac{1}{2}|\mathbf{U} - \mathbf{V}| \quad \forall \mathbf{U}, \mathbf{V} \in \mathbb{S}_d.$$

As in the scalar case, the minimum of two tensors is bounded above in the sense of Löwner ordering by one of its arguments

$$\min(\mathbf{U}, \mathbf{V}) = \tfrac{1}{2}\big((\mathbf{U} + \mathbf{V}) - \underbrace{|\mathbf{U} - \mathbf{V}|}_{\succcurlyeq \mathbf{V} - \mathbf{U}}\big) \preccurlyeq \tfrac{1}{2}\big((\mathbf{U} + \mathbf{V}) + (\mathbf{U} - \mathbf{V})\big) = \mathbf{U}$$
$$\forall \mathbf{U}, \mathbf{V} \in \mathbb{S}_d.$$

On the other hand, it is worth mentioning that a common lower bound of the two arguments is generally not a lower bound for the minimum

$$\mathbf{U}, \mathbf{V} \succcurlyeq \mathbf{W} \quad \not\Longrightarrow \quad \min(\mathbf{U}, \mathbf{V}) \succcurlyeq \mathbf{W},$$

which distinguishes the tensorial definition from its scalar counterpart. A simple counterexample in $\mathbb{S}_2$ is given by [Loh19, Section 2.3]

$$\mathbf{U} = \begin{pmatrix} 1 & 2 \\ 2 & 4 \end{pmatrix} \succcurlyeq \mathbf{0}, \quad \mathbf{V} = \begin{pmatrix} 4 & 2 \\ 2 & 1 \end{pmatrix} \succcurlyeq \mathbf{0}$$

$$\Longrightarrow \quad \min(\mathbf{U}, \mathbf{V}) = \tfrac{1}{2} \begin{pmatrix} 5 & 4 \\ 4 & 5 \end{pmatrix} - \tfrac{1}{2} \begin{pmatrix} 3 & 0 \\ 0 & 3 \end{pmatrix} = \begin{pmatrix} 1 & 2 \\ 2 & 1 \end{pmatrix} \not\succcurlyeq \mathbf{0}.$$

As in the scalar case, the above operators are Lipschitz continuous.

Theorem 5.3 ([Loh19]). *The tensorial extensions of* $|\cdot|$, $\min(\cdot, \cdot)$, *and* $\max(\cdot, \cdot)$ *are Lipschitz continuous and satisfy*

$$\left\| |\mathbf{V}| - |\bar{\mathbf{V}}| \right\|_{\mathrm{F}} \leqslant \|\mathbf{V} - \bar{\mathbf{V}}\|_{\mathrm{F}} \qquad \forall \mathbf{V}, \bar{\mathbf{V}} \in \mathbb{S}_d, \tag{5.13a}$$

$$\left\| \min(\mathbf{V}, \mathbf{U}) - \min(\bar{\mathbf{V}}, \bar{\mathbf{U}}) \right\|_{\mathrm{F}} \leqslant \|\mathbf{U} - \bar{\mathbf{U}}\|_{\mathrm{F}} + \|\mathbf{V} - \bar{\mathbf{V}}\|_{\mathrm{F}}$$
$$\forall \mathbf{U}, \bar{\mathbf{U}}, \mathbf{V}, \bar{\mathbf{V}} \in \mathbb{S}_d, \tag{5.13b}$$

$$\left\| \max(\mathbf{V}, \mathbf{U}) - \max(\bar{\mathbf{V}}, \bar{\mathbf{U}}) \right\|_{\mathrm{F}} \leqslant \|\mathbf{U} - \bar{\mathbf{U}}\|_{\mathrm{F}} + \|\mathbf{V} - \bar{\mathbf{V}}\|_{\mathrm{F}}$$
$$\forall \mathbf{U}, \bar{\mathbf{U}}, \mathbf{V}, \bar{\mathbf{V}} \in \mathbb{S}_d. \tag{5.13c}$$

Proof. The results are shown as in [Loh19].

Consider arbitrary tensors $\mathbf{V}, \bar{\mathbf{V}} \in \mathbb{S}_d$ and, without loss of generality, assume that $\mathbf{V} = \tilde{\mathbf{V}}$ is diagonal. Then straightforward algebraic manipulations lead to

$$(\mathbf{V}, \bar{\mathbf{V}})_{\mathrm{F}} = \sum_{k=1}^{d} v_{kk} \bar{v}_{kk} = \sum_{k=1}^{d} v_{kk} \sum_{\ell=1}^{d} \bar{q}_{k\ell} \bar{v}_{\ell} \bar{q}_{k\ell}$$

$$\leqslant \sum_{k=1}^{d} |v_{kk}| \sum_{\ell=1}^{d} |\bar{v}_{\ell}| \bar{q}_{k\ell}^2 = (|\mathbf{V}|, |\bar{\mathbf{V}}|)_{\mathrm{F}}$$

making use of the eigenvalue decomposition $\bar{\mathbf{V}} = \bar{\mathbf{Q}} \tilde{\mathbf{V}} \bar{\mathbf{Q}}^\top$. This implies the Lipschitz continuity of the tensorial absolute value because

$$\left\| |\mathbf{V}| - |\bar{\mathbf{V}}| \right\|_{\mathrm{F}}^2 = \||\mathbf{V}|\|_{\mathrm{F}}^2 + \||\bar{\mathbf{V}}|\|_{\mathrm{F}}^2 - 2(|\mathbf{V}|, |\bar{\mathbf{V}}|)_{\mathrm{F}}$$
$$\leqslant \|\mathbf{V}\|_{\mathrm{F}}^2 + \|\bar{\mathbf{V}}\|_{\mathrm{F}}^2 - 2(\mathbf{V}, \bar{\mathbf{V}})_{\mathrm{F}} = \|\mathbf{V} - \bar{\mathbf{V}}\|_{\mathrm{F}}^2.$$

Using this result, the statements for the minimum and maximum follow as in (4.124). □

For the sake of completeness, we define the positive and negative part of a symmetric tensor quantity as the tensorial extension of (2.22)

$$\mathbf{V} = \mathbf{V}_+ - \mathbf{V}_- := \max(\mathbf{V}, \mathbf{0}) - \max(-\mathbf{V}, \mathbf{0}) \qquad \forall \mathbf{V} \in \mathbb{S}_d. \tag{5.14}$$

5.3 Low order method

This section is devoted to the derivation of low order methods corresponding to the Galerkin discretization of (5.1) and (5.2). In view of the considerations presented in Section 5.1.2, the schemes to be constructed should satisfy discrete maximum principles for the eigenvalue range. These properties can be achieved by using the bound-preserving low order scheme of Section 4.3 to evolve each tensor component. Then tensorial extensions of Lemmas 4.16 and 4.24 can be proved and (local and global) DMPs for the range of eigenvalues hold in the case

$$\mathbf{F} = \mathbf{0}, \qquad c = 0 \tag{5.15}$$

provided that the low order system matrices are non-singular. Less restrictive assumptions may suffice to show the preservation of positive semidefiniteness. However, for the sake of simplicity, we restrict our analysis to situations in which the eigenvalue range of tensor quantities is preserved.

5.3.1 Steady problem

The steady advection-reaction equation for tensors (5.1) is discretized using (multi-)linear finite elements as described in Chapter 3. For this purpose, the Galerkin approximation $\mathbf{U}_h : \Omega \to \mathbb{S}_d$ with $\mathbf{U}_h \in V_h^{d \times d}$ is defined as a linear combination of the Lagrange basis functions $\varphi_1, \ldots, \varphi_N \in V_h$, i.e.,

$$\mathbf{U}_h(\mathbf{x}) = \sum_{i=1}^{N} \mathbf{U}_i \varphi_i(\mathbf{x}) \qquad \forall \mathbf{x} \in \Omega,$$

where the tensor-valued degrees of freedom $U = (\mathbf{U}_i)_{i=1}^{N} \subseteq \mathbb{S}_d^N$ satisfy the 'global' linear system of equations

$$\sum_{j=1}^{N} a_{ij}\mathbf{U}_j = \mathbf{G}_i \quad \forall i \in \{1,\dots,N\} \tag{5.16a}$$

$$\Longleftrightarrow \quad \sum_{j=1}^{N} a_{ij}u_{j,k\ell} = g_{i,k\ell} \quad \forall i \in \{1,\dots,N\}, \, k,\ell \in \{1,\dots,d\}. \tag{5.16b}$$

The coefficient matrix $\mathcal{A} = (a_{ij})_{i,j=1}^{N}$ is defined by (3.11b) and does not depend on the indices k, ℓ or values of the tensor components. Therefore, problem (5.16) admits a decomposition into d^2 independent linear systems of the form

$$\mathcal{A}(u_{i,k\ell})_{i=1}^{N} = (g_{i,k\ell})_{i=1}^{N} \quad \forall k,\ell \in \{1,\dots,d\}, \tag{5.17}$$

where the solution vector $(u_{i,k\ell})_{i=1}^{N}$ contains the unknown nodal values of the Galerkin approximation $u_{h,k\ell}$ at the vertices of the mesh. Recalling the symmetry of $\mathbf{U}_h$, the total number of segregated systems reduces to $\sum_{k=1}^{d} k = \frac{1}{2}d(d+1)$.

The application of discrete upwinding to each linear subsystem of (5.17) guarantees local and global DMP properties for every entry of the tensor-valued solution as discussed in Section 4.3.1. Using (4.49) to define the artificial diffusion operator $\mathcal{D} = (d_{ij})_{i,j=1}^{N}$ as before, we approximate (5.16) by the linear system

$$\sum_{j=1}^{N} (a_{ij} - d_{ij})\mathbf{U}_j^L = \mathbf{G}_i \quad \forall i \in \{1,\dots,N\}, \tag{5.18}$$

which yields the nodal values $U^L = (\mathbf{U}_i^L)_{i=1}^{N} \subseteq \mathbb{S}_d^N$ of a low order solution $\mathbf{U}_h^L \in V_h^{d \times d}$. Since this approximation forms the basis for the development of high order extensions, it is essential to verify if $\mathbf{U}_h^L$ possesses the desired properties.

To prove DMPs for the eigenvalues of U^L, we first multiply (5.18) from right and left by an arbitrary vector $\mathbf{x} \in \mathbb{R}^d$ and its transposed counterpart $\mathbf{x}^\top$

$$\sum_{j=1}^{N} (a_{ij} - d_{ij})(\mathbf{x}^\top \mathbf{U}_j^L \mathbf{x}) = (\mathbf{x}^\top \mathbf{G}_i \mathbf{x}) \quad \forall i \in \{1,\dots,N\}, \tag{5.19}$$

which is a linear system for the vector of scalar 'unknowns' $(\mathbf{x}^\top \mathbf{U}_i^\mathrm{L} \mathbf{x})_{i=1}^N$. Obviously, Lemma 4.16 remains valid if (5.19) is considered instead of (4.11). For instance, statement (4.14a) adapted to (5.19) reads

$$\max_{i \in I} \mathbf{x}^\top \mathbf{U}_i^\mathrm{L} \mathbf{x} \leqslant \max_{i \in (J \cap I) \cup \partial I} \mathbf{x}^\top \mathbf{U}_i^\mathrm{L} \mathbf{x}, \tag{5.20}$$

where

$$J := \Big\{ j \in \{1, \dots, N\} \ \Big| \ \mathbf{x}^\top \mathbf{G}_j \mathbf{x} > 0 \vee \sum_{l=1}^N a_{jl} > 0 \Big\}.$$

Even though this result is not very useful *per se*, it can be used to estimate the maximal eigenvalue of $\mathbf{U}_i^\mathrm{L}$ for any $i \in I$ by cycling through all eigenvectors $\mathbf{x} = \mathbf{q}_{i,d}$ corresponding to $u_{i,d}^\mathrm{L}$ and taking the maximum over all right hand sides of (5.20). An estimate obtained in this way leads to (5.21a) of the following lemma.

Lemma 5.4. *The discrete solution U^L to (5.18) satisfies DMPs for the range of eigenvalues corresponding to the following generalizations of Lemma 4.16:*
Consider the following index sets

$$J_1 := \Big\{ j \in \{1, \dots, N\} \ \Big| \ g_{j,d} > 0 \vee \sum_{l=1}^N a_{jl} > 0 \Big\},$$

$$J_2 := \Big\{ j \in \{1, \dots, N\} \ \Big| \ g_{j,1} < 0 \vee \sum_{l=1}^N a_{jl} > 0 \Big\},$$

$$J_3 := \Big\{ j \in \{1, \dots, N\} \ \Big| \ g_{j,d} > 0 \Big\},$$

$$J_4 := \Big\{ j \in \{1, \dots, N\} \ \Big| \ g_{j,1} < 0 \Big\}.$$

Let $\mathcal{A} - \mathcal{D} \in \mathbb{R}^{N \times N}$ be a non-singular matrix and $I \subseteq \{1, \dots, N\}$ be such that $(a_{ij} - d_{ij})_{i \in I, j \in \{1, \dots, N\}}$ is (i) a Z-matrix with (ii) nonnegative row sums. Define $\bar{I} := \bigcup_{j \in I} \mathcal{N}_j$ and $\partial I := \bar{I} \setminus I$. Then the eigenvalues of nodal tensors $(\mathbf{U}_i^\mathrm{L})_{i \in I}$ corresponding to the solution of (5.18) are bounded as follows:

$$\max_{i \in I} u_{i,d}^\mathrm{L} \leqslant \max_{i \in (J \cap I) \cup \partial I} u_{i,d}^\mathrm{L} \qquad\qquad J := J_1, \tag{5.21a}$$

$$\min_{i \in I} u_{i,1}^\mathrm{L} \geqslant \min_{i \in (J \cap I) \cup \partial I} u_{i,1}^\mathrm{L} \qquad\qquad J := J_2, \tag{5.21b}$$

$$\max_{i \in I} u_{i,d}^\mathrm{L} \leqslant \max\big(0, \max_{i \in (J \cap I) \cup \partial I} u_{i,d}^\mathrm{L}\big) \qquad J := J_3, \tag{5.21c}$$

$$\min_{i \in I} u_{i,1}^\mathrm{L} \geqslant \min\big(0, \min_{i \in (J \cap I) \cup \partial I} u_{i,1}^\mathrm{L}\big) \qquad J := J_4. \tag{5.21d}$$

If the invertible system matrix $\mathcal{A} - \mathcal{D}$ is (iii) a Z-matrix with (iv) non-negative row sums and

$$\sum_{j=1}^{N} a_{ij} = 0 \quad \Longrightarrow \quad \mathbf{G}_i = 0 \qquad \forall i \in \{1, \ldots, N\}, \tag{5.22}$$

then the following generalized maximum principles hold for the eigenvalues:

$$\max_{i \in \{1,\ldots,N\}} u_{i,d}^{\mathrm{L}} \leqslant \max_{i \in J} \left(\sum_{j=1}^{N} a_{ij} \right)^{-1} g_{i,d} \qquad J := J_1, \tag{5.23a}$$

$$\min_{i \in \{1,\ldots,N\}} u_{i,1}^{\mathrm{L}} \geqslant \min_{i \in J} \left(\sum_{j=1}^{N} a_{ij} \right)^{-1} g_{i,1} \qquad J := J_2. \tag{5.23b}$$

Note that all inequalities involving row sums of $\mathcal{A}$ also hold for row sums of $\mathcal{A} - \mathcal{D}$ since the artificial diffusion operator $\mathcal{D}$ has the zero row sum property.

Proof. Let us prove (5.21a). For any $\mathbf{x} \in \mathbb{R}^d$, the quadratic form $\mathbf{x}^\top \mathbf{U}_i^{\mathrm{L}} \mathbf{x}$ satisfies (5.20). Let $\iota \in I$ be arbitrary but fixed and $\mathbf{x} = \mathbf{q}_{\iota,d}^{\mathrm{L}}$ be the eigenvector of $\mathbf{U}_\iota^{\mathrm{L}}$ corresponding to the maximal eigenvalue $u_{\iota,d}^{\mathrm{L}}$. Then estimate (5.20) implies

$$u_{\iota,d} = \mathbf{q}_{\iota,d}^\top \mathbf{U}_\iota^{\mathrm{L}} \mathbf{q}_{\iota,d} \leqslant \max_{i \in (J' \cap I) \cup \partial I} \mathbf{q}_{\iota,d}^\top \mathbf{U}_i^{\mathrm{L}} \mathbf{q}_{\iota,d}$$

$$\leqslant \max_{i \in (J' \cap I) \cup \partial I} u_{i,d}^{\mathrm{L}} \leqslant \max_{i \in (J \cap I) \cup \partial I} u_{i,d}^{\mathrm{L}},$$

where

$$J' = \left\{ j \in \{1, \ldots, N\} \,\Big|\, \mathbf{q}_{\iota,d}^\top \mathbf{G}_j \mathbf{q}_{\iota,d} > 0 \vee \sum_{l=1}^{N} a_{jl} > 0 \right\} \subseteq J.$$

This shows (5.21a) because $\iota \in I$ was chosen arbitrarily.

The other results of Lemma 5.4 can be shown using similar arguments. $\square$

Therefore, as in the scalar case, local and global DMPs for the range of eigenvalues are satisfied if $\mathcal{A} - \mathcal{D}$ is non-singular (cf. condition (4.51)) and (5.15) holds.

5.3.2 Unsteady problem

The techniques of the previous section can also be used to derive a bound-preserving low order method for the time dependent advection-reaction equation (5.2). Since we are interested in obtaining two-sided estimates for the eigenvalue range, we assume that $c = 0$ and $\mathbf{F} = \mathbf{0}$, as we did in (4.62) and (5.15).

Proceeding as in Section 3.3.1, the time dependent degrees of freedom $\left(\mathbf{U}_i(t)\right)_{i=1}^N$ of the Galerkin approximation

$$\mathbf{U}_h(\mathbf{x},t) = \sum_{i=1}^N \mathbf{U}_i(t)\varphi_i(\mathbf{x}) \qquad \forall \mathbf{x} \in \Omega, \, t \in [0,T]$$

to the solution of (5.2) can be shown to satisfy the semi-discrete problem

$$\sum_{j=1}^N m_{ij}\frac{\mathrm{d}\mathbf{U}_j}{\mathrm{d}t} - \sum_{j=1}^N k_{ij}\mathbf{U}_j = \mathbf{G}_i \qquad \forall i \in \{1,\dots,N\}, \tag{5.24}$$

where $\mathcal{M} = (m_{ij})_{i,j=1}^N$, $\mathcal{K} = (k_{ij})_{i,j=1}^N$, and $(\mathbf{G}_i)_{i=1}^N$ are defined as above. Then the application of the discrete upwinding (and mass lumping) techniques of Section 4.3.2 leads to

$$m_i\frac{\mathrm{d}\mathbf{U}_i^{\mathrm{L}}}{\mathrm{d}t} - \sum_{j=1}^N (k_{ij} + d_{ij})\mathbf{U}_j^{\mathrm{L}} = \mathbf{G}_i \qquad \forall i \in \{1,\dots,N\}. \tag{5.25}$$

As in the stationary case, the subsystems associated with different tensor entries are decoupled and the scalar LED property of Theorem 4.48 holds true for each tensor component. Furthermore, the range of eigenvalues of the low order solution to (5.25) is preserved in the following sense.

Theorem 5.5. *Assume that condition (5.15) is satisfied. Let* $(\mathbf{U}_i^{\mathrm{L}}(t))_{i=1}^N$ *be the solution to problem (5.25) with differentiable eigenvalues and eigenvectors. Then* $(\mathbf{U}_i^{\mathrm{L}}(t))_{i=1}^N$ *is local extremum diminishing (LED) w.r.t. the range of eigenvalues. That is,*

$$\mathrm{d}_t u_{i,d}^{\mathrm{L}} \leqslant 0 \qquad \forall i \in \{1,\dots,M\} \text{ s.t. } u_{i,d}^{\mathrm{L}} = \max_{j\in\mathcal{N}_i} u_{j,d}^{\mathrm{L}}, \tag{5.26a}$$

$$\mathrm{d}_t u_{i,1}^{\mathrm{L}} \geqslant 0 \qquad \forall i \in \{1,\dots,M\} \text{ s.t. } u_{i,1}^{\mathrm{L}} = \min_{j\in\mathcal{N}_i} u_{j,1}^{\mathrm{L}}. \tag{5.26b}$$

Proof. For the sake of simplicity, we omit the superscript 'L' in this proof and denote the solution to problem (5.25) by $\left(\mathbf{U}_i(t)\right)_{i=1}^N$. Without loss of generality, we prove condition (5.26a) following [Loh17b].

The time derivative of $\tilde{\mathbf{U}}_i = \mathbf{Q}_i^\top \mathbf{U}_i \mathbf{Q}_i$ can be expressed as (cf. [VC03, Section 3.1])

$$
\begin{aligned}
\frac{d\tilde{\mathbf{U}}_i}{dt} &= \frac{d}{dt}(\mathbf{Q}_i^\top \mathbf{U}_i \mathbf{Q}_i) = (d_t \mathbf{Q}_i^\top)\mathbf{U}_i \mathbf{Q}_i + \mathbf{Q}_i^\top(d_t \mathbf{U}_i)\mathbf{Q}_i + \mathbf{Q}_i^\top \mathbf{U}_i(d_t \mathbf{Q}_i) \\
&= (d_t \mathbf{Q}_i^\top)\mathbf{Q}_i \tilde{\mathbf{U}}_i + \mathbf{Q}_i^\top(d_t \mathbf{U}_i)\mathbf{Q}_i + \tilde{\mathbf{U}}_i \mathbf{Q}_i^\top d_t \mathbf{Q}_i \\
&= \mathbf{H}_i \tilde{\mathbf{U}}_i + \mathbf{Q}_i^\top(d_t \mathbf{U}_i)\mathbf{Q}_i - \tilde{\mathbf{U}}_i \mathbf{H}_i,
\end{aligned}
\tag{5.27}
$$

where the auxiliary tensor $\mathbf{H}_i := (d_t \mathbf{Q}_i^\top)\mathbf{Q}_i$ is skew symmetric due to

$$
\begin{aligned}
\mathbf{H}_i &= (d_t \mathbf{Q}_i^\top)\mathbf{Q}_i = d_t(\mathbf{Q}_i^\top \mathbf{Q}_i) - \mathbf{Q}_i^\top(d_t \mathbf{Q}_i) \\
&= d_t \mathbf{I} - \mathbf{Q}_i^\top(d_t \mathbf{Q}_i) = -\mathbf{Q}_i^\top d_t \mathbf{Q}_i = -\mathbf{H}_i^\top.
\end{aligned}
$$

Consider a degree of freedom $\mathbf{U}_i$ associated with a non-Dirichlet node, i.e., $i \in \{1, \dots, M\}$. Then we have $\sum_j k_{ij} = 0$ and $\mathbf{G}_i = \mathbf{0}$ by (5.15) and, hence,

$$
m_i \frac{d\mathbf{U}_i}{dt} = \sum_{j=1}^N (k_{ij} + d_{ij})(\mathbf{U}_j - \mathbf{U}_i)
\tag{5.28}
$$

due to (5.25). Substituting (5.28) into (5.27) and assuming that the maximal eigenvalue of $\mathbf{U}_i$ is a local maximum, i.e., $u_{i,d} = \max_{j \in \mathcal{N}_i} u_{j,d}$, we obtain

$$
\begin{aligned}
\frac{du_{i,d}}{dt} &= \frac{d\tilde{u}_{i,dd}}{dt} = h_{i,dd}\tilde{u}_{i,dd} + \left(\mathbf{Q}_i^\top(d_t \mathbf{U}_i)\mathbf{Q}_i\right)_{dd} - \tilde{u}_{i,dd}h_{i,dd} \\
&= \mathbf{q}_{i,d}^\top(d_t \mathbf{U}_i)\mathbf{q}_{i,d} = \mathbf{q}_{i,d}^\top m_i^{-1} \sum_{j=1}^N (k_{ij} + d_{ij})(\mathbf{U}_j - \mathbf{U}_i)\mathbf{q}_{i,d} \\
&= m_i^{-1} \sum_{j \neq i} (k_{ij} + d_{ij})(\mathbf{q}_{i,d}^\top \mathbf{U}_j \mathbf{q}_{i,d} - u_{i,d}) \leqslant 0
\end{aligned}
$$

by using the fact that

$$
\begin{aligned}
k_{ij} + d_{ij} &\geqslant 0 && \forall i \in \{1, \dots, M\}, j \in \{1, \dots, N\}, j \neq i, \\
\mathbf{q}_{i,d}^\top \mathbf{U}_j \mathbf{q}_{i,d} &\leqslant u_{j,d} \leqslant u_{i,d} && \forall j \in \mathcal{N}_i, i \in \{1, \dots, M\}.
\end{aligned}
$$

This completes the proof of Theorem 5.5. $\square$

Theorem 5.5 shows that the solution of the semi-discrete low order problem (5.25) possesses the LED property for the range of eigenvalues. In particular, the eigenvalues of the linear finite element approximation $\mathbf{U}_h^{\mathrm{L}}(t)$

cannot grow arbitrarily and are bounded by the eigenvalues of the prescribed initial and inflow boundary data. Hereafter, we show that this property remains valid if (5.25) is discretized in time using the θ-scheme and a time step restriction is satisfied for $\theta \in [0, 1)$. For this purpose, we consider the fully discretized problem

$$\sum_{j=1}^{N} a_{ij} \mathbf{U}_j^{\mathrm{L},n+1} + \sum_{j=1}^{N} b_{ij} \mathbf{U}_j^{\mathrm{L},n} = \underline{\mathbf{G}}_i \qquad \forall i \in \{1, \dots, N\}, \tag{5.29}$$

where $\mathcal{A} = (a_{ij})_{i,j=1}^{N}$ and $\mathcal{B} = (b_{ij})_{i,j=1}^{N}$ are defined as in (4.65) and

$$\underline{\mathbf{G}} := \Delta t \big(\theta \mathbf{G}^{n+1} + (1-\theta) \mathbf{G}^n \big).$$

Proceeding as in Section 5.3.1, we formulate the following tensorial extension of Lemma 4.24.

Lemma 5.6. *Consider the following index sets*

$$J_1 := \big\{ j \in \{1, \dots, N\} \mid \underline{g}_{j,d} > 0 \vee \underline{p}_j \neq 0 \big\},$$
$$J_2 := \big\{ j \in \{1, \dots, N\} \mid \underline{g}_{j,1} < 0 \vee \underline{p}_j \neq 0 \big\},$$
$$J_3 := \big\{ j \in \{1, \dots, N\} \mid \underline{g}_{j,d} > 0 \vee \underline{p}_j < 0 \big\},$$
$$J_4 := \big\{ j \in \{1, \dots, N\} \mid \underline{g}_{j,1} < 0 \vee \underline{p}_j < 0 \big\}.$$

Let $I \subseteq \{1, \dots, N\}$ be such that (i) $(a_{ij})_{i \in I, j \in \{1,\dots,N\}}$ is a strictly diagonally dominant L-matrix and (ii) $(b_{ij})_{i \in I, j \in \{1,\dots,N\}}$ is nonpositive. Define $\bar{I} := \bigcup_{j \in I} \mathcal{N}_j$, and $\partial I := \bar{I} \setminus I$. Then the eigenvalues of nodal tensors $(\mathbf{U}_i^{\mathrm{L},n+1})_{i \in I}$ corresponding to the solution of (5.29) are bounded as follows:

$$\max_{i \in I} u_{i,d}^{\mathrm{L},n+1} \leqslant \max\big(\max_{i \in \bar{I}} u_{i,d}^{\mathrm{L},n}, \max_{i \in (J \cap I) \cup \partial I} u_{i,d}^{\mathrm{L},n+1} \big) \qquad J := J_1, \tag{5.30a}$$

$$\min_{i \in I} u_{i,1}^{\mathrm{L},n+1} \geqslant \min\big(\min_{i \in \bar{I}} u_{i,1}^{\mathrm{L},n}, \min_{i \in (J \cap I) \cup \partial I} u_{i,1}^{\mathrm{L},n+1} \big) \qquad J := J_2, \tag{5.30b}$$

$$\max_{i \in I} u_{i,d}^{\mathrm{L},n+1} \leqslant \max\big(0, \max_{i \in \bar{I}} u_{i,d}^{\mathrm{L},n}, \max_{i \in (J \cap I) \cup \partial I} u_{i,d}^{\mathrm{L},n+1} \big) \qquad J := J_3, \tag{5.30c}$$

$$\min_{i \in I} u_{i,1}^{\mathrm{L},n+1} \geqslant \min\big(0, \min_{i \in \bar{I}} u_{i,1}^{\mathrm{L},n}, \min_{i \in (J \cap I) \cup \partial I} u_{i,1}^{\mathrm{L},n+1} \big) \qquad J := J_4. \tag{5.30d}$$

If (iii) $\mathcal{A}$ is monotone, (iv) $\mathcal{B}$ is nonpositive, (v) $\underline{p}$ is nonnegative, and

$$\underline{p}_i = 0 \quad \Longrightarrow \quad \underline{\mathbf{G}}_i = \mathbf{0} \qquad \forall i \in \{1, \dots, N\}, \tag{5.31}$$

then the following generalized maximum principles hold for the eigenvalues:

$$\max_{i \in \{1,\dots,N\}} u_{i,d}^{L,n+1} \leqslant \max\Big(\max_{i \in \{1,\dots,N\}} u_{i,d}^{L,n}, \max_{i \in J} \underline{g}_{i,d} \underline{p}_i^{-1}\Big) \qquad J := J_1, \quad (5.32a)$$

$$\min_{i \in \{1,\dots,N\}} u_{i,1}^{L,n+1} \geqslant \min\Big(\min_{i \in \{1,\dots,N\}} u_{i,1}^{L,n}, \min_{i \in J} \underline{g}_{i,1} \underline{p}_i^{-1}\Big) \qquad J := J_2. \quad (5.32b)$$

Proof. This lemma can be proved following the arguments that we used in the proof of Lemma 5.4. $\qquad\qquad\qquad\qquad\qquad\qquad\qquad\qquad\qquad$ $\square$

As in the scalar case, Lemma 5.6 guarantees the validity of local and global DMPs for the range of eigenvalues if the CFL-like condition (4.66) holds for every $i \in \{1,\dots,N\}$. Furthermore, the eigenvalues of the low order solution are bounded by the eigenvalues of the prescribed initial and boundary data. The validity of these results can be shown as in the scalar case and a detailed proof is omitted.

5.4 Fractional step approach

Now that the definition and analysis of low order methods for the steady and unsteady advection-reaction equation are complete, this section deals with the tensorial extension of the algebraic correction technique presented in Section 4.4. To preserve the eigenvalue range in the process of antidiffusive correction, a scalar correction factor is applied to each component of the tensorial flux. As in the scalar case, the optimal value of this parameter is given by the solution of a local optimization problem. Even though the eigenvalue calculation is nonlinear, the computation of optimal correction factors can be performed in an efficient way.

This section summarizes the main results obtained in [Loh17b, Section 5], where an element-based FCT algorithm for a symmetric tensor quantity was proposed. For consistency reasons, we convert the original FCT scheme into an edge-based format, which exploits the localization technique presented in [Loh17a]. We remark that all algorithms to be considered in this section can be easily extended to their element-based counterparts.

In what follows, we do not distinguish between the treatment of a steady reaction dominated problem and the transient advection-reaction equation. The array U^L of nodal values to be corrected can be either the solution of (5.18) or the solution of (5.29) corresponding to a low order approximation at the time level t^{n+1}. The degrees of freedom $U = (\mathbf{U}_i)_{i=1}^N$ of the FCT solution $\mathbf{U}_h \in V_h^{d \times d}$ to be constrained are given by

$$\mathbf{U}_i = \mathbf{U}_i^L + m_i^{-1} \sum_{j \neq i} \alpha_{ij} \mathbf{F}_{ij} \qquad \forall i \in \{1, \ldots, N\}, \qquad (5.33)$$

where the entries of the antidiffusive flux $\mathbf{F}_{ij} = -\mathbf{F}_{ji} \in \mathbb{S}_d$ are defined as in Section 4.4 since the tensorial systems of high and low order consist of decoupled subsystems for individual components of the tensor field. The edge-based scalar correction factors $\alpha_{ij} = \alpha_{ji} \in [0,1]$ should be chosen as large as possible without violating the inequality constraints

$$u_i^{\min} \overset{!}{\leqslant} u_{i,1}, \quad u_{i,d} \overset{!}{\leqslant} u_i^{\max} \qquad \forall i \in \{1, \ldots, N\} \qquad (5.34)$$

for the range of eigenvalues (cf. (4.72)). The local bounds $u_i^{\min}$ and $u_i^{\max}$ should satisfy

$$u_i^{\min} \leqslant u_{i,1}^L, \quad u_{i,d}^L \leqslant u_i^{\max} \qquad \forall i \in \{1, \ldots, N\}$$

to guarantee the existence of a solution. We adopt the following definition which represents a consistent generalization of (4.79).

Definition 5.7. The *local bounds* for tensorial FCT constraints are defined by

$$u_i^{\max} := \max_{j \in \mathcal{N}_i} u_{j,d}^L, \qquad u_i^{\min} := \min_{j \in \mathcal{N}_i} u_{j,1}^L \qquad \forall i \in \{1, \ldots, N\}. \qquad (5.35)$$

The most straightforward way to define the correction factors α_{ij} is to adapt the ideas of Section 4.4. For instance, Algorithm 4.54 is based on separate estimates for the sums of positive and negative antidiffusive fluxes under worst case assumptions. The corresponding inequality constraints are summarized in (4.74). The simplified optimization problems are solved exactly to derive an upper bound for correction factors such that neither positive nor negative antidiffusive fluxes can violate the FCT constraints (4.72) even in the absence of fluxes that have the opposite sign. Invoking estimate (5.11b), eigenvalues of a tensor field can be constrained to satisfy

$$m_i u_i^{\min} \overset{!}{\leqslant} m_i u_{i,1}^L + \sum_{j \neq i} \alpha_{ij} f_{ij,1} \leqslant m_i u_{i,1} \qquad \forall i \in \{1, \ldots, N\}, \qquad (5.36a)$$

$$m_i u_i^{\max} \overset{!}{\geqslant} m_i u_{i,d}^L + \sum_{j \neq i} \alpha_{ij} f_{ij,d} \geqslant m_i u_{i,d} \qquad \forall i \in \{1, \ldots, N\}. \qquad (5.36b)$$

Then estimating each edge contribution by its positive or negative part, we find that the following conditions suffice to guarantee the validity of (5.36) in the worst case scenario:

$$m_i u_i^{\min} \overset{!}{\leqslant} m_i u_{i,1}^{\mathrm{L}} + \sum_{j \neq i} \alpha_{ij} \min(0, f_{ij,1}) \leqslant m_i u_{i,1} \qquad \forall i \in \{1, \ldots, N\},$$
$$\tag{5.37a}$$

$$m_i u_i^{\max} \overset{!}{\geqslant} m_i u_{i,d}^{\mathrm{L}} + \sum_{j \neq i} \alpha_{ij} \max(0, f_{ij,d}) \geqslant m_i u_{i,d} \qquad \forall i \in \{1, \ldots, N\}.$$
$$\tag{5.37b}$$

The optimal correction factors satisfying (5.37) can be computed using the following Zalesak-like eigenvalue range limiter.

Algorithm 5.8.

1. *Estimate the maximal/minimal eigenvalue of all increments to the maximal/minimal eigenvalue of* $\mathbf{U}_i^{\mathrm{L}}$

$$P_i^+ := \sum_{j \neq i} \max(0, f_{ij,d}), \qquad P_i^- := \sum_{j \neq i} \min(0, f_{ij,1}) \qquad \forall i \in \{1, \ldots, N\};$$
$$\tag{5.38}$$

2. *Determine the distance from the low order solution to the local maximum and minimum*

$$Q_i^+ := m_i(u_i^{\max} - u_{i,d}^{\mathrm{L}}), \qquad Q_i^- := m_i(u_i^{\min} - u_{i,1}^{\mathrm{L}}) \qquad \forall i \in \{1, \ldots, N\};$$
$$\tag{5.39}$$

3. *Calculate bounds for the nodal correction factors such that no violation of* (5.37) *occurs*

$$R_i^+ := \min\left(1, \frac{Q_i^+}{P_i^+}\right), \qquad R_i^- := \min\left(1, \frac{Q_i^-}{P_i^-}\right) \qquad \forall i \in \{1, \ldots, N\};$$
$$\tag{5.40}$$

4. *Limit all components of* $\mathbf{F}_{ij}$ *using the edge-based correction factors* α_{ij} *defined by*

$$S_{ij}^+ := \begin{cases} R_i^+ & : f_{ij,d} > 0, \\ 1 & : f_{ij,d} \leqslant 0, \end{cases} \qquad S_{ij}^- := \begin{cases} R_i^- & : f_{ij,1} < 0, \\ 1 & : f_{ij,1} \geqslant 0, \end{cases} \tag{5.41a}$$

$$\alpha_{ij} := \min(S_{ij}^-, S_{ij}^+, S_{ji}^-, S_{ji}^+) \qquad \forall i, j \in \{1, \ldots, N\}, j \neq i. \tag{5.41b}$$

Step 4 of Algorithm 5.8 is more involved than its scalar counterpart due to the possibility that the smallest and largest eigenvalue of $\mathbf{F}_{ij}$ may have different signs. Then the contribution of $\mathbf{F}_{ij}$ to (5.37a) and (5.37b) does not vanish and both requirements must be taken into consideration.

As in the scalar case, the inequality constraints (5.36) can be localized using the fact that $m_i = \sum_{j \neq i} \tilde{m}_{ij}$ (cf. (4.82))

$$\tilde{m}_{ij} u_i^{\min} \overset{!}{\leqslant} \tilde{m}_{ij} u_{i,1}^{L} + \alpha_{ij} f_{ij,1} \qquad \forall i,j \in \{1,\dots,N\}, \, j \neq i, \qquad (5.42a)$$

$$\tilde{m}_{ij} u_i^{\max} \overset{!}{\geqslant} \tilde{m}_{ij} u_{i,d}^{L} + \alpha_{ij} f_{ij,d} \qquad \forall i,j \in \{1,\dots,N\}, \, j \neq i. \qquad (5.42b)$$

Then step 1 of Algorithm 5.8 becomes redundant and the correction factors α_{ij} satisfying (5.42) can be computed using the following localized version of the edge-based eigenvalue range limiter.

Algorithm 5.9.

1. *Determine upper bounds for edge contributions to node i*

$$S_{ij}^{+} := \begin{cases} \min\big(1, \tilde{m}_{ij}(u_i^{\max} - u_{i,d}^{L})|f_{ij,d}|^{-1}\big) & : f_{ij,d} > 0, \\ 1 & : f_{ij,d} \leqslant 0, \end{cases} \qquad (5.43)$$

$$S_{ij}^{-} := \begin{cases} \min\big(1, \tilde{m}_{ij}(u_{i,1}^{L} - u_i^{\min})|f_{ij,1}|^{-1}\big) & : f_{ij,1} < 0, \\ 1 & : f_{ij,1} \geqslant 0. \end{cases} \qquad (5.44)$$

2. *Perform flux limiting using correction factors defined by*

$$\alpha_{ij} := \min(S_{ij}^{-}, S_{ij}^{+}, S_{ji}^{-}, S_{ji}^{+}) \qquad \forall i,j \in \{1,\dots,N\}, \, j \neq i. \qquad (5.45)$$

Both algorithms are based on criterion (5.36) in which estimate (5.11b) is exploited to approximate the eigenvalues of a sum of tensor quantities. As in the scalar case, each edge contribution to the antidiffusive correction term is considered separately and possible cancellation effects are neglected. This may lead to very pessimistic estimates of correction factors and quite diffusive results.

In what follows, we find a more general way to localize problem (5.34) without the need to make worst case assumptions like (5.36). The key ingredient of this approach is Löwner ordering as defined in Section 5.2. By definition, condition (5.34) is equivalent to

$$u_i^{\min} \mathbf{I} \overset{!}{\preccurlyeq} \mathbf{U}_i \overset{!}{\preccurlyeq} u_i^{\max} \mathbf{I} \qquad \forall i \in \{1,\dots,N\}. \qquad (5.46)$$

Positive and negative semidefinite parts of the edge-based antidiffusive fluxes $\mathbf{F}_{ij} = \mathbf{F}_{ij}^+ - \mathbf{F}_{ij}^-$, as defined in (5.14), can be readily used to localize this problem. The inequality constraints (5.46) hold if the limited antidiffusive tensor corrections satisfy (cf. (4.74))

$$m_i u_i^{\min} \mathbf{I} \overset{!}{\preccurlyeq} m_i \mathbf{U}_i^{\mathrm{L}} + \sum_{j \neq i} \alpha_{ij} \min(\mathbf{0}, \mathbf{F}_{ij}) \preccurlyeq m_i \mathbf{U}_i \qquad \forall i \in \{1, \ldots, N\},$$

(5.47a)

$$m_i u_i^{\max} \mathbf{I} \overset{!}{\succcurlyeq} m_i \mathbf{U}_i^{\mathrm{L}} + \sum_{j \neq i} \alpha_{ij} \max(\mathbf{0}, \mathbf{F}_{ij}) \succcurlyeq m_i \mathbf{U}_i \qquad \forall i \in \{1, \ldots, N\}.$$

(5.47b)

According to (5.12), the correction factors can be defined in terms of the largest nodal correction factors $R_i^+, R_i^- \in [0, 1]$ such that

$$m_i(u_i^{\min} \mathbf{I} - \mathbf{U}_i^{\mathrm{L}}) \overset{!}{\preccurlyeq} R_i^- \sum_{j \neq i} \min(\mathbf{0}, \mathbf{F}_{ij}) \qquad \forall i \in \{1, \ldots, N\}, \qquad (5.48a)$$

$$m_i(u_i^{\max} \mathbf{I} - \mathbf{U}_i^{\mathrm{L}}) \overset{!}{\succcurlyeq} R_i^+ \sum_{j \neq i} \max(\mathbf{0}, \mathbf{F}_{ij}) \qquad \forall i \in \{1, \ldots, N\}. \qquad (5.48b)$$

Explicit formulas for the calculation of R_i^- and R_i^+ are presented below in Section 5.4.1. Leaving $R_i^\pm$ unspecified for the time being, we consider the following alternative generalization of Zalesak's FCT limiter to tensor fields.

Algorithm 5.10.

1. *Compute the sums of positive and negative semidefinite unlimited fluxes into node i*

$$\mathbf{P}_i^+ := \sum_{j \neq i} \max(\mathbf{0}, \mathbf{F}_{ij}), \qquad \mathbf{P}_i^- := \sum_{j \neq i} \min(\mathbf{0}, \mathbf{F}_{ij}) \qquad \forall i \in \{1, \ldots, N\};$$

(5.49)

2. *Determine the distance from the low order solution to the local maximum and minimum*

$$\mathbf{Q}_i^+ := m_i(u_i^{\max} \mathbf{I} - \mathbf{U}_i^{\mathrm{L}}), \qquad \mathbf{Q}_i^- := m_i(u_i^{\min} \mathbf{I} - \mathbf{U}_i^{\mathrm{L}}) \qquad \forall i \in \{1, \ldots, N\};$$

(5.50)

3. *Calculate the largest nodal correction factors $R_i^+, R_i^- \in [0, 1]$, $i \in \{1, \ldots, N\}$ satisfying (5.48);*

4. *Define the correction factors depending on the eigenvalues of* $\mathbf{F}_{ij}$ *as follows:*

$$S_{ij}^+ := \begin{cases} R_i^+ & : f_{ij,d} > 0, \\ 1 & : f_{ij,d} \leqslant 0, \end{cases} \qquad S_{ij}^- := \begin{cases} R_i^- & : f_{ij,1} < 0, \\ 1 & : f_{ij,1} \geqslant 0, \end{cases} \qquad (5.51a)$$

$$\alpha_{ij} := \min(S_{ij}^-, S_{ij}^+, S_{ji}^-, S_{ji}^+) \qquad \forall i,j \in \{1,\dots,N\}, \, j \neq i. \qquad (5.51b)$$

In step 1, two tensor quantities must be stored for each degree of freedom, which requires a considerable amount of computer memory. This storage overhead can be avoided by exploiting the decomposition $m_i = \sum_{j \neq i} \tilde{m}_{ij}$ as in the case of Algorithm 4.58. The resulting localized edge-based limiter calculates the correction factors for (5.34) in the following manner.

Algorithm 5.11.

1. *Calculate the largest values of* $S_{ij}^-, S_{ij}^+ \in [0,1]$ *such that*

$$0 \overset{!}{\preccurlyeq} \tilde{m}_{ij}(\mathbf{U}_i^{\mathrm{L}} - u_i^{\min}\mathbf{I}) + S_{ij}^-\mathbf{F}_{ij} \qquad \forall i,j \in \{1,\dots,N\}, \, j \neq i, \quad (5.52a)$$

$$0 \overset{!}{\preccurlyeq} \tilde{m}_{ij}(u_i^{\max}\mathbf{I} - \mathbf{U}_i^{\mathrm{L}}) - S_{ij}^+\mathbf{F}_{ij} \qquad \forall i,j \in \{1,\dots,N\}, \, j \neq i. \quad (5.52b)$$

2. *Perform flux limiting using the correction factors defined by*

$$\alpha_{ij} := \min(S_{ij}^-, S_{ij}^+, S_{ji}^-, S_{ji}^+) \qquad \forall i,j \in \{1,\dots,N\}, \, j \neq i. \qquad (5.53)$$

The so defined eigenvalue range limiter guarantees the validity of the tensorial inequality constraints (5.46). For example, according to (5.12), (5.52a), and (5.53), we have

$$0 \preccurlyeq \tilde{m}_{ij}(\mathbf{U}_i^{\mathrm{L}} - u_i^{\min}\mathbf{I}) + \alpha_{ij}\mathbf{F}_{ij} \qquad \forall i,j \in \{1,\dots,N\}, \, j \neq i$$

and, hence,

$$m_i u_i^{\min}\mathbf{I} = \sum_{j \neq i} \tilde{m}_{ij} u_i^{\min}\mathbf{I} \preccurlyeq \sum_{j \neq i} (\tilde{m}_{ij}\mathbf{U}_i^{\mathrm{L}} + \alpha_{ij}\mathbf{F}_{ij}) = m_i\mathbf{U}_i \qquad \forall i \in \{1,\dots,N\}.$$

5.4.1 Semidefinite limiting strategies

Algorithms 5.10 and 5.11 require solution of local optimization problems with tensorial inequality constraints. This task is nontrivial if $d \in \{2,3\}$ due to the complexity of the nonlinear relationship (5.5) between the eigenvalues and components of a symmetric tensor. In what follows, we present different frame invariant strategies to find a usable approximation to the optimal lower bound α defined by [Loh17b, Eq. (12)]

$$\text{maximize } \alpha \in [0,1] \text{ s.t. } \quad \mathbf{0} \preccurlyeq \mathbf{A} + \alpha\mathbf{B}, \tag{5.54}$$

where $\mathbf{A} \in \mathbb{S}_{d,+}$ and $\mathbf{B} \in \mathbb{S}_d$. Note that optimization problems to be solved in Algorithms 5.10 and 5.11 can be written in this form. For instance the tensors $\mathbf{A}$ and $\mathbf{B}$ corresponding to (5.52a) are given by

$$\mathbf{A} = \tilde{m}_{ij}(\mathbf{U}_i^{\mathrm{L}} - u_i^{\min}\mathbf{I}), \qquad \mathbf{B} = \mathbf{F}_{ij},$$

where $\mathbf{A}$ is positive semidefinite by definition of the local bounds in (5.35).

In the literature on nonlinear constrained optimization, problem (5.54) belongs to the category of *semidefinite programming* and is closely related to the problem of finding so called 'positive semidefinite intervals' for matrix pencils (see, e.g., [Car+05]).

5.4.1.1 Min-min criterion

The most straightforward way to solve problem (5.54) is to invoke (5.11b) and estimate the smallest eigenvalue of the matrix pencil $\mathbf{A} + \alpha\mathbf{B}$ in terms of the minimal eigenvalues of $\mathbf{A}$ and $\mathbf{B}$. Then the maximal correction factor $\alpha \in [0,1]$ providing the positive semidefiniteness of $\mathbf{A} + \alpha\mathbf{B}$ can be determined by [Loh17b, Eq. (13)]

$$0 \overset{!}{\leqslant} a_1 + \alpha b_1 \leqslant \lambda_1(\mathbf{A} + \alpha\mathbf{B}) \quad \Longrightarrow \quad \alpha := \begin{cases} -a_1 b_1^{-1} & : b_1 < -a_1, \\ 1 & : b_1 \geqslant -a_1. \end{cases} \tag{5.55}$$

This formula is very simple but may lead to pessimistic results. For example, the problem [Loh17b, Eq. (14)]

$$\text{maximize } \alpha \in [0,1] \text{ s.t.}$$
$$\begin{pmatrix} 0 & 0 \\ 0 & 0 \end{pmatrix} \preccurlyeq \begin{pmatrix} 1 & 0 \\ 0 & 0 \end{pmatrix} + \alpha \begin{pmatrix} -3 & 0 \\ 0 & 0 \end{pmatrix} = \begin{pmatrix} 1 - 3\alpha & 0 \\ 0 & 0 \end{pmatrix} \tag{5.56}$$

possesses the exact solution $\alpha = \frac{1}{3}$ while (5.55) yields the lower bound $\alpha = 0$. This discrepancy is due to unexploited cancellation effects and caused by the simplicity of the formula. Unfortunately, vanishing correction factors occur whenever $a_1 = 0$ and $\mathbf{B}$ has a negative eigenvalue. The former condition holds at every local extremum, where, e.g., $u_i^{\min}$ coincides with the minimal eigenvalue of $\mathbf{U}_i^L$. Therefore, the FCT method utilizing this criterion is likely to produce less accurate results than an algorithm which solves the optimization problem (5.54) exactly. Note that Algorithm 5.11 equipped with the min-min criterion is equivalent to Algorithm 5.9, whereas Algorithm 5.10 using (5.55) is expected to be more accurate than Algorithm 5.8.

5.4.1.2 Regularized criterion

The criterion to be presented now does not ignore cancellation effects and overestimates the solution by regularizing the tensor $\mathbf{A}$. The main idea is based on the following result by Caron, Song, and Traynor [Car+05].

Theorem 5.12 ([Car+05, Theorem 3.1]). *Let* $\mathbf{A} \in \mathbb{S}_d$ *be positive definite and* $\mathbf{B} \in \mathbb{S}_d$. *Then* $\mathbf{A} + \beta\mathbf{B}$ *is positive semidefinite if and only if* $\beta \in [\underline{\beta}, \bar{\beta}]$, *where the bounds of the interval* $\underline{\beta} < \bar{\beta} \in \bar{\mathbb{R}}$ *are given by*

$$\underline{\beta} := \begin{cases} -\left(\lambda_d(\mathbf{A}^{-1}\mathbf{B})\right)^{-1} & : \lambda_d(\mathbf{A}^{-1}\mathbf{B}) > 0, \\ -\infty & : \lambda_d(\mathbf{A}^{-1}\mathbf{B}) \leqslant 0, \end{cases} \tag{5.57a}$$

$$\bar{\beta} := \begin{cases} -\left(\lambda_1(\mathbf{A}^{-1}\mathbf{B})\right)^{-1} & : \lambda_1(\mathbf{A}^{-1}\mathbf{B}) < 0, \\ +\infty & : \lambda_1(\mathbf{A}^{-1}\mathbf{B}) \geqslant 0. \end{cases} \tag{5.57b}$$

Proof. A proof of this theorem can be found in [Car+05]. □

This theorem leads to the following definition of the correction factor $\alpha \in [0, 1]$ [Loh17b, Eq. (15)]

$$\alpha := \begin{cases} -\left(\lambda_1(\mathbf{A}^{-1}\mathbf{B})\right)^{-1} & : \lambda_1(\mathbf{A}^{-1}\mathbf{B}) < -1, \\ 1 & : \lambda_1(\mathbf{A}^{-1}\mathbf{B}) \geqslant -1 \end{cases} \tag{5.58}$$

under the assumption that $\mathbf{A}$ is positive definite and, therefore, invertible. Unfortunately, as already mentioned, the minimal eigenvalue of $\mathbf{A}$ is equal to zero at every local extremum and (5.58) cannot be used to calculate the correction factors in these cases. Caron, Song, and Traynor [Car+05] also proposed formulas to calculate the bounds of the so called 'positive semidefinite interval' $[\underline{\beta}, \bar{\beta}]$ if $\mathbf{A}$ is only positive semidefinite (cf. [Car+05, Corollary 4.4]). However, these results are much more involved and, hence, not considered henceforth. Instead, we extend (5.58) to positive semidefinite tensors $\mathbf{A}$ by adding a small perturbation $\varepsilon\mathbf{I}$, $0 < \varepsilon \ll 1$ to $\mathbf{A}$. Substituting

$\bar{\mathbf{A}} := \mathbf{A} + \varepsilon \mathbf{I} \succ \mathbf{0}$ for $\mathbf{A}$ and using (5.58) to calculate the correction factor of the perturbed problem, we approximate the solution to (5.54) by

$$\alpha := \begin{cases} -\gamma^{-1} & : \gamma < -1, \\ 1 & : \gamma \geqslant -1, \end{cases}$$

$$\text{where} \quad \gamma := \lambda_1\big((\tilde{\mathbf{A}} + \varepsilon \mathbf{I})^{-1/2}(\mathbf{Q}^\top \mathbf{B} \mathbf{Q})(\tilde{\mathbf{A}} + \varepsilon \mathbf{I})^{-1/2}\big) \quad (5.59)$$

is defined using the spectral decomposition $\mathbf{A} = \mathbf{Q}\tilde{\mathbf{A}}\mathbf{Q}^\top$. This formula for α guarantees the positive semidefiniteness of $\mathbf{A} + \varepsilon \mathbf{I} + \alpha \mathbf{B}$ and overestimates the bound corresponding to the exact solution of (5.54). Consequently, each iteration of Algorithm 5.10 or 5.11 produces an FCT solution whose eigenvalues may violate the local bounds by ε. To illustrate this drawback, we solve problem (5.56) approximately using (5.59). The result is given by $\alpha = \frac{1}{3}(1 + \varepsilon)$, which follows from [Loh17b]

$$(\tilde{\mathbf{A}} + \varepsilon \mathbf{I})^{-1/2}(\mathbf{Q}^\top \mathbf{B} \mathbf{Q})(\tilde{\mathbf{A}} + \varepsilon \mathbf{I})^{-1/2}$$

$$= \begin{pmatrix} 1+\varepsilon & 0 \\ 0 & \varepsilon \end{pmatrix}^{-1/2} \begin{pmatrix} -3 & 0 \\ 0 & 0 \end{pmatrix} \begin{pmatrix} 1+\varepsilon & 0 \\ 0 & \varepsilon \end{pmatrix}^{-1/2} = \begin{pmatrix} -3(1+\varepsilon)^{-1} & 0 \\ 0 & 0 \end{pmatrix}.$$

In the limit of a vanishing perturbation $\varepsilon \mathbf{I} = \mathbf{0}$, we obtain the exact solution. However, for any $\varepsilon > 0$, we have $\alpha > \frac{1}{3}$ which is, strictly speaking, not admissible.

5.4.1.3 Invariant criterion

Both criteria presented so far involve the computation of eigenvalues and produce approximations to the exact solution of problem (5.54). A criterion that yields the exact solution without any eigenvalue calculation is based on the fact that a tensor is positive semidefinite if and only if all principal invariants $\mathrm{I}_1, \ldots, \mathrm{I}_d$ are nonnegative. Therefore, the exact solution to (5.54) coincides with [Loh17b]

$$\alpha := \min(\alpha_1, \ldots, \alpha_d), \tag{5.60a}$$

where the bound α_k, $k \in \{1, \ldots, d\}$, ensures that the k-th principal invariant of $\mathbf{A} + \beta \mathbf{B}$ is nonnegative for all $\beta \in [0, \alpha_k]$ and, hence, coincides with the first root of the polynomial $\mathrm{i}_k(\beta) := \mathrm{I}_k(\mathbf{A} + \beta \mathbf{B})$ in $[0, 1]$ where i_k changes its sign from positive to negative. That is,

$$\alpha_k = \inf\Big(\big\{\beta \geqslant 0 \mid \mathrm{I}_k(\mathbf{A} + \beta \mathbf{B}) < 0\big\} \cup \{1\}\Big) \qquad \forall k \in \{1, \ldots, d\}. \tag{5.60b}$$

If $\mathbf{A}$ is positive definite, this solution obviously coincides with the correction factor of the regularized criterion because both approaches are analytically exact. The interested reader is referred to [Loh17b, Section 5.3] for a detailed description of algorithms to calculate (or approximate) the auxiliary quantities $\alpha_1, \ldots, \alpha_d$.

5.4.1.4 Bisection criterion

Another very robust approach to calculating an arbitrarily sharp lower bound for the solution to (5.54) is based on the 'bisection method'. It exploits (5.12) and the fact that only roots in $[0, 1]$ are of interest.

Algorithm 5.13. *If all principal invariants of* $\mathbf{A} + \mathbf{B}$ *are nonnegative, then* $\alpha = 1$ *is the desired bound. Otherwise, initialize the search interval* $[\underline{\beta}, \bar{\beta}]$ *by setting* $\underline{\beta} = 0$ *and* $\bar{\beta} = 1$. *Then perform the following steps:*

1. *Check if all principal invariants of* $\mathbf{A} + \frac{1}{2}(\underline{\beta} + \bar{\beta})\mathbf{B}$ *are nonnegative. If so, set* $\underline{\beta} := \frac{1}{2}(\underline{\beta} + \bar{\beta})$. *Otherwise, set* $\bar{\beta} := \frac{1}{2}(\underline{\beta} + \bar{\beta})$.

2. *If* $\bar{\beta} - \underline{\beta} < \varepsilon$, *a guaranteed lower bound for the solution of* (5.54) *is given by* $\alpha = \underline{\beta}$. *Otherwise, repeat step 1 until this stopping criterion is satisfied.*

If $\alpha \in [0, 1]$ is the result of Algorithm 5.13, we have $\mathbf{A} \succcurlyeq \mathbf{0}$ and $\mathbf{A} + \alpha\mathbf{B} \succcurlyeq \mathbf{0}$. Consequently, the matrix pencil $\mathbf{A} + \beta\mathbf{B}$ is positive semidefinite for every $\beta \in [0, \alpha]$ by virtue of (5.12).

5.5 Monolithic approach

This section extends the monolithic limiting approach of Section 4.5 to the treatment of hyperbolic equations for symmetric tensor quantities. Similarly to the algorithms presented so far, this methodology constrains the range of eigenvalues in a manner that preserves the corresponding DMP property of the low order solution. In contrast to the FCT algorithms of the previous section, the monolithic approach to tensor limiting for steady problems makes it possible to design tensor-valued correction factors leading to improved approximations. The potentially significant gain of accuracy is achieved by using spectral decompositions and individual scalar correction factors for each eigenvalue of the tensor to be limited. In the case of unsteady problems, this approach seems to guarantee the DMP properties for the eigenvalue range only if fully implicit time integrators are used.

Most results of the next section, in which we consider the steady advection-reaction equation for symmetric tensors, were originally published in [Loh19].

5.5.1 Steady problem

To analyze more general limiting approaches and derive sufficient conditions for the validity of DMPs, we introduce limiting operators $\mathscr{A}_{ij} : \mathbb{S}_d \to \mathbb{S}_d$ which are applied to $\mathbf{U}_j - \mathbf{U}_i$. In general, $\mathscr{A}_{ij}$ depends on the solution $U = (\mathbf{U}_i)_{i=1}^N$ and returns $\mathbf{U}_j - \mathbf{U}_i$ in the best case scenario, in which the high order solution is recovered. If $\mathscr{A}_{ij}[\mathbf{U}_j - \mathbf{U}_i] = \mathbf{0}$ for all $i, j \in \{1, \dots, N\}$, $j \neq i$, the obtained solution coincides with its low order counterpart. The notation $\mathscr{A}_{ij}(U)[\mathbf{U}_j - \mathbf{U}_i]$ means the application of the solution dependent limiting operator $\mathscr{A}_{ij}(U)$ to $\mathbf{U}_j - \mathbf{U}_i$. For brevity, we will frequently abstain from indicating the dependence of $\mathscr{A}_{ij}$ on U and use the shorthand notation $\mathscr{A}_{ij}[\cdot]$ for $\mathscr{A}_{ij}(U)[\cdot]$. As in the case of scalar correction factors, we employ symmetric limiting operators for conservation reasons. That is,

$$\mathscr{A}_{ij}[\mathbf{U}_j - \mathbf{U}_i] = -\mathscr{A}_{ji}[\mathbf{U}_i - \mathbf{U}_j] \qquad \forall i, j \in \{1, \dots, N\}, j \neq i. \qquad (5.61)$$

Written in terms of these operators, the general AFC system for the steady advection-reaction equation (5.1) reads [Loh19, Eq. (14)]

$$\sum_{j=1}^N a_{ij}\mathbf{U}_j - \sum_{j \neq i} d_{ij}(\mathscr{I} - \mathscr{A}_{ij})[\mathbf{U}_j - \mathbf{U}_i] = \mathbf{G}_i \qquad \forall i \in \{1, \dots, N\}, \quad (5.62)$$

where $\mathscr{I} : \mathbb{S}_d \to \mathbb{S}_d$ is the identity operator. The next sections deal with the existence of a solution and furnish sufficient conditions for the validity of DMPs. Additionally, examples of limiting operators based on scalar and tensorial correction factors are presented below.

5.5.1.1 Existence of a solution

Let us now prove the existence of a solution to (5.62) exploiting Lemma 4.64 as in the scalar case (cf. Section 4.5.1.1). For presentation of the results in a compact and intuitive form, the following definition of the scalar product

$(\cdot,\cdot)_{2,\mathrm{F}} : \mathbb{S}_d^N \times \mathbb{S}_d^N \to \mathbb{R}$ and its naturally induced norm $\|\cdot\|_{2,\mathrm{F}} : \mathbb{S}_d^N \to \mathbb{R}_0^+$ is given:

$$(V,U)_{2,\mathrm{F}} := \sum_{i=1}^N (\mathbf{V}_i, \mathbf{U}_i)_{\mathrm{F}}, \qquad \forall V, U \in \mathbb{S}_d^N,$$

$$\|V\|_{2,\mathrm{F}}^2 := (V,V)_{2,\mathrm{F}} = \sum_{i=1}^N \|\mathbf{V}_i\|_{\mathrm{F}}^2 \qquad \forall V \in \mathbb{S}_d^N.$$

Theorem 5.14. *The AFC system* (5.62) *with an arbitrary right hand side* $G = (\mathbf{G}_i)_{i=1}^N$ *possesses at least one solution* $U = (\mathbf{U}_i)_{i=1}^N$ *with* $\|U\|_{2,\mathrm{F}} \leqslant C_M^{-1} \|G\|_{2,\mathrm{F}}$ *for* $C_M > 0$ *defined by* (3.12) *if the coercivity condition* (2.14) *holds and*

$$\big(\mathbf{V}_j - \mathbf{V}_i, (\mathscr{I} - \mathscr{A}_{ij})[\mathbf{V}_j - \mathbf{V}_i]\big)_{\mathrm{F}} \geqslant 0 \quad \forall i,j \in \{1,\dots,N\},\ j \neq i, \quad (5.63\mathrm{a})$$

$$\mathbb{S}_d^N \ni V \ \mapsto \ \mathscr{A}_{ij}(V)[\mathbf{V}_j - \mathbf{V}_i] \in \mathbb{S}_d \quad \text{is continuous}$$
$$\forall i,j \in \{1,\dots,N\},\ j \neq i. \qquad (5.63\mathrm{b})$$

Proof. To prove this result, we extend the proof of Theorem 4.65 to the treatment of the tensor version of the steady model problem. For this purpose, the operator $\mathscr{L} : \mathbb{S}_d^N \to \mathbb{S}_d^N$ of Lemma 4.64 is defined by

$$(\mathscr{L}V)_i := \sum_{j=1}^N a_{ij}\mathbf{V}_j - \sum_{j\neq i} d_{ij}\big(\mathscr{I} - \mathscr{A}_{ij}(V)\big)[\mathbf{V}_j - \mathbf{V}_i] - \mathbf{G}_i$$

$$\forall V \in \mathbb{S}_d^N,\ i \in \{1,\dots,N\}.$$

In view of (5.61) and (5.63a), we have

$$\sum_{j\neq i} d_{ij}\big(\mathbf{V}_i, (\mathscr{I} - \mathscr{A}_{ij})[\mathbf{V}_j - \mathbf{V}_i]\big)_{\mathrm{F}}$$

$$= \sum_{i,j,\,j<i} d_{ij}\Big(\big(\mathbf{V}_i, (\mathscr{I} - \mathscr{A}_{ij})[\mathbf{V}_j - \mathbf{V}_i]\big)_{\mathrm{F}}$$
$$+ \big(\mathbf{V}_j, (\mathscr{I} - \mathscr{A}_{ji})[\mathbf{V}_i - \mathbf{V}_j]\big)_{\mathrm{F}}\Big)$$

$$= \sum_{i,j,\,j<i} d_{ij}\big(\mathbf{V}_i - \mathbf{V}_j, (\mathscr{I} - \mathscr{A}_{ij})[\mathbf{V}_j - \mathbf{V}_i]\big)_{\mathrm{F}} \leqslant 0 \qquad \forall V \in \mathbb{S}_d^N.$$

Additionally, the definition of $C_M > 0$ (cf. (3.12)) implies

$$
\sum_{i,j=1}^{N} a_{ij} (\mathbf{V}_i, \mathbf{V}_j)_{\mathrm{F}} = \sum_{k,\ell=1}^{d} \sum_{i,j=1}^{N} a_{ij} v_{i,k\ell} v_{j,k\ell}
$$

$$
\geqslant \sum_{k,\ell=1}^{d} C_M \sum_{i=1}^{N} v_{i,k\ell}^2 = C_M \sum_{i=1}^{N} \sum_{k,\ell=1}^{d} v_{i,k\ell}^2
$$

$$
= C_M \sum_{i=1}^{N} \|\mathbf{V}_i\|_{\mathrm{F}}^2 = C_M \|V\|_{2,\mathrm{F}}^2 \qquad \forall V \in \mathbb{S}_d^N.
$$

Using these estimates, we follow the proof of Theorem 4.65 and obtain

$$
\left(\mathscr{L} V, V \right)_{2,\mathrm{F}} = \sum_{i,j=1}^{N} a_{ij} (\mathbf{V}_i, \mathbf{V}_j)_{\mathrm{F}} - \sum_{i=1}^{N} (\mathbf{G}_i, \mathbf{V}_i)_{\mathrm{F}}
$$

$$
- \sum_{i,j=1}^{N} d_{ij} \left(\mathbf{V}_i, (\mathscr{I} - \mathscr{A}_{ij}) [\mathbf{V}_j - \mathbf{V}_i] \right)_{\mathrm{F}}
$$

$$
\geqslant C_M \|V\|_{2,\mathrm{F}}^2 - c_2 \|V\|_{2,\mathrm{F}}^2 - (4c_2)^{-1} \|G\|_{2,\mathrm{F}}^2
$$

$$
\geqslant \frac{C_M}{2} \|V\|_{2,\mathrm{F}}^2 - (2C_M)^{-1} \|G\|_{2,\mathrm{F}}^2 > 0
$$

$$
\forall V \in \mathbb{S}_d^N \text{ s.t. } \|V\|_{2,\mathrm{F}}^2 = C_M^{-2} \|G\|_{2,\mathrm{F}}^2 + \varepsilon
$$

and all $\varepsilon > 0$, where $c_2 = \frac{C_M}{2}$. The desired result follows by Lemma 4.64 and assumption (5.63b). $\square$

5.5.1.2 Discrete maximum principles

While Theorem 5.14 guarantees the existence of a solution to problem (5.62) under appropriate assumptions, the next lemma shows that every solution inherits the discrete maximum principle properties of its low order counterpart. In contrast to [Loh19], the inequality constraints are shown under slightly different requirements on $\mathscr{A}_{ij}$, which can be formulated compactly and lead to a family of eigenvalue range preserving limiting operators to be presented below.

Lemma 5.15. *The assertions of Lemma 5.4 without requirements (i) and (iii) hold for the solution to problem (5.62) if*

$$\mathbf{q}_{i,d}^\top \mathscr{A}_{ij}[\mathbf{U}_j - \mathbf{U}_i]\mathbf{q}_{i,d} \geqslant \begin{cases} 0 & : a_{ij} > 0, \\ \mathbf{q}_{i,d}^\top \mathbf{U}_j \mathbf{q}_{i,d} - u_{i,d} & : a_{ij} \leqslant 0 \end{cases}$$

$$\forall i,j \in \{1,\dots,N\}, \; j \neq i \; s.t. \; u_{i,d} = \max_{l \in \mathcal{N}_i} u_{l,d}, \quad (5.64a)$$

$$\mathbf{q}_{i,1}^\top \mathscr{A}_{ij}[\mathbf{U}_j - \mathbf{U}_i]\mathbf{q}_{i,1} \leqslant \begin{cases} 0 & : a_{ij} > 0, \\ \mathbf{q}_{i,1}^\top \mathbf{U}_j \mathbf{q}_{i,1} - u_{i,1} & : a_{ij} \leqslant 0 \end{cases}$$

$$\forall i,j \in \{1,\dots,N\}, \; j \neq i \; s.t. \; u_{i,1} = \min_{l \in \mathcal{N}_i} u_{l,1}. \quad (5.64b)$$

Proof. To show that the statement of Lemma 5.15 is true, we slightly modify the proof given in [Loh19].

Suppose that (5.21a) does not hold. Then there exists an index $i \in I \setminus J$ such that

$$u_{i,d} = \max_{j \in I} u_{j,d} > \max_{j \in (J \cap I) \cup \partial I} u_{j,d}. \quad (5.65)$$

Furthermore, we define

$$I_* := \{l \in I \setminus J \mid \mathbf{q}_{i,d}^\top \mathbf{U}_l \mathbf{q}_{i,d} = u_{i,d}\} \subseteq I \setminus J.$$

This implies

$$u_{l,d} \geqslant \mathbf{q}_{i,d}^\top \mathbf{U}_l \mathbf{q}_{i,d} = u_{i,d} = \max_{j \in I} u_{j,d} \geqslant u_{l,d} \qquad \forall l \in I_* \subseteq I \quad (5.66)$$

and, hence,

$$u_{l,d} = u_{i,d} \geqslant \max_{j \in I} u_{j,d} \geqslant \max_{j \in \mathcal{N}_l} u_{j,d} \qquad \forall l \in I_*$$

by (5.65). Consequently, $u_{l,d}$ is a local maximum of maximal eigenvalues for each $l \in I_*$ and (5.64a) provides a lower bound for the limited quadratic form. Furthermore, according to (3.12) and (4.115),

$$\exists \iota \in I_* : \sum_{j \in I_*} a_{\iota j} > 0. \quad (5.67)$$

Multiplication of (5.62) for $i = \iota$ from right and left by $\mathbf{q}_{\iota,d} \in \mathbb{R}^d$ and $\mathbf{q}_{\iota,d}^\top$ yields

$$\sum_{j=1}^N a_{\iota j} \mathbf{q}_{\iota,d}^\top \mathbf{U}_j \mathbf{q}_{\iota,d} - \sum_{j \neq \iota} d_{\iota j} \mathbf{q}_{\iota,d}^\top (\mathscr{I} - \mathscr{A}_{\iota j})[\mathbf{U}_j - \mathbf{U}_\iota]\mathbf{q}_{\iota,d} = \mathbf{q}_{\iota,d}^\top \mathbf{G}_\iota \mathbf{q}_{\iota,d}$$

and, hence,

$$\sum_{j=1}^{N} a_{\iota j}\mathbf{q}_{\iota,d}^{\top}\mathbf{U}_j\mathbf{q}_{\iota,d} - \sum_{j,\,a_{\iota j}>0} d_{\iota j}(\mathbf{q}_{\iota,d}^{\top}\mathbf{U}_j\mathbf{q}_{\iota,d} - u_{\iota,d}) \leqslant \mathbf{q}_{\iota,d}^{\top}\mathbf{G}_\iota\mathbf{q}_{\iota,d} \leqslant g_{\iota,d} \leqslant 0$$

(5.68)

by (5.64a) and the fact that $\iota \in I_* \subseteq I \setminus J$. Invoking inequality (5.68), we find

$$\Big(\sum_{j\in I_*} a_{\iota j} + \sum_{j\notin I_*,\,a_{\iota j}>0} d_{\iota j}\Big)u_{\iota,d}$$

$$\leqslant \sum_{j\in I_*} a_{\iota j}(u_{\iota,d} - \mathbf{q}_{\iota,d}^{\top}\mathbf{U}_j\mathbf{q}_{\iota,d}) + \sum_{j\in I_*,\,a_{\iota j}>0} d_{\iota j}(\mathbf{q}_{\iota,d}^{\top}\mathbf{U}_j\mathbf{q}_{\iota,d} - u_{\iota,d})$$

$$- \sum_{j\notin I_*} a_{\iota j}\mathbf{q}_{\iota,d}^{\top}\mathbf{U}_j\mathbf{q}_{\iota,d} + \sum_{j\notin I_*,\,a_{\iota j}>0} d_{\iota j}\mathbf{q}_{\iota,d}^{\top}\mathbf{U}_j\mathbf{q}_{\iota,d}$$

$$= \sum_{j\in I_*,\,a_{\iota j}>0} (a_{\iota j} - d_{\iota j})(u_{\iota,d} - \mathbf{q}_{\iota,d}^{\top}\mathbf{U}_j\mathbf{q}_{\iota,d})$$

$$+ \sum_{j\in \bar{I}\setminus I_*,\,a_{\iota j}>0} (-a_{\iota j} + d_{\iota j})\mathbf{q}_{\iota,d}^{\top}\mathbf{U}_j\mathbf{q}_{\iota,d}$$

$$+ \sum_{j\in I_*,\,a_{\iota j}\leqslant 0} a_{\iota j}(u_{\iota,d} - \mathbf{q}_{\iota,d}^{\top}\mathbf{U}_j\mathbf{q}_{\iota,d}) + \sum_{j\in \bar{I}\setminus I_*,\,a_{\iota j}\leqslant 0} (-a_{\iota j})\mathbf{q}_{\iota,d}^{\top}\mathbf{U}_j\mathbf{q}_{\iota,d}$$

$$\leqslant \sum_{j\in \bar{I}\setminus I_*,\,a_{\iota j}>0} (-a_{\iota j} + d_{\iota j})\max_{j\in \bar{I}\setminus I_*} \mathbf{q}_{\iota,d}^{\top}\mathbf{U}_j\mathbf{q}_{\iota,d}$$

$$+ \sum_{j\in \bar{I}\setminus I_*,\,a_{\iota j}\leqslant 0} (-a_{\iota j})\max_{j\in \bar{I}\setminus I_*} \mathbf{q}_{\iota,d}^{\top}\mathbf{U}_j\mathbf{q}_{\iota,d}$$

$$= \Big(\sum_{j\in I_*} a_{\iota j} + \sum_{j\notin I_*,\,a_{\iota j}>0} d_{\iota j}\Big)\max_{j\in \bar{I}\setminus I_*} \mathbf{q}_{\iota,d}^{\top}\mathbf{U}_j\mathbf{q}_{\iota,d}$$

(5.69)

by the fact that $\mathbf{q}_{\iota,d}^{\top}\mathbf{U}_j\mathbf{q}_{\iota,d} \leqslant u_{j,d} = u_{\iota,d}$ for all $j \in I_* \subset I \setminus J$. Recalling (5.67), we arrive at

$$\max_{j\in(I\setminus J)\setminus I_*} \mathbf{q}_{\iota,d}^{\top}\mathbf{U}_j\mathbf{q}_{\iota,d} < u_{\iota,d}$$

$$\leqslant \max_{j\in \bar{I}\setminus I_*} \mathbf{q}_{\iota,d}^{\top}\mathbf{U}_j\mathbf{q}_{\iota,d} = \max\Big(\max_{j\in(I\setminus J)\setminus I_*} \mathbf{q}_{\iota,d}^{\top}\mathbf{U}_j\mathbf{q}_{\iota,d}, \max_{j\in(J\cap I)\cup\partial I} \mathbf{q}_{\iota,d}^{\top}\mathbf{U}_j\mathbf{q}_{\iota,d}\Big)$$

because $(I \setminus J) \setminus I_* \cup (J \cap I) \cup \partial I = (I \setminus I_*) \cup \partial I = \bar{I} \setminus I_*$. This leads to

$$u_{\iota,d} \leqslant \max_{j\in(J\cap I)\cup\partial I} \mathbf{q}_{\iota,d}^{\top}\mathbf{U}_j\mathbf{q}_{\iota,d} \leqslant \max_{j\in(J\cap I)\cup\partial I} u_{j,d}$$

in contradiction to assumption (5.65) and shows the bound (5.21a).

Condition (5.21c) can be shown in the same way, whereby the last identity of (5.69) needs to be replaced by an inequality with a nonnegative maximum.

To show the validity of (5.23a), we first notice that there exists an $i \in J$ such that $u_{i,d} = \max_j u_{j,d}$ by (5.21a). In particular, $u_{i,d}$ is a local maximum of maximal eigenvalues and, hence, (5.64a) holds. The same argumentation as above yields

$$
\begin{aligned}
g_{i,d} &\geqslant \sum_{j=1}^{N} a_{ij} \mathbf{q}_{i,d}^{\top} \mathbf{U}_j \mathbf{q}_{i,d} - \sum_{j \neq i,\, a_{ij}>0} d_{ij}(\mathbf{q}_{i,d}^{\top} \mathbf{U}_j \mathbf{q}_{i,d} - u_{i,d}) \\
&= \big(a_{ii} + \sum_{j \neq i,\, a_{ij}>0} d_{ij}\big) u_{i,d} + \sum_{j \neq i,\, a_{ij}\leqslant 0} a_{ij} \mathbf{q}_{i,d}^{\top} \mathbf{U}_j \mathbf{q}_{i,d} \\
&\quad + \sum_{j \neq i,\, a_{ij}>0} (a_{ij} - d_{ij}) \mathbf{q}_{i,d}^{\top} \mathbf{U}_j \mathbf{q}_{i,d} \\
&\geqslant \big(a_{ii} + \sum_{j \neq i,\, a_{ij}>0} d_{ij}\big) u_{i,d} + \sum_{j \neq i,\, a_{ij}\leqslant 0} a_{ij} u_{i,d} \\
&\quad + \sum_{j \neq i,\, a_{ij}>0} (a_{ij} - d_{ij}) u_{i,d} = \sum_{j=1}^{N} a_{ij} u_{i,d},
\end{aligned}
$$

which completes the proof.

The lower bounds of Lemma 5.4 can be shown similarly. $\qquad\square$

Condition (5.64) is more involved than its scalar counterpart (4.112) because a general limiting operator $\mathscr{A}_{ij}$ is used to limit $\mathbf{U}_j - \mathbf{U}_i$. This operator can be more sophisticated than multiplication of its argument by a scalar correction factor and additional conditions must be met even if $a_{ij} \leqslant 0$.

5.5.1.3 Examples

To devise examples of limiting operators $\mathscr{A}_{ij}$ satisfying the above requirements, we proceed as in the scalar case and first prove a tensorial extension of Lemma 4.76. The next lemma shows the Lipschitz continuity of relevant terms for tensor-valued counterparts $\boldsymbol{\xi}_l$ of the scalar functions ξ_l. Due to the lack of commutativity of tensor products, it is worthwhile to consider the case in which $\mathbf{V}_j - \mathbf{V}_i$ is multiplied from left and right by different tensors $\boldsymbol{\xi}_1$ and $\boldsymbol{\xi}_2$. The norms of the tensor quantities under investigation are intentionally left unspecified and can be chosen arbitrarily due to their equivalence in finite dimensional spaces.

Lemma 5.16. *Let* $i, j \in \{1, \dots, N\}$, $j \neq i$, *be arbitrary and* $\boldsymbol{\xi}_1, \boldsymbol{\xi}_2 : \mathbb{S}_d^N \to \mathbb{S}_d$ *such that*

$$\left\| \boldsymbol{\xi}_l(V) \right\| \leqslant C \qquad\qquad \forall V \in \mathbb{S}_d^N, \qquad (5.70a)$$

$$\left\| \boldsymbol{\xi}_l(V) - \boldsymbol{\xi}_l(\bar{V}) \right\| \left\| \mathbf{V}_j - \mathbf{V}_i \right\| \leqslant C \| V - \bar{V} \| \qquad \forall V, \bar{V} \in V_* \qquad (5.70b)$$

for $l \in \{1, 2\}$, *where* $C > 0$ *and* $V_* := \{V \in \mathbb{S}_d^N \mid \mathbf{V}_i \neq \mathbf{V}_j\}$. *Then*

$$\left\| \boldsymbol{\xi}_1(V)(\mathbf{V}_j - \mathbf{V}_i)\boldsymbol{\xi}_2(V) - \boldsymbol{\xi}_1(\bar{V})(\bar{\mathbf{V}}_j - \bar{\mathbf{V}}_i)\boldsymbol{\xi}_2(\bar{V}) \right\| \leqslant \| V - \bar{V} \| \qquad \forall V, \bar{V} \in \mathbb{S}_d^N. \qquad (5.71)$$

Examples of $\boldsymbol{\xi}(V)$ *satisfying* (5.70) *include*

E1. *Tensors* $\boldsymbol{\xi}$ *with* $\| \boldsymbol{\xi} \| \leqslant C$ *which can be written as*

$$\boldsymbol{\xi} = \frac{\mathbf{A}(V)}{\| \mathbf{V}_j - \mathbf{V}_i \| + B(V)} \qquad \forall v \in V_*, \qquad (5.72)$$

where $\mathbf{A} : V_* \to \mathbb{S}_d$, $B : V_* \to \mathbb{R}_0^+$ *are Lipschitz continuous functions;*

E2. *The minimum of two tensors* $\boldsymbol{\xi}_1$ *and* $\boldsymbol{\xi}_2$ *satisfying condition* (5.70);

E3. *The product of tensors* $\boldsymbol{\xi}_l$, $l \in \{1, \dots, \zeta\}$, *satisfying condition* (5.70);

E4. *A polynomial in tensors* $\boldsymbol{\xi}_l$, $l \in \{1, \dots, \zeta\}$, *satisfying condition* (5.70).

Proof. Let us first show that (5.71) holds for arbitrary $V, \bar{V} \in \mathbb{S}_d^N$ if $\boldsymbol{\xi}_1$ and $\boldsymbol{\xi}_2$ satisfy (5.70). If $\mathbf{V}_i = \mathbf{V}_j$ (similarly for $\bar{\mathbf{V}}_i = \bar{\mathbf{V}}_j$), we have

$$\left\| \boldsymbol{\xi}_1(V)(\mathbf{V}_j - \mathbf{V}_i)\boldsymbol{\xi}_2(V) - \boldsymbol{\xi}_1(\bar{V})(\bar{\mathbf{V}}_j - \bar{\mathbf{V}}_i)\boldsymbol{\xi}_2(\bar{V}) \right\|$$
$$\leqslant \left\| \boldsymbol{\xi}_1(\bar{V})(\bar{\mathbf{V}}_j - \bar{\mathbf{V}}_i)\boldsymbol{\xi}_2(\bar{V}) \right\| \leqslant C \left\| \boldsymbol{\xi}_1(\bar{V}) \right\| \left\| \bar{\mathbf{V}}_j - \bar{\mathbf{V}}_i \right\| \left\| \boldsymbol{\xi}_2(\bar{V}) \right\|$$
$$\leqslant C \left\| (\mathbf{V}_j - \mathbf{V}_i) - (\bar{\mathbf{V}}_j - \bar{\mathbf{V}}_i) \right\| \leqslant C \| V - \bar{V} \|.$$

Otherwise, we obtain

$$\left\| \boldsymbol{\xi}_1(V)(\mathbf{V}_j - \mathbf{V}_i)\boldsymbol{\xi}_2(V) - \boldsymbol{\xi}_1(\bar{V})(\bar{\mathbf{V}}_j - \bar{\mathbf{V}}_i)\boldsymbol{\xi}_2(\bar{V}) \right\|$$
$$\leqslant C \left\| \boldsymbol{\xi}_1(V) - \boldsymbol{\xi}_1(\bar{V}) \right\| \left\| \mathbf{V}_j - \mathbf{V}_i \right\| \left\| \boldsymbol{\xi}_2(V) \right\|$$
$$+ C \left\| \boldsymbol{\xi}_1(\bar{V}) \right\| \left\| (\mathbf{V}_j - \mathbf{V}_i) - (\bar{\mathbf{V}}_j - \bar{\mathbf{V}}_i) \right\| \left\| \boldsymbol{\xi}_2(V) \right\|$$
$$+ C \left\| \boldsymbol{\xi}_1(\bar{V}) \right\| \left\| \bar{\mathbf{V}}_j - \bar{\mathbf{V}}_i \right\| \left\| \boldsymbol{\xi}_2(V) - \boldsymbol{\xi}_2(\bar{V}) \right\|$$
$$\leqslant C \| V - \bar{V} \|.$$

To prove (5.70) for the examples E1–E4, it suffices to verify (5.70b) because condition (5.70a) is satisfied trivially.

E1. Assume that $\boldsymbol{\xi}$ can be written in the form (5.72). Then, for all $V, \bar{V} \in V_*$, we have

$$\boldsymbol{\xi}(V) - \boldsymbol{\xi}(\bar{V}) - \frac{\mathbf{A}(V) - \mathbf{A}(\bar{V})}{\|\mathbf{V}_j - \mathbf{V}_i\| + B(V)}$$

$$= \frac{\mathbf{A}(\bar{V})}{\|\mathbf{V}_j - \mathbf{V}_i\| + B(V)} - \frac{\mathbf{A}(\bar{V})}{\|\bar{\mathbf{V}}_j - \bar{\mathbf{V}}_i\| + B(\bar{V})}$$

$$= \mathbf{A}(\bar{V}) \frac{\left(\|\bar{\mathbf{V}}_j - \bar{\mathbf{V}}_i\| + B(\bar{V})\right) - \left(\|\mathbf{V}_j - \mathbf{V}_i\| + B(V)\right)}{\left(\|\mathbf{V}_j - \mathbf{V}_i\| + B(V)\right)\left(\|\bar{\mathbf{V}}_j - \bar{\mathbf{V}}_i\| + B(\bar{V})\right)}$$

$$= \boldsymbol{\xi}(\bar{V}) \frac{\left(B(\bar{V}) - B(V)\right) + \left(\|\bar{\mathbf{V}}_j - \bar{\mathbf{V}}_i\| - \|\mathbf{V}_j - \mathbf{V}_i\|\right)}{\|\mathbf{V}_j - \mathbf{V}_i\| + B(V)}.$$

Therefore, (5.70b) follows from the fact that $B \geqslant 0$, $\mathbf{A}$ and B are Lipschitz continuous, and $\|\boldsymbol{\xi}\| \leqslant C$.

E2. Using to the Lipschitz continuity property (5.13b) of the minimum, the validity of condition (5.70b) can be shown as follows:

$$\left\| \min\big(\boldsymbol{\xi}_1(V), \boldsymbol{\xi}_2(V)\big) - \min\big(\boldsymbol{\xi}_1(\bar{V}), \boldsymbol{\xi}_2(\bar{V})\big) \right\| \|\mathbf{V}_j - \mathbf{V}_i\|$$

$$\leqslant \left\| \min\big(\boldsymbol{\xi}_1(V), \boldsymbol{\xi}_2(V)\big) - \min\big(\boldsymbol{\xi}_1(\bar{V}), \boldsymbol{\xi}_2(V)\big) \right\| \|\mathbf{V}_j - \mathbf{V}_i\|$$

$$+ \left\| \min\big(\boldsymbol{\xi}_1(\bar{V}), \boldsymbol{\xi}_2(V)\big) - \min\big(\boldsymbol{\xi}_1(\bar{V}), \boldsymbol{\xi}_2(\bar{V})\big) \right\| \|\mathbf{V}_j - \mathbf{V}_i\|$$

$$\leqslant \left\| \boldsymbol{\xi}_1(V) - \boldsymbol{\xi}_1(\bar{V}) \right\| \|\mathbf{V}_j - \mathbf{V}_i\| + \left\| \boldsymbol{\xi}_2(V) - \boldsymbol{\xi}_2(\bar{V}) \right\| \|\mathbf{V}_j - \mathbf{V}_i\|$$

$$\leqslant C \|V - \bar{V}\| \qquad \forall V, \bar{V} \in V_*.$$

E3. As in the scalar case, it is sufficient to prove the statement for $\boldsymbol{\xi}_1 \boldsymbol{\xi}_2$. Then the more general result follows by induction. The result for the product of two quantities can be shown in exactly the same way as for the minimum exploiting the boundedness of $\|\boldsymbol{\xi}_l\| \leqslant C$ instead of (5.13b).

E4. Condition (5.70b) for this statement can be easily shown by virtue of the result for the product and the triangle inequality. $\qquad \square$

The most straightforward way to define a Lipschitz continuous limiting operator $\mathscr{A}_{ij}$, which guarantees that the solution to (5.62) possesses the DMP properties for the range of eigenvalues, is given by an extension of Algorithm 4.79. As before, scalar correction factors are used to limit the antidiffusive fluxes in this algorithm. The below limiting operator $\mathscr{A}_{ij}$ is designed to return $\mathbf{0}$ if $a_{ij} > 0$ (resp. $a_{ji} > 0$) and some eigenvalue of $\mathbf{U}_i$ (resp. $\mathbf{U}_j$) attains a local extremum.

Algorithm 5.17 ([Loh19]). *For each pair of nodes i and $j \neq i$, calculate the auxiliary quantities*

$$\beta_{ij} := \begin{cases} 1 & : a_{ij} \leqslant 0, \\ 1 - \left(1 - \min\left(1, \frac{q(u_i^{\max} - u_{i,d})}{\|\mathbf{U}_j - \mathbf{U}_i\|_F + \varepsilon}, \frac{q(u_{i,1} - u_i^{\min})}{\|\mathbf{U}_j - \mathbf{U}_i\|_F + \varepsilon}\right)\right)^p & : a_{ij} > 0 \end{cases} \quad (5.73)$$

and define the limiting operators $\mathscr{A}_{ij}$ as follows:

$$\mathscr{A}_{ij}[\mathbf{U}_j - \mathbf{U}_i] := \min(\beta_{ij}, \beta_{ji})(\mathbf{U}_j - \mathbf{U}_i) \qquad \forall i, j \in \{1, \dots, N\}, \; j \neq i. \quad (5.74)$$

The local extrema $u_i^{\max}$ and $u_i^{\min}$ are defined by (5.35).

The Frobenius norm in the denominator is used because it can be calculated efficiently in a practical implementation. Due to the equivalence of norms in finite dimensional spaces, it can be replaced by any other tensor norm. We will come back to this issue below when discussing the accuracy of the obtained solution (cf. Remark 5.19).

Using $s_{ij} := \min\left(q(u_i^{\max} - u_{i,d}), q(u_{i,1} - u_i^{\min})\right)$, the application of $\mathscr{A}_{ij}$ is equivalent to

$$\mathbf{S}_{ij} := \begin{cases} \mathbf{I} & : a_{ij} \leqslant 0, \\ \mathbf{I} - \left(\mathbf{I} - \frac{\min\left(\|\mathbf{U}_j - \mathbf{U}_i\|_F + \varepsilon, s_{ij}\right)\mathbf{I}}{\|\mathbf{U}_j - \mathbf{U}_i\|_F + \varepsilon}\right)^p & : a_{ij} > 0, \end{cases}$$

$$\mathscr{A}_{ij}[\mathbf{U}_j - \mathbf{U}_i] = \min(\mathbf{S}_{ij}, \mathbf{S}_{ji})(\mathbf{U}_j - \mathbf{U}_i) \qquad \forall i, j \in \{1, \dots, N\}, \; j \neq i,$$

which can be exploited to show its Lipschitz continuity using Lemma 5.16. For this purpose, we first find that

$$\boldsymbol{\xi} := \frac{\min\left(\|\mathbf{U}_j - \mathbf{U}_i\|_F + \varepsilon, q(u_i^{\max} - u_{i,d}), q(u_{i,1} - u_i^{\min})\right)\mathbf{I}}{\|\mathbf{U}_j - \mathbf{U}_i\|_F + \varepsilon}$$

has the form of (5.72), where $B := \varepsilon \geqslant 0$ and

$$\mathbf{A} := \min\left(\|\mathbf{U}_j - \mathbf{U}_i\|_F + \varepsilon, q(u_i^{\max} - u_{i,d}), q(u_{i,1} - u_i^{\min})\right)\mathbf{I}$$

are Lipschitz continuous for every $q \geqslant 0$ by virtue of Lemma 4.75 and Theorem 5.2. Therefore, $\mathbf{I} - (\mathbf{I} - \boldsymbol{\xi})^p$ satisfies (5.70) in view of the result E4 of Lemma 5.16. The Lipschitz continuity of $\mathscr{A}_{ij}[\mathbf{U}_j - \mathbf{U}_i]$ can be readily shown using property E2 again.

The limiting operator defined by Algorithm 5.17 scales each tensor entry using the same correction factor $\alpha_{ij} := \min(\beta_{ij}, \beta_{ji}) \leqslant 1$. Therefore, condition (5.63a) holds due to

$$\big(\mathbf{U}_j - \mathbf{U}_i, (\mathscr{I} - \mathscr{A}_{ij})[\mathbf{U}_j - \mathbf{U}_i]\big)_{\mathrm{F}} = (1 - \alpha_{ij})\|\mathbf{U}_j - \mathbf{U}_i\|_{\mathrm{F}}^2 \geqslant 0$$

$$\forall i, j \in \{1, \ldots, N\}, j \neq i.$$

However, a disadvantage of $\mathscr{A}_{ij}[\mathbf{U}_j - \mathbf{U}_i]$ being a multiple of $\mathbf{U}_j - \mathbf{U}_i$ is the fact that the so defined limited antidiffusive contribution to (5.62) vanishes whenever *any* eigenvalue of $\mathbf{U}_i$ (resp. $\mathbf{U}_j$) is a local extremum and $a_{ij} > 0$ (resp. $a_{ji} > 0$) holds. Therefore, the resulting AFC solution coincides with its low order counterpart if the minimal or maximal eigenvalue stays constant no matter how the other eigenvalues may behave (cf. Section 5.6.1). A less restrictive spectral limiting operator avoids this drawback by using individual correction factors for each eigenvalue of the solution.

Algorithm 5.18 ([Loh19]). *For each pair of nodes i and $j \neq i$, calculate the auxiliary quantities*

$$\mathbf{S}_{ij} := \begin{cases} \mathbf{I} & : a_{ij} \leqslant 0, \\ \mathbf{I} - \Big(\mathbf{I} - \min\Big(\mathbf{I}, \min\big(\frac{q(u_i^{\max}\mathbf{I} - \mathbf{U}_i)}{\|\mathbf{U}_j - \mathbf{U}_i\|_{\mathrm{F}} + \varepsilon}, \frac{q(\mathbf{U}_i - u_i^{\min}\mathbf{I})}{\|\mathbf{U}_j - \mathbf{U}_i\|_{\mathrm{F}} + \varepsilon}\big)\Big)\Big)^p & : a_{ij} > 0 \end{cases}$$

$$(5.75)$$

and define the limiting operators $\mathscr{A}_{ij}$ as follows:

$$\mathscr{A}_{ij}[\mathbf{U}_j - \mathbf{U}_i] := \frac{1}{2}\big(\mathbf{S}_{ij}(\mathbf{U}_j - \mathbf{U}_i)\mathbf{S}_{ji} + \mathbf{S}_{ji}(\mathbf{U}_j - \mathbf{U}_i)\mathbf{S}_{ij}\big)$$

$$\forall i, j \in \{1, \ldots, N\}, j \neq i. \quad (5.76)$$

The local extrema $u_i^{\max}$ and $u_i^{\min}$ are defined by (5.35).

Note that the computation of $\mathbf{S}_{ij}$ requires the spectral decomposition of $\mathbf{U}_i$ due to the definition of $\min(\cdot, \cdot)$ for two tensor quantities. The Lipschitz continuity of $\mathscr{A}_{ij}[\mathbf{U}_j - \mathbf{U}_i]$ was already shown in [Loh19]. A more compact proof can be constructed using Lemma 5.16 and the above argumentation.

To show the DMP properties of the solution to (5.62) for limiting operators defined by Algorithm 5.18, we prove (5.64a). The validity of (5.64b) can be shown similarly. We first notice that $\mathbf{S}_{ij}$ has the same eigenvectors as $\mathbf{U}_i$ and the diagonal entries

$$(\tilde{\mathbf{S}}_{ij})_{kk} := \begin{cases} 1 & : a_{ij} \leqslant 0, \\ 1 - \left(1 - \min\left(1, \frac{q(u_i^{\max} - u_{i,k})}{\|\mathbf{U}_j - \mathbf{U}_i\|_F + \varepsilon}, \frac{q(u_{i,k} - u_i^{\min})}{\|\mathbf{U}_j - \mathbf{U}_i\|_F + \varepsilon}\right)\right)^p & : a_{ij} > 0 \end{cases}$$
$$\forall k \in \{1, \ldots, d\}$$

of $\tilde{\mathbf{S}}_{ij}$ correspond to scalar correction factors that prevent local overshoots and undershoots in $u_{i,k}$. Indeed, the entry $(\tilde{\mathbf{S}}_{ij})_{dd}$ vanishes if $u_{i,d} = u_i^{\max}$ (or $u_{i,d} = u_i^{\min}$) and $a_{ij} > 0$. Consequently, condition (5.64a) holds due to

$$\mathbf{q}_{i,d}^\top \mathscr{A}_{ij}[\mathbf{U}_j - \mathbf{U}_i]\mathbf{q}_{i,d} = \mathbf{q}_{i,d}^\top \mathbf{S}_{ij}(\mathbf{U}_j - \mathbf{U}_i)\mathbf{S}_{ji}\mathbf{q}_{i,d}$$
$$= (\tilde{\mathbf{S}}_{ij})_{dd}\mathbf{q}_{i,d}^\top(\mathbf{U}_j - \mathbf{U}_i)\mathbf{S}_{ji}\mathbf{q}_{i,d} = 0$$

if $a_{ij} > 0$. Otherwise, we have

$$\mathbf{q}_{i,d}^\top \mathscr{A}_{ij}[\mathbf{U}_j - \mathbf{U}_i]\mathbf{q}_{i,d} = \mathbf{q}_{i,d}^\top \mathbf{S}_{ij}(\mathbf{U}_j - \mathbf{U}_i)\mathbf{S}_{ji}\mathbf{q}_{i,d}$$
$$= (\tilde{\mathbf{S}}_{ij})_{dd}\mathbf{q}_{i,d}^\top(\mathbf{U}_j - u_{i,d}\mathbf{I})\mathbf{S}_{ji}\mathbf{q}_{i,d}$$
$$= (\tilde{\mathbf{S}}_{ij})_{dd} \sum_{k=1}^N (\mathbf{q}_{i,d}^\top \mathbf{q}_{j,k})^2 (u_{j,k} - u_{i,d})(\tilde{\mathbf{S}}_{ij})_{kk}$$
$$\geqslant \sum_{k=1}^N (\mathbf{q}_{i,d}^\top \mathbf{q}_{j,k})^2 (u_{j,k} - u_{i,d}) = \mathbf{q}_{i,d}^\top \mathbf{U}_j \mathbf{q}_{i,d} - u_{i,d}$$

in view of the fact that $u_{j,k} \leqslant u_{i,d}$ for all $k \in \{1, \ldots, d\}$.

Furthermore, requirement (5.63a) of Theorem 5.14 holds

$$\left(\mathbf{U}_j - \mathbf{U}_i, (\mathscr{I} - \mathscr{A}_{ij})[\mathbf{U}_j - \mathbf{U}_i]\right)_F$$
$$= \|\mathbf{U}_j - \mathbf{U}_i\|_F^2 - \tfrac{1}{2}\left(\mathbf{U}_j - \mathbf{U}_i, \mathbf{S}_{ij}(\mathbf{U}_j - \mathbf{U}_i)\mathbf{S}_{ji}\right)_F$$
$$- \tfrac{1}{2}\left(\mathbf{U}_j - \mathbf{U}_i, \mathbf{S}_{ji}(\mathbf{U}_j - \mathbf{U}_i)\mathbf{S}_{ij}\right)_F$$
$$= \|\mathbf{U}_j - \mathbf{U}_i\|_F^2 - \left(\mathbf{U}_j - \mathbf{U}_i, \mathbf{S}_{ij}(\mathbf{U}_j - \mathbf{U}_i)\mathbf{S}_{ji}\right)_F$$
$$\geqslant \|\mathbf{U}_j - \mathbf{U}_i\|_F^2 - \|\mathbf{U}_j - \mathbf{U}_i\|_F\|\mathbf{S}_{ij}(\mathbf{U}_j - \mathbf{U}_i)\mathbf{S}_{ji}\|_F$$
$$\geqslant (1 - \|\mathbf{S}_{ij}\|_2\|\mathbf{S}_{ji}\|_2)\|\mathbf{U}_j - \mathbf{U}_i\|_F^2 \geqslant 0 \qquad \forall i,j \in \{1, \ldots, N\}, j \neq i$$

by virtue of (5.10).

In the following remark, we discuss the properties of AFC schemes equipped with tensorial limiting operators, as compared to the possibility of solving system (4.98) for individual tensor components using Algorithm 4.79 to calculate the correction factors.

Remark 5.19. Let us first consider the special case $d = 1$, in which tensor quantities become scalars. Then a bound-preserving approximation to the steady advection-reaction equation can be obtained, for example, by solving
1. AFC system (4.98) with Algorithm 4.79;
2. AFC system (5.62) with Algorithm 5.17;
3. AFC system (5.62) with Algorithm 5.18.

In this case, all eigenvalues of the tensorial degrees of freedom coincide with the nodal values of the scalar field. Therefore, the first two approaches produce the same results as long as the same parameters $p \in \mathbb{N}$ and $q > 0$ are employed in both limiters. The solution to problem (5.62) obtained with Algorithm 5.18 is likely to be less accurate because the argument of the limiting operator is multiplied by both correction tensors and not by their minimum.

For $d > 1$, we extend the scalar problem to a tensorial one by multiplying the right hand side function f and the inflow boundary function u_{in} by the identity tensor $\mathbf{I} \in \mathbb{S}_d$. Then we deduce that q must be chosen $\sqrt{d}$ times larger for Algorithm 5.17 to produce the same result as in the scalar case (multiplied by $\mathbf{I}$). To avoid the dependence on d, the Frobenius norm in the denominator of Algorithm 5.17 can be replaced by the spectral norm.

For artificially constructed test problems that we consider in Section 5.6.1, the AFC scheme based on Algorithm 5.18 produces more accurate numerical solution than the one based on Algorithm 5.17. However, the application of the latter limiting operator is less expensive and does not involve the computation of eigenvectors. The simplicity of scaling each tensor entry with the same scalar correction factor will be exploited in the following section to prove DMPs for the AFC discretization of the transient problem (5.2).

5.5.2 Unsteady problem

After presenting the AFC methodology for a steady tensorial problem, we now extend the monolithic limiting approach to the treatment of evolving tensor fields. Even though discrete maximum principles seem to be provable only for limiting operators based on scalar correction factors, we use general limiting operators $\mathscr{A}_{ij}$ and $\dot{\mathscr{A}}_{ij}$ for our theoretical analysis. Similarly to the scalar case, these operators are applied to $\mathbf{U}_j - \mathbf{U}_i$ and $\dot{\mathbf{U}}_j - \dot{\mathbf{U}}_i$, where $\dot{U} = (\dot{\mathbf{U}}_i)_{i=1}^N$ is an approximation of $\mathrm{d}_t U = (\mathrm{d}_t \mathbf{U}_i)_{i=1}^N$ that can be defined as in the previous chapter. For example, the basic high and low order approximations are given by (cf. (4.140))

$$\dot{\mathbf{U}}_i^{\mathrm{H}} := \sum_{j=1}^{N} (\mathcal{M}^{-1})_{ij} \Big(\mathbf{G}_j + \sum_{l=1}^{N} k_{jl} \mathbf{U}_l \Big), \qquad \forall i \in \{1, \dots, N\},$$

$$\dot{\mathbf{U}}_i^{\mathrm{L}} := m_i^{-1} \Big(\mathbf{G}_i + \sum_{j=1}^{N} (k_{ij} + d_{ij}) \mathbf{U}_j \Big) \qquad \forall i \in \{1, \dots, N\}.$$

The semi-discrete AFC problem representing a tensorial counterpart of (4.139) reads

$$m_i \frac{\mathrm{d}\mathbf{U}_i}{\mathrm{d}t} = \mathbf{G}_i + \mathbf{Z}_i(U) \qquad \forall i \in \{1, \dots, N\}, \tag{5.77a}$$

where the right hand side $\mathbf{Z}$ is given by

$$\mathbf{Z}_i(U) := \sum_{j=1}^{N} k_{ij} \mathbf{U}_j + \sum_{j \neq i} d_{ij} (\mathscr{I} - \mathscr{A}_{ij}) [\mathbf{U}_j - \mathbf{U}_i] - \sum_{j \neq i} m_{ij} \dot{\mathscr{A}}_{ij} [\dot{\mathbf{U}}_j - \dot{\mathbf{U}}_i]$$
$$\forall i \in \{1, \dots, N\}. \tag{5.77b}$$

As before, the above discretization reduces to its low order counterpart (5.25) if every limiting operator returns $\mathbf{0}$. If $\mathscr{A}_{ij}$ and $\dot{\mathscr{A}}_{ij}$ are identity mappings and $\dot{U} = \dot{U}^{\mathrm{H}}$ holds, the Galerkin approximation defined by (5.24) solves problem (5.77). Discretization of this AFC system in time using the θ-scheme results in

$$m_i \mathbf{U}_i^{n+1} = \underline{\mathbf{G}}_i + m_i \mathbf{U}_i^n + \Delta t \theta \mathbf{Z}_i^{n+1}(U^{n+1}) + \Delta t (1 - \theta) \mathbf{Z}_i^n(U^n)$$
$$\forall i \in \{1, \dots, N\}, \tag{5.78}$$

where $U^{n+1} = (\mathbf{U}_i^{n+1})_{i=1}^{N}$ are the nodal function values of the AFC solution $\mathbf{U}_h^{n+1}$. The degrees of freedom possess the DMP properties for the range of eigenvalues under assumptions that resemble DMP conditions for the scalar case (cf. Lemma 4.82).

Lemma 5.20. *Let $\mathscr{A}_{ij}$ and $\dot{\mathscr{A}}_{ij}$ be limiting operators satisfying the design criteria*

$$\mathbf{q}_{i,d}^{\top}\mathscr{A}_{ij}[\mathbf{U}_j - \mathbf{U}_i]\mathbf{q}_{i,d} \geqslant \begin{cases} 0 & : k_{ij} < 0, \\ \mathbf{q}_{i,d}^{\top}\mathbf{U}_j\mathbf{q}_{i,d} - u_{i,d} & : k_{ij} \geqslant 0 \end{cases}$$
$$\forall i,j \in \{1,\ldots,N\},\ j \neq i,\ s.t.\ u_{i,d} = \max_{l \in \mathcal{N}_i} u_{l,d}, \quad (5.79\text{a})$$

$$\mathbf{q}_{i,1}^{\top}\mathscr{A}_{ij}[\mathbf{U}_j - \mathbf{U}_i]\mathbf{q}_{i,1} \leqslant \begin{cases} 0 & : k_{ij} < 0, \\ \mathbf{q}_{i,1}^{\top}\mathbf{U}_j\mathbf{q}_{i,1} - u_{i,1} & : k_{ij} \geqslant 0 \end{cases}$$
$$\forall i,j \in \{1,\ldots,N\},\ j \neq i,\ s.t.\ u_{i,1} = \min_{l \in \mathcal{N}_i} u_{l,1}, \quad (5.79\text{b})$$

$$\mathbf{q}_{i,d}^{\top}\dot{\mathscr{A}}_{ij}[\dot{\mathbf{U}}_j - \dot{\mathbf{U}}_i]\mathbf{q}_{i,d} \geqslant 0$$
$$\forall i,j \in \{1,\ldots,N\},\ j \neq i,\ s.t.\ u_{i,d} = \max_{l \in \mathcal{N}_i} u_{l,d}, \quad (5.79\text{c})$$

$$\mathbf{q}_{i,1}^{\top}\dot{\mathscr{A}}_{ij}[\dot{\mathbf{U}}_j - \dot{\mathbf{U}}_i]\mathbf{q}_{i,1} \leqslant 0$$
$$\forall i,j \in \{1,\ldots,N\},\ j \neq i,\ s.t.\ u_{i,1} = \min_{l \in \mathcal{N}_i} u_{l,1}. \quad (5.79\text{d})$$

For $\theta \in [0,1)$, we additionally require the existence of constants $L_1, L_2 > 0$ such that

$$\mathscr{A}_{ij}[\mathbf{U}_j - \mathbf{U}_i] \succcurlyeq \begin{cases} L_1(\mathbf{U}_i - u_i^{\max}\mathbf{I}) & : k_{ij} < 0, \\ \mathbf{U}_j - u_i^{\max}\mathbf{I} & : k_{ij} \geqslant 0 \end{cases} \quad \forall i,j \in \{1,\ldots,N\},\ j \neq i,$$
$$(5.80\text{a})$$

$$\mathscr{A}_{ij}[\mathbf{U}_j - \mathbf{U}_i] \preccurlyeq \begin{cases} L_1(\mathbf{U}_i - u_i^{\min}\mathbf{I}) & : k_{ij} < 0, \\ \mathbf{U}_j - u_i^{\min}\mathbf{I} & : k_{ij} \geqslant 0 \end{cases} \quad \forall i,j \in \{1,\ldots,N\},\ j \neq i,$$
$$(5.80\text{b})$$

$$\dot{\mathscr{A}}_{ij}[\dot{\mathbf{U}}_j - \dot{\mathbf{U}}_i] \succcurlyeq L_2(\mathbf{U}_i - u_i^{\max}\mathbf{I}) \qquad \forall i,j \in \{1,\ldots,N\},\ j \neq i, \quad (5.80\text{c})$$

$$\dot{\mathscr{A}}_{ij}[\dot{\mathbf{U}}_j - \dot{\mathbf{U}}_i] \preccurlyeq L_2(\mathbf{U}_i - u_i^{\min}\mathbf{I}) \qquad \forall i,j \in \{1,\ldots,N\},\ j \neq i. \quad (5.80\text{d})$$

For $\theta = 1$, the constants $L_1, L_2 > 0$ can be chosen arbitrarily. Assume that the CFL-like condition (4.145) is satisfied for every $i \in \{1,\ldots,N\}$. Then the statements of Lemma 5.6 without requirements (i)–(iv) hold for the solution $U^{n+1} = (\mathbf{U}_i^{n+1})_{i=1}^N$ of (5.78).

Proof. The assertion of this lemma is verified by adapting the proof of Lemma 4.82.

We first consider the case $\theta \in [0,1)$ in which

$$((k_{ij}^n + d_{ij}^n)\mathscr{I} - d_{ij}^n\mathscr{A}_{ij}^n)[\mathbf{U}_j^n - \mathbf{U}_i^n] \preccurlyeq (k_{ij}^n + (1+L_1)d_{ij}^n)(u_i^{\max,n}\mathbf{I} - \mathbf{U}_i^n)$$
$$\forall i,j \in \{1,\ldots,N\},\, j \neq i. \quad (5.81)$$

Indeed, if $k_{ij}^n < 0$, we have

$$\begin{aligned}
&((k_{ij}^n + d_{ij}^n)\mathscr{I} - d_{ij}^n\mathscr{A}_{ij}^n)[\mathbf{U}_j^n - \mathbf{U}_i^n] \\
&= (k_{ij}^n + d_{ij}^n)(\mathbf{U}_j^n - \mathbf{U}_i^n) - d_{ij}^n\mathscr{A}_{ij}^n[\mathbf{U}_j^n - \mathbf{U}_i^n] \\
&\preccurlyeq (k_{ij}^n + d_{ij}^n)(u_i^{\max,n}\mathbf{I} - \mathbf{U}_i^n) + d_{ij}^n L_1(u_i^{\max,n}\mathbf{I} - \mathbf{U}_i^n)
\end{aligned}$$

by virtue of (5.80a) and the fact that $\mathbf{U}_j^n \preccurlyeq u_i^{\max,n}\mathbf{I}$. In the case of $k_{ij}^n \geqslant 0$, we obtain

$$\begin{aligned}
&((k_{ij}^n + d_{ij}^n)\mathscr{I} - d_{ij}^n\mathscr{A}_{ij}^n)[\mathbf{U}_j^n - \mathbf{U}_i^n] \\
&= k_{ij}^n(\mathbf{U}_j^n - \mathbf{U}_i^n) + d_{ij}^n(\mathscr{I} - \mathscr{A}_{ij}^n)[\mathbf{U}_j^n - \mathbf{U}_i^n] \\
&\preccurlyeq k_{ij}^n(u_i^{\max,n}\mathbf{I} - \mathbf{U}_i^n) + d_{ij}^n(u_i^{\max,n}\mathbf{I} - \mathbf{U}_i^n)
\end{aligned}$$

by (5.80a). According to (5.81) and the assumption that $\theta \in [0,1)$, we deduce

$$\begin{aligned}
&\mathbf{Z}_i^n(U^n) - \Big(\sum_{j=1}^N k_{ij}^n\Big)\mathbf{U}_i^n \\
&= \sum_{j \neq i}((k_{ij}^n + d_{ij}^n)\mathscr{I} - d_{ij}^n\mathscr{A}_{ij}^n)[\mathbf{U}_j^n - \mathbf{U}_i^n] - \sum_{j \neq i} m_{ij}\mathscr{A}_{ij}^n[\dot{\mathbf{U}}_j^n - \dot{\mathbf{U}}_i^n] \\
&\preccurlyeq \Big(\sum_{j \neq i}(k_{ij}^n + (1+L_1)d_{ij}^n) + \sum_{j \neq i} m_{ij}L_2\Big)(u_i^{\max,n}\mathbf{I} - \mathbf{U}_i^n) \\
&= \Big(L_2(m_i - m_{ii}) + (1+L_1)|d_{ii}^n| + \sum_{j \neq i} k_{ij}^n\Big)(u_i^{\max,n}\mathbf{I} - \mathbf{U}_i^n) \\
&= \Big(L_2(m_i - m_{ii}) + (1+L_1)|d_{ii}^n| - k_{ii}^n\Big)(u_i^{\max,n}\mathbf{I} - \mathbf{U}_i^n) \\
&\quad + \Big(\sum_{j=1}^N k_{ij}^n\Big)(u_i^{\max,n}\mathbf{I} - \mathbf{U}_i^n) \qquad \forall i \in \{1,\ldots,N\}
\end{aligned}$$

by (5.80c) and the definition of $\mathcal{D}^n$. If (4.145) holds as well, this leads to

$$m_i\mathbf{U}_i^n + \Delta t(1-\theta)\mathbf{Z}_i^n(U^n) \preccurlyeq \Big(m_i + \Delta t(1-\theta)(\sum_{j=1}^N k_{ij}^n)\Big)u_i^{\max,n}\mathbf{I}$$
$$\forall i \in \{1,\ldots,N\} \quad (5.82)$$

regardless of the choice of $\theta \in [0, 1]$. Furthermore, we have

$$m_i + \Delta t (1 - \theta) \sum_{j=1}^{N} k_{ij}^n = m_i > 0 \qquad \forall i \in \{1, \ldots, N\}, \, \theta = 1 \qquad (5.83a)$$

and

$$m_i + \Delta t (1 - \theta) \sum_{j=1}^{N} k_{ij}^n$$

$$= m_i + \Delta t (1 - \theta) \big(k_{ii}^n + d_{ii}^n \big) + \Delta t (1 - \theta) \sum_{j \neq i} (k_{ij}^n + d_{ij}^n)$$

$$\geqslant \Delta t (1 - \theta) \big(L_2 (m_i - m_{ii}) + L_1 |d_{ii}^n| \big) > 0 \quad \forall i \in \{1, \ldots, N\}, \, \theta \in [0, 1)$$
$$(5.83b)$$

according to (4.145) and in view of the fact that $k_{ij}^n + d_{ij}^n \geqslant 0$ for all $j \neq i$.

To prove (5.30a), let us suppose that there exists an index $i \in I \setminus J$ such that

$$u_{i,d}^{n+1} = \max_{j \in I \setminus J} u_{j,d}^{n+1} > c := \max \big(\max_{i \in \bar{I}} u_{i,d}^n, \max_{i \in (J \cap I) \cup \partial I} u_{i,d}^{n+1} \big). \qquad (5.84)$$

If there is no such index i, then condition (5.30a) is satisfied and we are done. According to (5.84), we have $u_{i,d}^{n+1} = u_{i,d}^{\max, n+1}$. Furthermore, we define $\hat{\mathcal{K}} \in \mathbb{R}^{N \times N}$ by

$$\hat{k}_{ij}^{n+1} := \begin{cases} k_{ij}^{n+1} + d_{ij}^{n+1} & : k_{ij}^{n+1} < 0, \\ k_{ij}^{n+1} & : k_{ij}^{n+1} \geqslant 0 \end{cases} \qquad \forall j \in \{1, \ldots, N\}, \, j \neq i.$$

It follows that for $\mathbf{x} = \mathbf{q}_{i,d}^{n+1}$ the vector $(\mathbf{x}^\top \mathbf{U}_j^{n+1}\mathbf{x})_{j=1}^N$ satisfies

$$\sum_{j=1}^N \hat{a}_{ij}\mathbf{x}^\top\mathbf{U}_j^{n+1}\mathbf{x} := \Big(m_i - \Delta t\theta(\sum_{j=1}^N k_{ij}^{n+1}) + \Delta t\theta\sum_{j\neq i}\hat{k}_{ij}^{n+1}\Big)u_{i,d}^{n+1}$$
$$- \Delta t\theta\sum_{j\neq i}\hat{k}_{ij}^{n+1}\mathbf{x}^\top\mathbf{U}_j^{n+1}\mathbf{x}$$

$$= m_i u_{i,d}^{n+1} - \Delta t\theta\sum_{j=1}^N k_{ij}^{n+1}\mathbf{x}^\top\mathbf{U}_j^{n+1}\mathbf{x}$$
$$- \Delta t\theta\sum_{j\neq i,\, k_{ij}^{n+1}<0} d_{ij}^{n+1}\mathbf{x}^\top(\mathbf{U}_j^{n+1} - \mathbf{U}_i^{n+1})\mathbf{x}$$

$$\leqslant m_i u_{i,d}^{n+1} - \Delta t\theta\sum_{j=1}^N k_{ij}^{n+1}\mathbf{x}^\top\mathbf{U}_j^{n+1}\mathbf{x}$$
$$- \Delta t\theta\sum_{j\neq i} d_{ij}^{n+1}\mathbf{x}^\top(\mathscr{I} - \mathscr{A}_{ij}^{n+1})[\mathbf{U}_j^{n+1} - \mathbf{U}_i^{n+1}]\mathbf{x}$$
$$+ \Delta t\theta\sum_{j\neq i} m_{ij}\mathbf{x}^\top\mathscr{A}_{ij}^{n+1}[\dot{\mathbf{U}}_j^{n+1} - \dot{\mathbf{U}}_i^{n+1}]\mathbf{x}$$

$$= m_i u_{i,d}^{n+1} - \Delta t\theta\mathbf{x}^\top\mathbf{Z}_i^{n+1}(U^{n+1})\mathbf{x}$$
$$= \mathbf{x}^\top\underline{\mathbf{G}}_i\mathbf{x} + m_i\mathbf{x}^\top\mathbf{U}_i^n\mathbf{x} + \Delta t(1-\theta)\mathbf{x}^\top\mathbf{Z}_i^n(U^n)\mathbf{x}$$

by (5.78), (5.79a), and (5.79c), where $\hat{a}_{ij} \leqslant 0$ if $j \neq i$. Recalling (5.82) and the fact that $i \in I \setminus J$, we find that

$$\Big(m_i - \Delta t\theta\sum_{j=1}^N k_{ij}^{n+1}\Big)u_{i,d}^{n+1} = \Big(\sum_{j=1}^N \hat{a}_{ij}\Big)u_{i,d}^{n+1}$$

$$\leqslant \sum_{j=1}^N \hat{a}_{ij}\mathbf{x}^\top\mathbf{U}_j^{n+1}\mathbf{x} \leqslant \mathbf{x}^\top\underline{\mathbf{G}}_i\mathbf{x} + \mathbf{x}^\top\big(m_i\mathbf{U}_i^n + \Delta t(1-\theta)\mathbf{Z}_i^n(U^n)\big)\mathbf{x}$$

$$\leqslant \Big(m_i + \Delta t(1-\theta)(\sum_{j=1}^N k_{ij}^n)\Big)u_i^{\max,n} < \Big(m_i + \Delta t(1-\theta)\sum_{j=1}^N k_{ij}^n\Big)u_{i,d}^{n+1}$$

by virtue of (5.83) and (5.84). Therefore, we have $\underline{p}_i u_{i,d}^{n+1} < 0$, which cannot be the case since $\underline{p}_i = 0$ for $i \in I \setminus J$. This contradiction completes the proof of (5.30a).

In the proof of (5.30c), we redefine $c \geqslant 0$ as the right hand side of (5.30c). The same argumentation leads to $p_i u_{i,d}^{n+1} < 0$ again. In this case, the statement follows from a contradiction to $p_i \geqslant 0$ and $u_{i,d}^{n+1} > c \geqslant 0$.

To show (5.32a), we note that (5.30a) with $I = \{1, \ldots, N\} \setminus J$ implies the existence of an index $i \in J$ such that

$$u_{i,d}^{n+1} = \max_{j \in \{1,\ldots,N\}} u_{j,d}^{n+1}.$$

By (5.31) and the assumption that $\underline{p} \geqslant 0$, we have $\underline{p}_i > 0$. Without loss of generality, let us suppose that

$$u_{i,d}^{n+1} = \max_{j \in \{1,\ldots,N\}} u_{j,d}^{n+1} > \max_{j \in \{1,\ldots,N\}} u_{j,d}^{n}.$$

Otherwise, the result follows immediately. The same arguments as above yield

$$\left(m_i - \Delta t \theta \sum_{j=1}^{N} k_{ij}^{n+1}\right) u_{i,d}^{n+1} \leqslant \underline{g}_{i,d} + \left(m_i + \Delta t (1-\theta) \sum_{j=1}^{N} k_{ij}^{n}\right) u_{i,d}^{n+1}$$

and the result follows by rearranging.

The lower bounds of Lemma 5.6 can be proved similarly. $\qquad\square$

Limiting operators satisfying the conditions of Lemma 5.20 can be defined by combining the ideas of Algorithms 4.83 and 5.17.

Algorithm 5.21. *For each pair of nodes i and $j \neq i$, calculate the auxiliary quantities*

$$\beta_{ij} := \begin{cases} 1 & : k_{ij} \geqslant 0, \\ 1 - \left(1 - \min\left(1, \dfrac{q_1(u_i^{\max} - u_{i,d})}{\|\mathbf{U}_j - \mathbf{U}_i\|_F + \varepsilon}, \dfrac{q_1(u_{i,1} - u_i^{\min})}{\|\mathbf{U}_j - \mathbf{U}_i\|_F + \varepsilon}\right)\right)^{p_1} & : k_{ij} < 0, \end{cases}$$

$$\tag{5.85a}$$

$$\dot{\beta}_{ij} := 1 - \left(1 - \min\left(1, \frac{q_2(u_i^{\max} - u_{i,d})}{\|\dot{\mathbf{U}}_j - \dot{\mathbf{U}}_i\|_F + \varepsilon}, \frac{q_2(u_{i,1} - u_i^{\min})}{\|\dot{\mathbf{U}}_j - \dot{\mathbf{U}}_i\|_F + \varepsilon}\right)\right)^{p_2}, \tag{5.85b}$$

and limit the antidiffusive fluxes using

$$\mathscr{A}_{ij}[\mathbf{U}_j - \mathbf{U}_i] := \min(\beta_{ij}, \beta_{ji})(\mathbf{U}_j - \mathbf{U}_i) \qquad \forall i, j \in \{1, \ldots, N\},\ j \neq i, \tag{5.86a}$$

$$\mathscr{A}_{ij}[\dot{\mathbf{U}}_j - \dot{\mathbf{U}}_i] := \min(\dot{\beta}_{ij}, \dot{\beta}_{ji})(\dot{\mathbf{U}}_j - \dot{\mathbf{U}}_i) \qquad \forall i, j \in \{1, \ldots, N\},\ j \neq i. \tag{5.86b}$$

Remark 5.22. As in the scalar case, the limiting operators defined in Algorithm 5.21 satisfy (5.79) and (5.80) with $L_1 := p_1 q_1$ and $L_2 := p_2 q_2$. While the validity of condition (5.79) follows directly form the definition of β_{ij} and $\dot{\beta}_{ij}$, condition (5.80) holds by Bernoulli's inequality (4.152). For example, for (5.80c), we have

$$
\begin{aligned}
0 &\leqslant \min(\dot{\beta}_{ij}, \dot{\beta}_{ji}) \\
&\leqslant \dot{\beta}_{ij} = 1 - \left(1 - \min\left(1, \frac{q_2(u_i^{\max} - u_{i,d})}{\|\dot{\mathbf{U}}_j - \dot{\mathbf{U}}_i\|_{\mathrm{F}} + \varepsilon}, \frac{q_2(u_{i,1} - u_i^{\min})}{\|\dot{\mathbf{U}}_j - \dot{\mathbf{U}}_i\|_{\mathrm{F}} + \varepsilon}\right)\right)^{p_2} \\
&\leqslant 1 - \left(1 - p_2 \min\left(1, \frac{q_2(u_i^{\max} - u_{i,d})}{\|\dot{\mathbf{U}}_j - \dot{\mathbf{U}}_i\|_{\mathrm{F}} + \varepsilon}, \frac{q_2(u_{i,1} - u_i^{\min})}{\|\dot{\mathbf{U}}_j - \dot{\mathbf{U}}_i\|_{\mathrm{F}} + \varepsilon}\right)\right) \\
&= p_2 \min\left(1, \frac{q_2(u_i^{\max} - u_{i,d})}{\|\dot{\mathbf{U}}_j - \dot{\mathbf{U}}_i\|_{\mathrm{F}} + \varepsilon}, \frac{q_2(u_{i,1} - u_i^{\min})}{\|\dot{\mathbf{U}}_j - \dot{\mathbf{U}}_i\|_{\mathrm{F}} + \varepsilon}\right) \leqslant p_2 q_2 \frac{u_i^{\max} - u_{i,d}}{\|\dot{\mathbf{U}}_j - \dot{\mathbf{U}}_i\|_{\mathrm{F}} + \varepsilon}
\end{aligned}
$$

and, hence,

$$
\begin{aligned}
\mathscr{A}_{ij}[\dot{\mathbf{U}}_j - \dot{\mathbf{U}}_i] &= \min(\dot{\beta}_{ij}, \dot{\beta}_{ji})(\dot{\mathbf{U}}_j - \dot{\mathbf{U}}_i) \\
&\succcurlyeq -\min(\dot{\beta}_{ij}, \dot{\beta}_{ji})\|\dot{\mathbf{U}}_j - \dot{\mathbf{U}}_i\|_2 \mathbf{I} \\
&\succcurlyeq -p_2 q_2 \frac{u_i^{\max} - u_{i,d}}{\|\dot{\mathbf{U}}_j - \dot{\mathbf{U}}_i\|_{\mathrm{F}} + \varepsilon}\|\dot{\mathbf{U}}_j - \dot{\mathbf{U}}_i\|_2 \mathbf{I} \\
&\succcurlyeq -p_2 q_2(u_i^{\max} - u_{i,d})\mathbf{I} \succcurlyeq p_2 q_1(\mathbf{U}_i - u_i^{\max}\mathbf{I}).
\end{aligned}
$$

As in the scalar case, the choice of $q_2 = \Delta t^{-1}$, $p_2 = 1$ produces a dimensionless value of $\dot{\beta}_{ij}$.

Unfortunately, the tensorial limiting approach based on Algorithm 5.18 cannot be adapted to the transient case in a way that guarantees the validity of condition (5.80) for $L_1, L_2 < \infty$. Hence, discrete maximum principles can only be proved in the case $\theta = 1$.

5.6 Numerical examples

This section is devoted to a comparison of bound-preserving approaches and assessment of their accuracy in applications to numerical benchmarks. Unless stated otherwise, we restrict ourselves to illustrating the benefits and drawbacks of different limiting strategies for fixed parameter settings. The potential influence of adjustable constants like p and q, of the time increment Δt, and of different triangulations $\mathcal{T}_h$ is discussed selectively.

More advanced studies concerning these issues were already performed in Section 4.6. In the case of the FCT methodology, only the localized limiting strategy summarized in Algorithm 5.11 is considered. The corresponding optimization problems of the form (5.54) are solved using the min-min or bisection criterion. Other criteria presented in Section 5.4.1 perform similarly to the bisection criterion and, hence, the corresponding results are omitted. In contrast to the assumption of equal dimension d for the domain and the tensor quantity under investigation, we consider only two dimensional flow patterns and seek numerical approximations to three dimensional symmetric tensor fields $\mathbf{U}_h \in V_h^{3 \times 3}$ defined on a domain $\Omega \subset \mathbb{R}^2$.

5.6.1 Steady problem

The first test problem of this section is a tensorial extension of the stationary circular convection benchmark, which was originally introduced in [Loh19, Section 8.1]. As before, the inflow boundary profile $\mathbf{U}_{in}$ is rotated around the origin of the domain $\Omega = (0, 1)^2$ using the velocity field $\mathbf{v} = (-x_2, x_1)^\top$. Therefore, the exact solution depends only on the distance $r = \|\mathbf{x}\|_2$ and is given by

$$\mathbf{U}(\mathbf{x}) = \begin{cases} \mathbf{U}_{in}(\|\mathbf{x}\|_2, 0) & : \|\mathbf{x}\|_2 \leqslant 1, \\ \mathbf{U}_{in}(1, \sqrt{\|\mathbf{x}\|_2^2 - 1}) & : \|\mathbf{x}\|_2 > 1 \end{cases} \quad \forall \mathbf{x} \in \Omega. \quad (5.87)$$

As in Section 4.6.1, the coercivity condition (2.14) is violated for the solenoidal velocity field $\mathbf{v}$ and the vanishing reactivity parameter $c = 0$. Therefore, Theorem 5.14 is not applicable to guarantee the existence of a solution to the AFC problem (5.62). To investigate the numerical behavior of different limiting strategies, we choose the inflow boundary data so that the exact solution of the problem at hand is given by a piecewise constant function [Loh19]

$$\mathbf{U}(\mathbf{x}) = \begin{cases} \mathbf{U}^{(1)} & : 0 \leqslant \|\mathbf{x}\|_2 < \frac{1}{5}, \\ \mathbf{U}^{(2)} & : \frac{1}{5} \leqslant \|\mathbf{x}\|_2 < \frac{2}{5}, \\ \mathbf{U}^{(3)} & : \frac{2}{5} \leqslant \|\mathbf{x}\|_2 < \frac{3}{5}, \\ \mathbf{U}^{(4)} & : \frac{3}{5} \leqslant \|\mathbf{x}\|_2 < \frac{4}{5}, \\ \mathbf{U}^{(5)} & : \frac{4}{5} \leqslant \|\mathbf{x}\|_2 \leqslant \sqrt{2}, \end{cases}$$

where the tensors corresponding to the five states admit the spectral decompositions

$$\mathbf{U}^{(1)} = \tfrac{1}{3} \begin{pmatrix} 1 & 0 & 0 \\ 0 & 1 & 0 \\ 0 & 0 & 1 \end{pmatrix},$$

$$\mathbf{U}^{(2)} = \tfrac{1}{5} \begin{pmatrix} 4 & 3 & 0 \\ 3 & -4 & 0 \\ 0 & 0 & 5 \end{pmatrix} \begin{pmatrix} \tfrac{2}{3} & 0 & 0 \\ 0 & 0 & 0 \\ 0 & 0 & \tfrac{1}{3} \end{pmatrix} \tfrac{1}{5} \begin{pmatrix} 4 & 3 & 0 \\ 3 & -4 & 0 \\ 0 & 0 & 5 \end{pmatrix},$$

$$\mathbf{U}^{(3)} = \tfrac{1}{2} \begin{pmatrix} \sqrt{2} & \sqrt{2} & 0 \\ \sqrt{2} & -\sqrt{2} & 0 \\ 0 & 0 & 2 \end{pmatrix} \begin{pmatrix} 0 & 0 & 0 \\ 0 & \tfrac{2}{3} & 0 \\ 0 & 0 & \tfrac{1}{3} \end{pmatrix} \tfrac{1}{2} \begin{pmatrix} \sqrt{2} & \sqrt{2} & 0 \\ \sqrt{2} & -\sqrt{2} & 0 \\ 0 & 0 & 2 \end{pmatrix},$$

$$\mathbf{U}^{(4)} = \tfrac{1}{2} \begin{pmatrix} \sqrt{2} & \sqrt{2} & 0 \\ \sqrt{2} & -\sqrt{2} & 0 \\ 0 & 0 & 2 \end{pmatrix} \begin{pmatrix} \tfrac{2}{3} & 0 & 0 \\ 0 & 0 & 0 \\ 0 & 0 & \tfrac{1}{3} \end{pmatrix} \tfrac{1}{2} \begin{pmatrix} \sqrt{2} & \sqrt{2} & 0 \\ \sqrt{2} & -\sqrt{2} & 0 \\ 0 & 0 & 2 \end{pmatrix},$$

$$\mathbf{U}^{(5)} = \tfrac{\sqrt{6}}{6} \begin{pmatrix} \sqrt{2} & \sqrt{3} & 1 \\ \sqrt{2} & -\sqrt{3} & 1 \\ \sqrt{2} & 0 & -2 \end{pmatrix} \begin{pmatrix} 1 & 0 & 0 \\ 0 & 0 & 0 \\ 0 & 0 & 0 \end{pmatrix} \tfrac{\sqrt{6}}{6} \begin{pmatrix} \sqrt{2} & \sqrt{2} & \sqrt{2} \\ \sqrt{3} & -\sqrt{3} & 0 \\ 1 & 1 & -2 \end{pmatrix}.$$

These data are defined so that the trace of the exact solution is identically 1 in the whole domain and the eigenvalues are globally bounded above and below by 1 and 0, respectively. Furthermore, $\mathbf{U}$ exhibits different kinds of discontinuities that may occur in tensor solutions. For $\{\mathbf{x} \mid 0 \leqslant \|\mathbf{x}\|_2 < \tfrac{4}{5}\}$, one eigenvector is given by $(0, 0, 1)^\top$ and corresponds to the intermediate eigenvalue $\tfrac{1}{3}$. At the discontinuities separating the constant states inside this subdomain, the eigenvalues $\tfrac{2}{3}$ and 0 are either replaced by $\tfrac{1}{3}$ (as in the jump from $\mathbf{U}^{(2)}$ to $\mathbf{U}^{(1)}$) or interchanged. At the interface between $\mathbf{U}^{(2)}$ and $\mathbf{U}^{(3)}$ the associated eigenvectors are discontinuous as well. The jump at $\|\mathbf{x}\|_2 = \tfrac{4}{5}$ is characterized by the fact that $\mathbf{U}^{(4)}$ and $\mathbf{U}^{(5)}$ have the same eigenvector $(1, -1, 0)^\top$ corresponding to the minimal eigenvalue 0.

As before, the unstable Galerkin method produces a finite element approximation which is corrupted by spurious oscillations and violates the global bounds for the range of eigenvalues (cf. Fig. 5.1). Interestingly enough, the intermediate eigenvalue oscillates only for $x_2 > \tfrac{4}{5}$. This eigenvalue is not affected by other tensor entries because $u_{h,2} = u_{h,33}$ is computed independently from the other components (cf. (5.17)). In contrast to this, the eigenvalues of the low order approximation on the cutline $x_1 = 0$ do not oscillate and are bounded by 0 and 1. While every eigenvalue of the exact solution is constant on the interval $(\tfrac{1}{5}, \tfrac{4}{5})$, the eigenvalue range of the low order solution becomes particularly narrow near the discontinuities at $x_2 = \frac{2}{5}$ and $x_2 = \frac{3}{5}$. This behavior is caused by the discontinuities of the involved eigenvectors and occurs for every linear combination of the tensor quantities $\mathbf{U}^{(2)}, \ldots, \mathbf{U}^{(4)}$.

If the FCT methodology using the min-min criterion is applied to this problem using a pseudo time stepping approach, a great amount of artificial diffusion is removed. However, the accuracy of the solution does not improve as Δt is refined and there is hardly any difference between the numerical solutions presented in Fig. 5.1. This behavior matches that of the scalar FCT limiter at the discontinuities in Fig. 4.6a. Near the last discontinuity at $x_2 = \frac{4}{5}$ of the tensorial circular convection benchmark, the FCT solutions are still very diffusive because the minimal eigenvalue and the corresponding eigenvector of the inflow boundary condition are not affected by the discontinuity and stay constant. Then the tensor $\mathbf{A}$ of the optimization problem (5.54) possesses an eigenvalue close to 0 which implies small correction factors due to (5.55).

This drawback of the min-min criterion can be avoided by invoking the bisection criterion, which makes it possible to determine optimal correction factors without any worst case assumptions. However, this FCT approach does not produce a steady state solution even for very small pseudo time increments Δt and, hence, is not suitable for calculating bound-preserving solutions to this benchmark.

In the case of the monolithic limiting approach of Section 5.5, we use $p = q = 1$ in Algorithms 5.17 and 5.18. The results are less accurate than those corresponding to larger values of p, q. To allow a fair comparison of the different limiting techniques, we refrain from using other parameter settings in this study. However, more accurate results can be obtained by adjusting the values of free parameters as in the scalar case.

Regardless of the adopted limiting strategy, the AFC solution seems to be more accurate if the spectral norm $\| \cdot \|_2$ rather than the Frobenius norm $\| \cdot \|_F$ is used in the denominator of the correction factors (cf. Fig. 5.2). This confirms the considerations of Remark 5.19 but the gain of accuracy due to the use of the spectral norm comes at the expense of performance because the Frobenius norm can be computed more efficiently without any eigenvalue calculation. At the first three discontinuities, the eigenvalues of the AFC solutions obtained using Algorithm 5.17 nearly coincide with those of the solutions produced by Algorithm 5.18. In this case, the potentially beneficial use of tensorial correction factors does not pay off because limiters using scalar correction factors are more efficient.

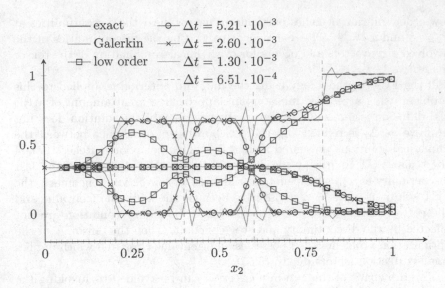

Figure 5.1: Tensorial circular convection: Cutline $x_1 = 0$ profiles (cf. Fig. 3.3)
of FCT solution with min-min criterion and different pseudo time
increments Δt on level 6 of $\mathcal{T}_h^3$.

However, as in the case of the FCT solution, the AFC method based
on Algorithm 5.17 produces an approximation which is very diffusive near
the discontinuity located at $x_2 = \frac{4}{5}$. The reason for this is the fact that
$(1, -1, 0)^\top$ is an eigenvector corresponding to the minimal eigenvalue 0
of both involved tensors, $\mathbf{U}^{(4)}$ and $\mathbf{U}^{(5)}$. This jump constellation forces
Algorithm 5.17 to use vanishing scalar correction factors and a diffusive
approximation is obtained regardless of the tensor norm in the denominator
of (5.73). For such jumps, Algorithm 5.18 performs much better since only
one eigenvalue of the tensorial correction factor $\mathbf{S}_{ij}$ is set to zero due to the
constant minimal eigenvalue.

Unfortunately, the decoupled limiting of individual eigenvalues in Algorithm 5.18 has the drawback that the intermediate eigenvalue is hardly
controlled by the limiting operator. Indeed, the admissible range is determined by the local bounds for the maximal and minimal eigenvalues.
As a consequence, the eigenvalue $u_{h,2}$ of the AFC solution obtained using
Algorithm 5.18 is polluted by oscillations in the interval $(\frac{3}{5}, \frac{4}{5})$ where even
the intermediate eigenvalue of the Galerkin approximation is smooth. This
drawback can be avoided by incorporating high order stabilization into

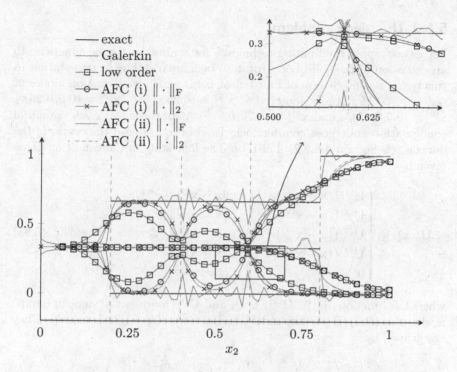

Figure 5.2: Tensorial circular convection: Cutline $x_1 = 0$ profiles (cf. Fig. 3.3) of AFC solutions with $p = q = 1$ on level 6 of $\mathcal{T}_h^3$ using (i) Algorithm 5.17 and (ii) Algorithm 5.18 as well as different norms in denominator.

the definition of the antidiffusive fluxes. For example, the AFC scheme presented in Lohmann, Christoph [Loh19, Section 8.1.1] reduces the amount of spurious oscillations by using the SUPG discretization as the high order target method. The scalar limiting technique of Algorithm 5.17 does not need such a stabilization for this benchmark. Indeed, the multiplication of the antidiffusive fluxes by synchronized correction factors leads to an algorithm that preserves the constant trace of the exact solution. Therefore, the imposition of local bounds on the maximal and minimal eigenvalues is sufficient to prevent the formation of spurious ripples in the distribution of the intermediate eigenvalue as well.

5.6.2 Unsteady problem

Let us now apply the limiting techniques for symmetric tensor quantities to an extension of the solid body rotation benchmark. The exact solution to this test is a composition of four tensor fields which we define on circles of radius $r = 0.15$ initially centered at $\mathbf{x}^{(1)} = (0.25, 0.5)^\top$, $\mathbf{x}^{(2)} = (0.5, 0.25)^\top$, $\mathbf{x}^{(3)} = (0.5, 0.75)^\top$, and $\mathbf{x}^{(4)} = (0.75, 0.5)^\top$. As in the scalar case, the initial configuration undergoes counterclockwise rotation around the center of the domain (cf. Section 4.6.2.2) [Loh17b]. The four-state initial condition $\mathbf{U}_0$ is given by

$$\mathbf{U}_0(\mathbf{x}) := \begin{cases} \mathbf{U}^{(1)}\big((\mathbf{x} - \mathbf{x}^{(1)})r^{-1}\big) & : \|\mathbf{x} - \mathbf{x}^{(1)}\|_2 \leqslant r, \\ \mathbf{U}^{(2)}\big((\mathbf{x} - \mathbf{x}^{(2)})r^{-1}\big) & : \|\mathbf{x} - \mathbf{x}^{(2)}\|_2 \leqslant r, \\ \mathbf{U}^{(3)}\big((\mathbf{x} - \mathbf{x}^{(3)})r^{-1}\big) & : \|\mathbf{x} - \mathbf{x}^{(3)}\|_2 \leqslant r, \qquad \forall \mathbf{x} \in \Omega, \quad (5.88) \\ \mathbf{U}^{(4)}\big((\mathbf{x} - \mathbf{x}^{(4)})r^{-1}\big) & : \|\mathbf{x} - \mathbf{x}^{(4)}\|_2 \leqslant r, \\ \mathbf{0} & : \text{otherwise} \end{cases}$$

where the functions $\mathbf{U}^{(1)}$, $\mathbf{U}^{(2)}$, $\mathbf{U}^{(3)}$, and $\mathbf{U}^{(4)}$ represent a 'smooth hump', a 'sharp cone', a 'slotted cylinder', and a 'semi-ellipse', respectively. They are defined by

$$\mathbf{U}^{(1)}(\mathbf{x}) := \|\mathbf{x}\|_2^{-2} \begin{pmatrix} 1 & 0 & 0 \\ 0 & \cos\phi & \sin\phi \\ 0 & \sin\phi & -\cos\phi \end{pmatrix} \begin{pmatrix} x_1 & x_2 & 0 \\ x_2 & -x_1 & 0 \\ 0 & 0 & \|\mathbf{x}\|_2 \end{pmatrix}$$

$$\cdot \begin{pmatrix} u_1^{(1)} & 0 & 0 \\ 0 & u_2^{(1)} & 0 \\ 0 & 0 & u_3^{(1)} \end{pmatrix} \begin{pmatrix} x_1 & x_2 & 0 \\ x_2 & -x_1 & 0 \\ 0 & 0 & \|\mathbf{x}\|_2 \end{pmatrix} \begin{pmatrix} 1 & 0 & 0 \\ 0 & \cos\phi & \sin\phi \\ 0 & \sin\phi & -\cos\phi \end{pmatrix},$$

where $u_1^{(1)} = \big(\tfrac{1}{2}(1 + \cos(\pi\|\mathbf{x}\|_2))\big)^3$, $u_2^{(1)} = \big(\tfrac{1}{2}(1 + \cos(\pi\|\mathbf{x}\|_2))\big)^2$, $u_3^{(1)} = \tfrac{1}{2}\big(1 + \cos(\pi\|\mathbf{x}\|_2)\big)$, and $\phi = \tfrac{1}{2}\arctan2(x_1, x_2)$,

$$\mathbf{U}^{(2)}(\mathbf{x}) := \big(10\|\mathbf{x}\|_2\big)^{-2} \begin{pmatrix} 10 & 0 & 0 \\ 0 & 8 & 6 \\ 0 & 6 & -8 \end{pmatrix} \begin{pmatrix} x_1 & x_2 & 0 \\ x_2 & -x_1 & 0 \\ 0 & 0 & \|\mathbf{x}\|_2 \end{pmatrix}$$

$$\cdot \begin{pmatrix} u_1^{(2)} & 0 & 0 \\ 0 & u_a^{(2)} & 0 \\ 0 & 0 & u_b^{(2)} \end{pmatrix} \begin{pmatrix} x_1 & x_2 & 0 \\ x_2 & -x_1 & 0 \\ 0 & 0 & \|\mathbf{x}\|_2 \end{pmatrix} \begin{pmatrix} 10 & 0 & 0 \\ 0 & 8 & 6 \\ 0 & 6 & -8 \end{pmatrix},$$

where $u_1^{(2)} = \frac{1}{2} - \frac{1}{2}\|\mathbf{x}\|_2$, $u_a^{(2)} = \frac{1}{2} - \frac{1}{2}|x_1|$, and $u_b^{(2)} = 1 - \|\mathbf{x}\|_2$,

$$
\mathbf{U}^{(3)}(\mathbf{x}) := \begin{cases}
10^{-2}\begin{pmatrix} -8 & 6 & 0 \\ 6 & 8 & 0 \\ 0 & 0 & 10 \end{pmatrix}\begin{pmatrix} u_3^{(3)} & 0 & 0 \\ 0 & u_1^{(3)} & 0 \\ 0 & 0 & u_2^{(3)} \end{pmatrix}\begin{pmatrix} -8 & 6 & 0 \\ 6 & 8 & 0 \\ 0 & 0 & 10 \end{pmatrix} \\
\qquad\qquad : \left(|x_1| \geqslant \frac{1}{6} \vee x_2 > \frac{2}{3}\right) \wedge x_1 > 0, \\[2mm]
10^{-2}\begin{pmatrix} -8 & 6 & 0 \\ 6 & 8 & 0 \\ 0 & 0 & 10 \end{pmatrix}\begin{pmatrix} u_1^{(3)} & 0 & 0 \\ 0 & u_3^{(3)} & 0 \\ 0 & 0 & u_2^{(3)} \end{pmatrix}\begin{pmatrix} -8 & 6 & 0 \\ 6 & 8 & 0 \\ 0 & 0 & 10 \end{pmatrix} \\
\qquad\qquad : \left(|x_1| \geqslant \frac{1}{6} \vee x_2 > \frac{2}{3}\right) \wedge x_1 < 0, \\[2mm]
0 \qquad\qquad\qquad\qquad : \text{otherwise}
\end{cases},
$$

where $u_1^{(3)} = 0.1$, $u_2^{(3)} = 0.45$, and $u_3^{(3)} = 1$,

$$
\mathbf{U}^{(4)}(\mathbf{x}) := \begin{pmatrix} u_1^{(4)} & 0 & 0 \\ 0 & u_1^{(4)} & 0 \\ 0 & 0 & u_1^{(4)} \end{pmatrix},
$$

where $u_1^{(4)} = u_2^{(4)} = u_3^{(4)} = \sqrt{1 - \|\mathbf{x}\|_2^2}$. The intermediate and largest eigenvalue of $\mathbf{U}^{(2)}$ are given by

$$
u_2^{(2)} = \begin{cases} u_a^{(2)} & : |x_1| \geqslant 2r - 1, \\ u_b^{(2)} & : |x_1| < 2r - 1, \end{cases}
\qquad
u_3^{(2)} = \begin{cases} u_b^{(2)} & : |x_1| \geqslant 2r - 1, \\ u_a^{(2)} & : |x_1| < 2r - 1. \end{cases}
$$

The initial condition $\mathbf{U}_{0,h}$ for all numerical methods under investigation is computed on level 7 of $\mathcal{T}_h^3$ using the lumped L^2-projection. This low order method guarantees that the range of eigenvalues is globally bounded by the extremal eigenvalues of the exact solution 0 and 1 (cf. Fig. 5.3). However, the minimal and maximal eigenvalues of this approximation possess a peak in the discontinuity of the 'slotted cylinder'. This is not a contradiction to the above discrete maximum principles and the overshoot and undershoot are still locally bounded by the eigenvalues of the exact solution.

As in the scalar case, the FCT methodology using Algorithm 5.11 and the bisection criterion produces spurious ripples if the largest time increment $\Delta t = 3.92 \cdot 10^{-3}$ satisfying the CFL-like condition (4.66) is used (cf. Fig. 5.4). However, local and global DMPs for the range of eigenvalues are satisfied and, in particular, the tensor solution remains positive semidefinite. As the time increment decreases to $\Delta t = 1.31 \cdot 10^{-3}$, spurious deformations vanish

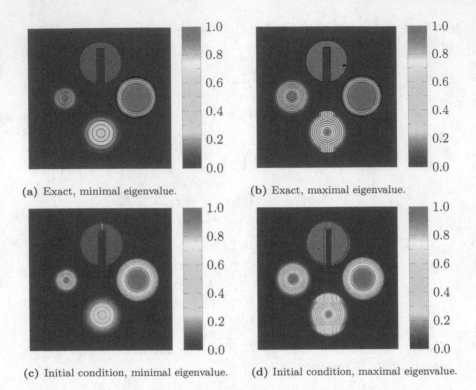

(a) Exact, minimal eigenvalue.   (b) Exact, maximal eigenvalue.

(c) Initial condition, minimal eigenvalue. (d) Initial condition, maximal eigenvalue.

Figure 5.3: Tensorial solid body rotation: Maximal and minimal eigenvalues of exact solution and projected initial condition on level 7 of $\mathcal{T}_h^3$.

but the symmetry of the tensor object located at $\mathbf{x}^{(1)} = (0.25, 0.5)^\top$ is still not reproduced exactly. The 'smooth hump' becomes rotationally symmetric when the time step size $\Delta t = 1.95 \cdot 10^{-4}$ is used. However, regardless of the time increment, the maximal eigenvalue of the obtained FCT solution is polluted by dispersive errors at the base of the 'smooth hump' (compare the distance between the corresponding contour lines).

In the remaining figures, we do not present the minimal eigenvalue of the finite element approximations under consideration because it behaves like the maximal eigenvalue and does not provide further insights regarding the quality of the limiting procedures.

The FCT solution obtained using the time increment $\Delta t = 1.31 \cdot 10^{-3}$ reveals that the min-min criterion produces less pronounced peaks of the 'slotted cylinder' and 'smooth hump' than the bisection criterion (cf. Fig. 5.5).

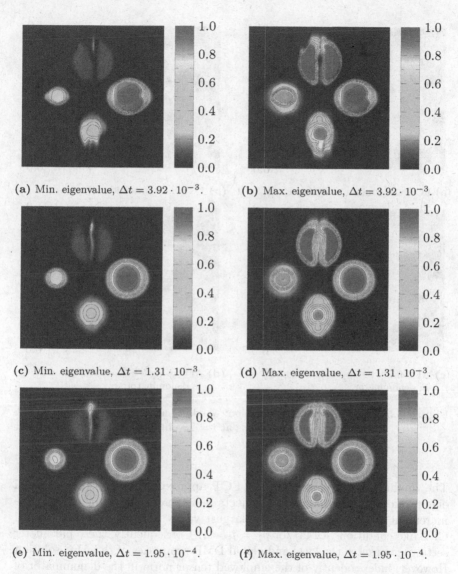

(a) Min. eigenvalue, $\Delta t = 3.92 \cdot 10^{-3}$.

(b) Max. eigenvalue, $\Delta t = 3.92 \cdot 10^{-3}$.

(c) Min. eigenvalue, $\Delta t = 1.31 \cdot 10^{-3}$.

(d) Max. eigenvalue, $\Delta t = 1.31 \cdot 10^{-3}$.

(e) Min. eigenvalue, $\Delta t = 1.95 \cdot 10^{-4}$.

(f) Max. eigenvalue, $\Delta t = 1.95 \cdot 10^{-4}$.

Figure 5.4: Tensorial solid body rotation: Extremal eigenvalues of FCT solution at $t = 2\pi$ obtained using bisection criterion and different time increments Δt.

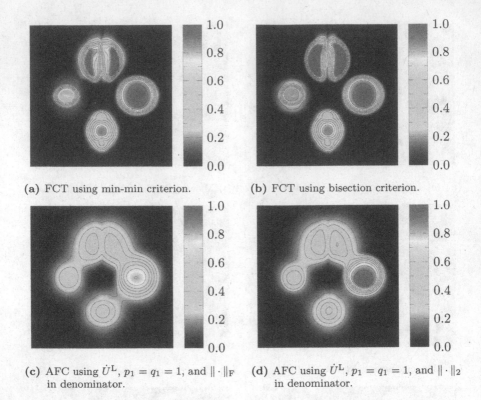

(a) FCT using min-min criterion.

(b) FCT using bisection criterion.

(c) AFC using $\dot{U}^{L}$, $p_1 = q_1 = 1$, and $\| \cdot \|_F$ in denominator.

(d) AFC using $\dot{U}^{L}$, $p_1 = q_1 = 1$, and $\| \cdot \|_2$ in denominator.

Figure 5.5: Tensorial solid body rotation: Maximal eigenvalues of different bound-preserving approximations at $t = 2\pi$ obtained using time increment $\Delta t = 1.31 \cdot 10^{-3}$.

The quality of the corresponding FCT approximations to the maximal eigenvalue of the other tensor objects is virtually the same. The time increment $\Delta t = 1.31 \cdot 10^{-3}$ is the largest value satisfying the node-based CFL-like condition (4.145) for $p_1 = q_1 = 1$. Consequently, these parameter settings guarantee the local and global DMP properties of the AFC solution. However, independently of the employed tensor norm in the denominator of the correction factors, the solution obtained using $\dot{U}^{L}$ is very diffusive and the initial configuration $\mathbf{U}_0$ is completely smeared.

The accuracy of the AFC approximations improves as p_1 and q_1 are increased while the time increment Δt is refined according to condition (4.145) (cf. Figs. 5.6 and 5.7). However, the approximation $\dot{U}^* = (\dot{\mathbf{U}}_i^*)_{i=1}^N$ defined by

$$\dot{\mathbf{U}}_i^* = m_i^{-1}\Big(\mathbf{G}_i + \sum_{j=1}^N k_{ij}\mathbf{U}_j + \sum_{j\neq i} d_{ij}(\mathscr{I} - \mathscr{A}_{ij})[\mathbf{U}_j - \mathbf{U}_i]\Big) \qquad \forall i \in \{1,\ldots,N\}$$

yields numerical solutions that are polluted by phase errors (cf. Figs. 5.6e and 5.6f). To avoid spurious deformations of this kind, the antidiffusive fluxes should be redefined to recover a stabilized high order target scheme instead of the Galerkin method in situations when no limiting is performed. A particularly simple way to do so is to exploit the smoothing effect of using $\dot{U}^L$ when it comes to the calculation of fluxes depending on the nodal time derivatives. The so defined AFC method produces bound-preserving approximations, which are comparable to the FCT results. In this case, the most accurate results are obtained when the parameter q_1 is maximized subject to condition (4.145). However, even in the case of $q_1 = 16$ and $p_1 = 1$, the maximal eigenvalue of the slotted cylinder is not preserved as well as in the FCT solutions. Only a further reduction of the time increment to $\Delta t = 1.02 \cdot 10^{-4}$ and the use of $q_1 = 32$, $p_1 = 1$ lead to AFC solutions which nearly coincide with their FCT counterparts (cf. Fig. 5.7). In contrast to the FCT solution obtained using the bisection criterion, the maximal eigenvalue distributions produced by other schemes are not distorted by dispersive effects at the base of the 'smooth hump'.

As discussed above, limiting techniques based on the use of scalar correction factors have the drawback that every entry of the antidiffusive flux is scaled by the same value. Therefore, the solution can be very diffusive, e.g., if the minimal eigenvalue of the initial data is (nearly) constant, while the other eigenvalues exhibit considerable spatial variations. To illustrate this behavior of the presented bound-preserving schemes for transient problems, we modify the initial (and exact) solution by setting the intermediate eigenvalue of the 'slotted cylinder' equal to zero, i.e., $u_2^{(3)} = 0$. This eigenvalue can only become positive due to fluxes from regions occupied by the other tensor objects. Therefore, the correction factors defined by the min-min criterion or Algorithm 5.17 are close to zero and produce very diffusive 'slotted cylinders' (cf. Fig. 5.8). In contrast to these approaches, the bisection criterion may deliver correction factors greater than 0. However, the outcome of limiting is highly sensitive to small perturbations like round off errors with the consequence that the resulting FCT solution may be corrupted by spurious artifacts. This deficiency is also responsible for the lack of convergence

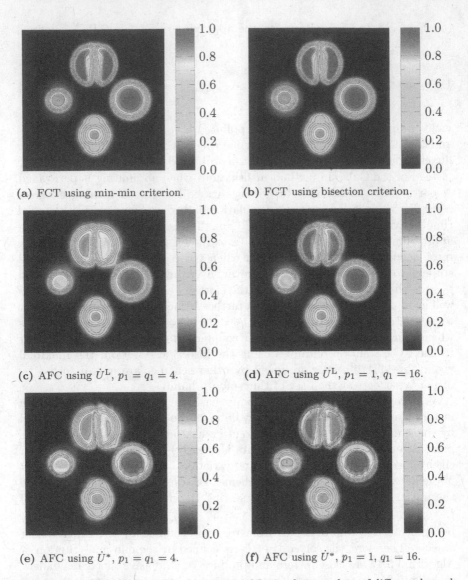

(a) FCT using min-min criterion.

(b) FCT using bisection criterion.

(c) AFC using $\dot{U}^{L}$, $p_1 = q_1 = 4$.

(d) AFC using $\dot{U}^{L}$, $p_1 = 1$, $q_1 = 16$.

(e) AFC using $\dot{U}^{*}$, $p_1 = q_1 = 4$.

(f) AFC using $\dot{U}^{*}$, $p_1 = 1$, $q_1 = 16$.

Figure 5.6: Tensorial solid body rotation: Maximal eigenvalues of different bound-preserving approximations at $t = 2\pi$ obtained using time increment $\Delta t = 1.95 \cdot 10^{-4}$ and $\|\cdot\|_2$ in denominator of AFC correction tensors.

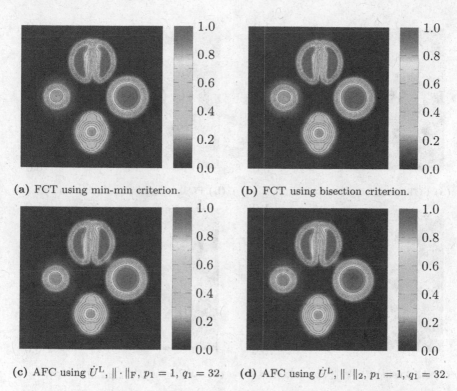

(a) FCT using min-min criterion.

(b) FCT using bisection criterion.

(c) AFC using $\dot{U}^{\mathrm{L}}$, $\|\cdot\|_{\mathrm{F}}$, $p_1 = 1$, $q_1 = 32$.

(d) AFC using $\dot{U}^{\mathrm{L}}$, $\|\cdot\|_2$, $p_1 = 1$, $q_1 = 32$.

Figure 5.7: Tensorial solid body rotation: Maximal eigenvalues of different bound-preserving approximations at $t = 2\pi$ obtained using time increment $\Delta t = 1.02 \cdot 10^{-4}$.

to steady state solutions which we observed while using the pseudo time stepping approach in Section 5.6.1.

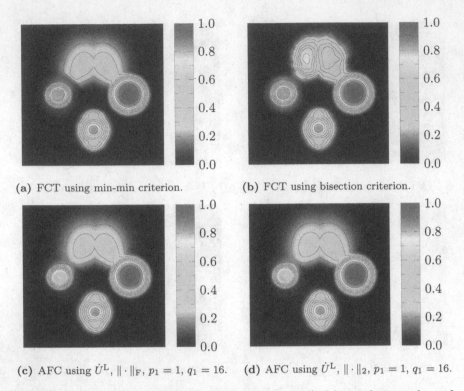

(a) FCT using min-min criterion.

(b) FCT using bisection criterion.

(c) AFC using $\dot{U}^{\mathrm{L}}$, $\|\cdot\|_{\mathrm{F}}$, $p_1 = 1$, $q_1 = 16$.

(d) AFC using $\dot{U}^{\mathrm{L}}$, $\|\cdot\|_2$, $p_1 = 1$, $q_1 = 16$.

Figure 5.8: Tensorial solid body rotation (disturbed): Maximal eigenvalues of different bound-preserving approximations at $t = 2\pi$ obtained using time increment $\Delta t = 1.95 \cdot 10^{-4}$.

6 Simulation of fiber suspensions

In the previous chapter, a family of eigenvalue range preserving numerical methods was developed for hyperbolic equations governing the evolution of symmetric tensor fields. Since the benchmarks considered in Section 5.6 are artificially constructed test problems, further analysis and additional numerical experiments are performed in this chapter to study the behavior of the proposed methods in practical applications. To motivate the need for using eigenvalue range preserving limiters, we apply them to orientation tensors that determine the rheological behavior of fiber suspension flows. In Section 6.1, we briefly introduce the system of PDEs which models the evolution of the fluid-fiber mixture at hand. Discretizations of the subproblems for the carrier fluid and the fiber orientation model are described in Sections 6.2 and 6.3, respectively. In the context of the latter subproblem, positive semidefiniteness of the evolving orientation tensor must be preserved to obtain physically meaningful results. Therefore, sufficient conditions of eigenvalue range preservation are derived for selected models of orientation dynamics. This case study concludes with a numerical experiment in which we use the limiting techniques of the previous chapter.

6.1 Modeling

Mixtures of a Newtonian solvent and rigid slender particles are commonly used as a simplified model for the simulation of fiber suspension flows occurring, e.g., in the papermaking process. In the headbox of a paper machine, millions of fibers move within a carrier fluid and strongly influence the macroscopic behavior of the flow. Numerical simulations can significantly reduce the costs of research and development aimed at shape optimization and obtaining desirable fiber distributions. Lagrangian models of fiber suspensions require massive computing power due to the need to translate and rotate a large number of particles. For that reason, macroscopic Eulerian models of the fluid-particle mixture are frequently employed. The effective stress of a fiber suspension depends on moments of a function $\psi : \Omega \times \partial B^d \times [0, T] \to \mathbb{R}_0^+$, $\partial B^d := \left\{ \mathbf{x} \in \mathbb{R}^d \mid \|\mathbf{x}\|_2 = 1 \right\}$. By definition, $\psi(\mathbf{x}, \mathbf{p}, t)$

© Springer Fachmedien Wiesbaden GmbH, part of Springer Nature 2019
C. Lohmann, *Physics-Compatible Finite Element Methods for Scalar and Tensorial Advection Problems*, https://doi.org/10.1007/978-3-658-27737-6_6

is the probability that a fiber occupying the space location $\mathbf{x} \in \Omega$ will have orientation $\mathbf{p} \in \partial B^d$ at time $t \in [0, T]$. The so defined *orientation distribution function* satisfies

$$\psi(\mathbf{x}, \mathbf{p}, t) = \psi(\mathbf{x}, -\mathbf{p}, t), \quad \psi(\mathbf{x}, \mathbf{p}, t) \geqslant 0 \qquad \forall \mathbf{x} \in \Omega, \, \mathbf{p} \in \partial B^d, \, t \in [0, T],$$

$$\oint_{\partial B^d} \psi(\mathbf{x}, \mathbf{p}, t) \, d\mathbf{p} = 1 \qquad \forall \mathbf{x} \in \Omega, \, t \in [0, T].$$

The first property holds because both ends of a rigid fiber are supposed to be indistinguishable form each other and, therefore, a unit vector $\mathbf{p}$ parallel to the fiber symmetry axis describes the same orientation state if its sign is reversed. The other properties hold for any probability density distribution. The evolution of ψ is governed by the *Fokker-Planck equation*

$$\frac{d\psi}{dt} = \frac{\partial \psi}{\partial t} + \mathbf{v} \cdot \mathrm{grad}(\psi) = - \mathrm{div}_{\mathbf{p}}(\psi \dot{\mathbf{p}}), \tag{6.1}$$

where $\mathrm{div}_{\mathbf{p}}$ denotes the divergence operator on ∂B^d, $\mathbf{v}$ is the velocity field of the carrier fluid, and $\dot{\mathbf{p}}$ stands for the change of orientation for a single fiber under the influence of velocity gradients (at a fixed space location in the Eulerian reference frame or along a characteristic in the Lagrangian reference frame). The corresponding rate of rotation can be approximated by the *Jeffery equation* [Jef22]

$$\dot{\mathbf{p}} = \mathbf{W}\mathbf{p} + \lambda\Big(\mathbf{D}\mathbf{p} - (\mathbf{D} : (\mathbf{p} \otimes \mathbf{p}))\mathbf{p}\Big), \tag{6.2}$$

where $\mathbf{D} = \frac{1}{2}(\nabla \mathbf{v} + \nabla \mathbf{v}^\top)$ and $\mathbf{W} = \frac{1}{2}(\nabla \mathbf{v} - \nabla \mathbf{v}^\top)$ are the strain rate and spin tensors, respectively, and $\lambda = (r_e^2 - 1)/(r_e^2 + 1)$ depends on the aspect ratio $r_e = \frac{l}{d}$ of the fibers with length l and diameter d.

The first order PDE model (6.1) is based on the assumption that the particles evolve independently of each other and possible fiber-fiber interactions are neglected. However, this simplification is only suitable for dilute suspensions. As the volume fraction of fibers increases, the influence of fiber-fiber interactions on the orientation dynamics becomes more pronounced and must be taken into account. To that end, Brownian diffusion terms are commonly incorporated into the equation for ψ. For example, Hinch and Leal [HL73] generalized (6.1) to the case of semi-dilute fiber suspensions as follows:

$$\frac{d\psi}{dt} = \frac{\partial \psi}{\partial t} + \mathbf{v} \cdot \mathrm{grad}(\psi) = - \mathrm{div}_{\mathbf{p}}\big(\dot{\mathbf{p}}\psi - D_r \, \mathrm{grad}_{\mathbf{p}}(\psi)\big), \tag{6.3}$$

where $\mathrm{grad}_{\mathbf{p}}$ denotes the gradient operator on ∂B^d and $D_r \geqslant 0$ is the rotary diffusion coefficient. In the model of Folgar and Tucker [FT84], the intensity of Brownian motion is given by $D_r = C_I \dot{\gamma}$, where $\dot{\gamma} := \sqrt{2}\|\mathbf{D}\|_F$ is the effective shear rate and C_I is a material constant known as the interaction coefficient.

While the evolution equation for ψ depends on the velocity $\mathbf{v}$ of the carrier fluid, the presence of fibers changes the rheology of the mixture by producing additional stresses that depend on the local orientation pattern. Hence, there is an intricate two-way coupling between the subproblems for the fluid and the fibers. The non-Newtonian overall behavior of the mixture can be described by the Navier-Stokes equations (2.44) equipped with the effective stress tensor [Tuc91, Eq. (2.2)]

$$\tau = 2\mu\mathbf{D} + 2\mu\phi\big(c_1\underline{\underline{\mathbf{A}}} : \mathbf{D} + c_2(\mathbf{D}\mathbf{A} + \mathbf{A}\mathbf{D}) + c_3\mathbf{D} + 2c_4 D_r \mathbf{A}\big), \qquad (6.4)$$

where ϕ is the volume fraction of the fibers, $c_1, \ldots, c_4$ are nonnegative material coefficients, and the inner product $\underline{\underline{\mathbf{A}}} : \mathbf{D}$ between a fourth and second order tensor quantity is a second order tensor with entries $(\underline{\underline{\mathbf{A}}} : \mathbf{D})_{k\ell} := \sum_{mn} \underline{\underline{a}}_{k\ell mn} d_{mn}$. The *orientation tensors* $\mathbf{A} \in \mathbb{S}_d$ and $\underline{\underline{\mathbf{A}}} \in \mathbb{S}_d^{d \times d}$ are defined by

$$\mathbf{A} = (a_{k\ell})_{k,\ell=1}^d, \qquad a_{k\ell} := \oint_{\partial B^d} p_k p_\ell \psi \, \mathrm{d}\mathbf{p}, \qquad (6.5a)$$

$$\underline{\underline{\mathbf{A}}} = (\underline{\underline{a}}_{k\ell mn})_{k,\ell,m,n=1}^d, \qquad \underline{\underline{a}}_{k\ell mn} := \oint_{\partial B^d} p_k p_\ell p_m p_n \psi \, \mathrm{d}\mathbf{p} \qquad (6.5b)$$

and represent the second and fourth order moments of ψ.

If the contribution of the Brownian motion of particles to the stress tensor τ is removed by setting $c_4 = 0$, the constitutive law (6.4) becomes equivalent to

$$\tau = 2\mu_I\big(\mathbf{D} + N_p \underline{\underline{\mathbf{A}}} : \mathbf{D} + N_s(\mathbf{D}\mathbf{A} + \mathbf{A}\mathbf{D})\big),$$

where $\mu_I := \mu(1 + \phi c_3)$ contains all isotropic contributions to the viscosity. The particle number N_p and the shear number N_s are defined by

$$N_p := \frac{\phi c_1}{1 + \phi c_3}, \qquad N_s := \frac{\phi c_2}{1 + \phi c_3}.$$

For slender fibers, the shear number is much smaller than N_p and the term proportional to N_s can be neglected [Tuc91].

In summary, the full system of PDEs to be considered in this chapter is given by

$$\rho\frac{\partial \mathbf{v}}{\partial t} + \rho(\mathbf{v}\cdot\nabla)\mathbf{v} = \rho\mathbf{g} - \text{grad}(p) + \nabla\cdot\boldsymbol{\tau}, \qquad \text{div}(\mathbf{v}) = 0, \qquad (6.6a)$$

$$\frac{\partial \psi}{\partial t} + \mathbf{v}\cdot\text{grad}(\psi) = -\text{div}_{\mathbf{p}}\big(\dot{\mathbf{p}}\psi - D_r\,\text{grad}_{\mathbf{p}}(\psi)\big), \qquad (6.6b)$$

$$\boldsymbol{\tau} = 2\mu_I(\mathbf{D} + N_p\underline{\underline{\mathbf{A}}}:\mathbf{D}), \qquad \dot{\mathbf{p}} = \mathbf{W}\mathbf{p} + \lambda\Big(\mathbf{D}\mathbf{p} - (\mathbf{D}:(\mathbf{p}\otimes\mathbf{p}))\mathbf{p}\Big). \quad (6.6c)$$

While the momentum equation of (6.6a) depends on the fourth order orientation tensor $\underline{\underline{\mathbf{A}}}$ by definition of the stress tensor $\boldsymbol{\tau}$, the evolution of the orientation distribution function ψ is affected by the velocity field $\mathbf{v}$. In what follows, we present numerical methods to approximate the solutions $(\mathbf{v}, p)$ of (6.6a) and ψ of (6.6b) independently by assuming that the orientation tensor $\underline{\underline{\mathbf{A}}}$ and the velocity field $\mathbf{v}$ are given parameters. Then the solution $(\mathbf{v}, p, \psi)$ to the coupled problem (6.6) can be approximated using an operator splitting approach.

6.2 Fluid dynamics

The time dependent flow behavior of an incompressible Newtonian fluid or a fiber suspension is governed by the unsteady Navier-Stokes equations with a suitably defined stress tensor $\boldsymbol{\tau}$. Leaving $\boldsymbol{\tau}$ unspecified for the time being, we consider the initial-boundary value problem

$$\rho\frac{\partial \mathbf{v}}{\partial t} + \rho(\mathbf{v}\cdot\nabla)\mathbf{v} = \rho\mathbf{g} - \text{grad}(p) + \nabla\cdot\boldsymbol{\tau} \qquad \text{in } \Omega\times(0,T), \qquad (6.7a)$$

$$\text{div}(\mathbf{v}) = 0 \qquad \text{in } \Omega\times(0,T), \qquad (6.7b)$$

$$\mathbf{v}(\cdot, 0) = \mathbf{v}_0 \qquad \text{on } \Omega, \qquad (6.7c)$$

$$\mathbf{v} = \mathbf{v}_D \qquad \text{on } \Gamma_D\times(0,T), \qquad (6.7d)$$

$$-p\mathbf{n} + \boldsymbol{\tau}\mathbf{n} = \rho\mathbf{h} \qquad \text{on } \Gamma_N\times(0,T), \qquad (6.7e)$$

where $\Gamma_D \cup \Gamma_N = \partial\Omega$ is the decomposition of the boundary $\partial\Omega$ into its Dirichlet and Neumann parts, $\mathbf{v}_0 : \Omega \to \mathbb{R}^d$ is the initial velocity field, $\mathbf{v}_D : \Gamma_D \times (0,T) \to \mathbb{R}^d$ denotes the Dirichlet boundary data, and $\mathbf{h} : \Gamma_N \times (0,T) \to \mathbb{R}^d$ is the prescribed traction, i.e., the force acting on the fluid in the outward normal direction [KH14, Chapter 8]. For the problem at hand to be well-defined, the compatibility conditions

$$\operatorname{div}(\mathbf{v}_0) = 0 \qquad \text{on } \Omega,$$
$$\mathbf{n} \cdot \mathbf{v}_0 = \mathbf{n} \cdot \mathbf{v}_D(\cdot, 0) \qquad \text{on } \Gamma_D$$

must be satisfied. Note that there is no initial condition for p and, in contrast to the thermodynamic pressure of a compressible fluid, p is not required to be nonnegative. In fact, the pressure field p can be interpreted as a distributed Lagrange multiplier for the incompressibility constraint (6.7b) [KH14, Chapter 8].

In the special case $\Gamma_D = \partial\Omega$ (and $\Gamma_N = \emptyset$), only spatial derivatives of p occur in (6.7) and, hence, the pressure is defined up to an arbitrary constant. The lack of uniqueness can be cured, e.g., by imposing the zero-mean constraint $\int_\Omega p = 0$. Integrating (6.7b) over the domain Ω and using Theorem 2.1, we find that the Dirichlet boundary data $\mathbf{v}|_{\partial\Omega} = \mathbf{v}_D$ must satisfy

$$0 = \int_\Omega \operatorname{div}(\mathbf{v}) = \int_{\partial\Omega} \mathbf{v} \cdot \mathbf{n} = \int_{\partial\Omega} \mathbf{v}_D \cdot \mathbf{n}.$$

For simplicity, we restrict further discussion to the case $\Gamma_N \neq \emptyset$ in which the pressure is uniquely determined by (6.7e) and no integral compatibility conditions apply to the normal components of $\mathbf{v}_D$.

In what follows, we consider a finite element discretization of problem (6.7), briefly discuss its stability properties, and summarize a numerical solution algorithm. This part is mainly based on [Tur99; KH14, Chapter 8].

6.2.1 Discretization

Let $\mathbf{W}_D := \left\{ \mathbf{w} \in H^1(\Omega)^d \,\middle|\, \mathbf{w}|_{\Gamma_D} = \mathbf{v}_D \right\}$ and $Q := L^2(\Omega)$ denote the function spaces in which a weak solution $(\mathbf{v}, p) \in \mathbf{W}_D \times Q$ of the incompressible

Navier-Stokes equations is supposed to exist. The variational formulation of the momentum equation (6.7a) is given by

$$\rho \int_\Omega \mathbf{w} \cdot \frac{\partial \mathbf{v}}{\partial t} + \rho \int_\Omega \mathbf{w} \cdot ((\mathbf{v} \cdot \nabla)\mathbf{v})$$

$$= \rho \int_\Omega \mathbf{w} \cdot \mathbf{g} - \int_\Omega \nabla \mathbf{w} : (-p\mathbf{I} + \boldsymbol{\tau}) + \rho \int_{\Gamma_N} \mathbf{w} \cdot \mathbf{h}$$

$$+ \int_{\Gamma_D} \mathbf{w} \cdot ((-p\mathbf{I} + \boldsymbol{\tau})\mathbf{n}) \tag{6.8a}$$

$$= \rho \int_\Omega \mathbf{w} \cdot \mathbf{g} + \int_\Omega p \operatorname{div}(\mathbf{w}) - \int_\Omega \nabla \mathbf{w} : \boldsymbol{\tau} + \rho \int_{\Gamma_N} \mathbf{w} \cdot \mathbf{h}$$

$$+ \int_{\Gamma_D} \mathbf{w} \cdot ((-p\mathbf{I} + \boldsymbol{\tau})\mathbf{n}) \qquad \forall \mathbf{w} \in \mathbf{W}_0, t \in (0,T],$$

where $\mathbf{W}_0 := \{\mathbf{w} \in H^1(\Omega)^d \mid \mathbf{w}|_{\Gamma_D} = \mathbf{0}\}$ and integration by parts is performed to incorporate the Neumann boundary condition (6.7e). The weak formulation of the continuity equation (6.7b) reads

$$\int_\Omega q \operatorname{div}(\mathbf{v}) = 0 \qquad \forall q \in Q, t \in (0,T]. \tag{6.8b}$$

To discretize problem (6.8) in space, we introduce finite dimensional subspaces $\mathbf{W}_h \subseteq \mathbf{W} := H^1(\Omega)^d$ and $Q_h \subseteq Q$ spanned by the basis functions $\boldsymbol{\psi}_1, \ldots, \boldsymbol{\psi}_{N_v}$ and $\varphi_1, \ldots, \varphi_{N_p}$, respectively. As in Section 3.3.1, we approximate the velocity $\mathbf{v}$ and the pressure p by

$$\mathbf{v}_h := \sum_{j=1}^{N_v} v_j(t) \boldsymbol{\psi}_j(\mathbf{x}), \quad p_h := \sum_{l=1}^{N_p} p_l(t) \varphi_l(\mathbf{x}) \qquad \forall \mathbf{x} \in \Omega, t \in [0,T],$$

where $v = v(t) = (v_j(t))_{j=1}^{N_v}$ and $p = p(t) = (p_l(t))_{l=1}^{N_p}$ are the vectors of time dependent degrees of freedom. Then the semi-discrete counterparts of (6.8a) and (6.8b) can be written as

$$\rho \int_\Omega \mathbf{w}_h \cdot \frac{\partial \mathbf{v}_h}{\partial t} + \rho \int_\Omega \mathbf{w}_h \cdot ((\mathbf{v}_h \cdot \nabla)\mathbf{v}_h) + \int_\Omega \nabla \mathbf{w}_h : \boldsymbol{\tau}(\mathbf{v}_h) - \int_\Omega p_h \operatorname{div}(\mathbf{w}_h)$$

$$= \rho \int_\Omega \mathbf{w}_h \cdot \mathbf{g} + \rho \int_{\Gamma_N} \mathbf{w}_h \cdot \mathbf{h} + \int_{\Gamma_D} \mathbf{w}_h \cdot \left((-p_h\mathbf{I} + \boldsymbol{\tau}(\mathbf{v}_h))\mathbf{n} \right)$$

$$\forall \boldsymbol{\psi}_h \in \mathbf{W}_h \cap \mathbf{W}_0, t \in (0,T], \tag{6.9a}$$

$$\int_\Omega q_h \operatorname{div}(\mathbf{v}_h) = 0 \qquad \forall \varphi_h \in Q_h, t \in (0,T]. \tag{6.9b}$$

Discretization in time using the θ-scheme yields

$$a(\mathbf{v}_h^{n+1}; \mathbf{v}_h^{n+1}, \mathbf{w}_h) + b(p_h^{n+1}, \mathbf{w}_h)$$

$$+ \Delta t \int_{\Gamma_D} \mathbf{w}_h \cdot \mathbf{n} p_h^{n+1} - \Delta t \theta \int_{\Gamma_D} \mathbf{w}_h \cdot (\boldsymbol{\tau}(\mathbf{v}_h^{n+1})\mathbf{n})$$

$$= \underline{g}(\mathbf{w}_h) + \Delta t(1-\theta) \int_{\Gamma_D} \mathbf{w}_h \cdot (\boldsymbol{\tau}(\mathbf{v}_h^n)\mathbf{n}) \qquad \forall \psi_h \in \mathbf{W}_h \cap \mathbf{W}_0, \quad (6.10a)$$

$$b(\varphi_h, \mathbf{v}_h^{n+1}) = 0 \qquad \forall \varphi_h \in Q_h, \tag{6.10b}$$

where

$$a(\mathbf{v}_h; \mathbf{u}_h, \mathbf{w}_h) := \rho \int_\Omega \mathbf{w}_h \cdot \mathbf{u}_h$$

$$+ \Delta t \theta \Big(\rho \int_\Omega \mathbf{w}_h \cdot ((\mathbf{v}_h \cdot \nabla)\mathbf{u}_h) + \int_\Omega \nabla \mathbf{w}_h : \boldsymbol{\tau}(\mathbf{u}_h) \Big),$$

$$b(q_h, \mathbf{w}_h) := -\Delta t \int_\Omega q_h \operatorname{div}(\mathbf{w}_h),$$

$$\underline{g}(\mathbf{w}_h) := \rho \int_\Omega \mathbf{w}_h \cdot \mathbf{v}_h^n + \rho \Delta t \theta \int_\Omega \mathbf{w}_h \cdot \mathbf{g}^{n+1}$$

$$+ \rho \Delta t (1-\theta) \int_\Omega \mathbf{w}_h \cdot \mathbf{g}^n + \rho \Delta t \theta \int_{\Gamma_N} \mathbf{w}_h \cdot \mathbf{h}^{n+1}$$

$$+ \rho \Delta t (1-\theta) \Big(\int_{\Gamma_N} \mathbf{w}_h \cdot \mathbf{h}^n - \int_\Omega \mathbf{w}_h \cdot ((\mathbf{v}_h^n \cdot \nabla)\mathbf{v}_h^n) \Big)$$

$$- \Delta t (1-\theta) \int_\Omega \nabla \mathbf{w}_h : \boldsymbol{\tau}(\mathbf{u}_h^n).$$

Note that all terms depending on the pressure are treated implicitly no matter how θ is chosen. In the case $\theta \in (0,1)$, the pressure can also be approximated using a θ-weighed convex combination of p^{n+1} and p^n. However, the fully implicit treatment does not affect the accuracy of the pressure approximation if appropriate post-processing is performed [Tur99].

To guarantee that $a(\mathbf{v}_h; \cdot, \cdot)$ is bilinear, we assume that the stress tensor $\boldsymbol{\tau}(\mathbf{v}_h)$ is linear in $\mathbf{v}_h$. Furthermore, let all indices $i \in \{1, \dots, M_\mathbf{v}\}$ correspond to degrees of freedom that are not determined by the Dirichlet boundary conditions (cf. Section 3.2.3). Then system (6.10) with appropriate boundary conditions $\mathbf{v}_h(\cdot, t^{n+1}) \approx \mathbf{v}_D(\cdot, t^{n+1})$ on Γ_D is equivalent to the *saddle point problem*

$$\begin{bmatrix} \tilde{\mathcal{S}} & \tilde{\mathcal{B}} \\ \mathcal{B}^\top & \mathbf{0} \end{bmatrix} \begin{bmatrix} v^{n+1} \\ p^{n+1} \end{bmatrix} = \begin{bmatrix} g \\ 0 \end{bmatrix} \tag{6.11}$$

for the degrees of freedom of the velocity $\mathbf{v}_h^{n+1}$ and the pressure p_h^{n+1} approximating the solution to (6.7) at t^{n+1}. The entries of the matrix blocks $\tilde{S} = \tilde{S}(v^{n+1}) \in \mathbb{R}^{N_{\mathbf{v}} \times N_{\mathbf{v}}}$, $\mathcal{B}, \tilde{\mathcal{B}} \in \mathbb{R}^{N_{\mathbf{v}} \times N_p}$ and the components of the right hand side vector $\underline{g} \in \mathbb{R}^{N_{\mathbf{v}}}$ are given by

$$\tilde{s}_{ij} := \begin{cases} a(\mathbf{v}_h^{n+1}; \boldsymbol{\psi}_j, \boldsymbol{\psi}_i) & : i \leqslant M_{\mathbf{v}}, \\ \delta_{ij} & : i > M_{\mathbf{v}} \end{cases} \qquad \forall i, j \in \{1, \dots, N_{\mathbf{v}}\},$$

$$b_{il} := b(\varphi_l, \boldsymbol{\psi}_i) \qquad\qquad \forall i \in \{1, \dots, N_{\mathbf{v}}\}, \, l \in \{1, \dots, N_p\},$$

$$\tilde{b}_{il} := \begin{cases} b_{il} & : i \leqslant M_{\mathbf{v}}, \\ 0 & : i > M_{\mathbf{v}} \end{cases} \qquad \forall i \in \{1, \dots, N_{\mathbf{v}}\}, \, l \in \{1, \dots, N_p\},$$

$$\underline{g}_i := \underline{g}(\boldsymbol{\psi}_i) \qquad\qquad \forall i \in \{1, \dots, M_{\mathbf{v}}\},$$

where $\underline{g}_{M_{\mathbf{v}}+1}, \dots \underline{g}_{N_{\mathbf{v}}}$ are determined by the Dirichlet boundary data $\mathbf{v}_{\mathrm{D}}(\cdot, t^{n+1})$.

6.2.2 Stability

Due to the saddle point structure of (6.11), the finite dimensional subspaces $\mathbf{W}_h$ and Q_h must be chosen appropriately to guarantee the existence of a unique solution. In particular, the number of degrees of freedom for the velocity field may not be smaller than the number of pressure unknowns, that is, $N_{\mathbf{v}} \geqslant N_p$ must hold. Otherwise, the columns of $\mathcal{B}$ are linearly dependent and the system matrix is singular.

In the literature on finite element methods for incompressible fluid dynamics, the stability analysis for different combinations of $\mathbf{W}_h$ and Q_h is typically restricted to discretizations of the linear Stokes problem

$$\begin{aligned} -\mu \Delta \mathbf{v} + \mathrm{grad}(p) &= \mathbf{0} && \text{in } \Omega, \\ \mathrm{div}(\mathbf{v}) &= 0 && \text{in } \Omega, \\ \mathbf{v} &= \mathbf{0} && \text{on } \partial\Omega, \end{aligned}$$

where $\mathbf{W} = H_0^1(\Omega)^d$ and $Q = L_0^2(\Omega) := \{ q \in L^2(\Omega) \mid \int_\Omega q = 0 \}$. A discretely divergence-free finite element approximation $\mathbf{v}_h \in \mathbf{W}_h$ and a pressure approximation $p_h \in Q_h$ exist and are unique if the discrete *Ladyzhenskaya-Babuška-Brezzi (LBB) condition* [GR86]

$$\min_{q_h \in Q_h \setminus \{0\}} \max_{\mathbf{w}_h \in \mathbf{W}_h \setminus \{0\}} \frac{(q_h, \mathrm{div}(\mathbf{w}_h))_{L^2(\Omega)}}{\|q_h\|_{L^2(\Omega)} \|\nabla \mathbf{w}_h\|_{L^2(\Omega)}} \geqslant \gamma > 0$$

holds for a mesh-independent constant γ. In the two dimensional case, examples of stable finite element pairs include the nonconforming Crouzeix-Raviart $(\tilde{P}_1/P_0)$ and Rannacher-Turek $(\tilde{Q}_1/Q_0)$ as well as the continuous Taylor-Hood (P_2/P_1) and Bercovier-Pironneau $(P_1\text{-iso-}P_2/P_1)$ elements [CR73b; RT92; QV94, Section 9.3]. In the three dimensional numerical examples below, we employ continuous piecewise triquadratic velocity approximations on a hexahedral mesh, while the pressure p_h is discontinuous and piecewise linear [BF91].

6.2.3 Iterative solver

There exists a great variety of (iterative) solution techniques which are readily applicable to the coupled nonlinear problem (6.11). However, only solvers that exploit the saddle point structure of the system matrix are likely to deliver sufficiently high performance in practical applications. In what follows, we present a discrete projection scheme which is based on the *pressure Schur complement (PSC) equation*

$$\mathcal{A}p^{n+1} := \mathcal{B}^\top \tilde{\mathcal{S}}^{-1} \tilde{\mathcal{B}} p^{n+1} = \mathcal{B}^\top \tilde{\mathcal{S}}^{-1} \underline{g}. \qquad (6.12)$$

If the matrix $\mathcal{A} = \mathcal{B}^\top \tilde{\mathcal{S}}^{-1} \tilde{\mathcal{B}}$ is invertible, then the unique solution of (6.11) is formally given by

$$p^{n+1} = \mathcal{A}^{-1} \mathcal{B}^\top \tilde{\mathcal{S}}^{-1} \underline{g}, \qquad v^{n+1} = \tilde{\mathcal{S}}^{-1}(\underline{g} - \tilde{\mathcal{B}} p^{n+1}).$$

However, a stand-alone computation of p^{n+1} is impossible due to the dependence of $\tilde{\mathcal{S}}$ on the unknown vector of degrees of freedom v^{n+1}. For that reason, the solution p^{n+1} has to be computed iteratively, e.g., using fixed point iterations of the form

$$p^{n+1,s+1} = p^{n+1,s} + \omega \mathcal{C}^{-1}(\mathcal{B}^\top \tilde{\mathcal{S}}^{-1} \underline{g} - \mathcal{B}^\top \tilde{\mathcal{S}}^{-1} \tilde{\mathcal{B}} \tilde{p}^{n+1,s}) \quad s = 0, 1, \dots, \quad (6.13)$$

where $\omega \in (0, 2]$ is a relaxation parameter and $\mathcal{C} \in \mathbb{R}^{N_p \times N_p}$ denotes a suitable approximation to $\mathcal{A}$ acting as a preconditioner. In each cycle, the velocity degrees of freedom must also be updated by solving the system $\tilde{\mathcal{S}} \tilde{v}^{n+1,s+1} = \underline{g} - \tilde{\mathcal{B}} p^{n+1,s}$ with a solution dependent matrix $\tilde{\mathcal{S}}$. Then the preconditioned Richardson iteration (6.13) can be written as

$$p^{n+1,s+1} = p^{n+1,s} + \omega \mathcal{C}^{-1}(\mathcal{B}^\top \tilde{v}^{n+1,s+1}) \qquad s = 0, 1, \dots$$

Using the sparse Schur complement preconditioner $\mathcal{C} := \mathcal{B}^\top \mathcal{M}_L^{-1} \tilde{\mathcal{B}}$, where $\mathcal{M}_L \in \mathbb{R}^{N_v \times N_v}$ is the lumped velocity mass matrix with diagonal entries

$$m_i = \sum_{j=1}^{N_v} \int_\Omega \boldsymbol{\psi}_i \cdot \boldsymbol{\psi}_j \qquad \forall i \in \{1, \dots, N_v\},$$

the velocity and pressure unknowns can be updated in the following way.

Algorithm 6.1 ([Tur97]). *Let $p^{n+1,0} \in \mathbb{R}^{N_p}$ be a suitable initial approximation to the vector p^{n+1} of pressure unknowns. For $s = 0, 1, \ldots$, perform the following steps:*

1. *Solve the discrete nonlinear Burgers equation with the system matrix $\tilde{\mathcal{S}} = \tilde{\mathcal{S}}(\tilde{v}^{n+1,s+1})$*

$$\tilde{\mathcal{S}}\tilde{v}^{n+1,s+1} = \underline{g} - \tilde{\mathcal{B}}p^{n+1,s}. \tag{6.14}$$

2. *Calculate the pressure correction $q^{n+1,s+1}$ by solving*

$$\mathcal{B}^{\top}\mathcal{M}_{\mathrm{L}}^{-1}\tilde{\mathcal{B}}q^{n+1,s+1} = \mathcal{B}^{\top}\tilde{v}^{n+1,s+1}. \tag{6.15}$$

3. *Correct the pressure*

$$p^{n+1,s+1} = p^{n+1,s} + \omega q^{n+1,s+1}. \tag{6.16}$$

Finally, project the velocity

$$v^{n+1} = \tilde{v}^{n+1,s+1} - \mathcal{M}_{\mathrm{L}}^{-1}\tilde{\mathcal{B}}q^{n+1,s+1}. \tag{6.17}$$

The nonlinear subproblem (6.14) can be solved, e.g., using a fixed point iteration or a quasi-Newton method with the initial guess $\tilde{v}^{n+1,s+1} = \tilde{v}^{n+1,s}$. At the first outer iteration, the initial guess $\tilde{v}^{n+1,0} = v^n$ or $\tilde{v}^{n+1,0} = 2v^n - v^{n-1}$ may be employed.

The solution $\tilde{v}^{n+1,s+1}$ of (6.14) can be computed independently of the pressure $p^{n+1,s+1}$ but the associated velocity field may violate the discretized incompressibility condition (6.10b). To rectify this, the optional step 6.17 projects the tentative velocity into the space of 'discretely divergence-free' functions. Note that (6.15) is the Schur complement equation of the saddle point problem

$$\begin{aligned} \mathcal{M}_{\mathrm{L}}(v^{n+1} - \tilde{v}^{n+1,s+1}) + \tilde{\mathcal{B}}q^{n+1,s+1} &= 0, \\ \mathcal{B}^{\top}v^{n+1} &= 0 \end{aligned} \tag{6.18}$$

such that v^{n+1} calculated in (6.17) satisfies

$$\mathcal{B}^{\top}v^{n+1} = \mathcal{B}^{\top}\tilde{v}^{n+1,s+1} - \mathcal{B}^{\top}\mathcal{M}_{\mathrm{L}}^{-1}\tilde{\mathcal{B}}q^{n+1,s+1} = 0.$$

Problem (6.18) corresponds to a mixed finite element approximation of the Laplacian operator. Therefore, the subproblem to be solved in step 2 can be interpreted as a "discrete pressure Poisson" equation. Especially in the case of small time increments Δt, in which $\mathcal{S}$ behaves like a mass matrix, the above choice of $\mathcal{C}$ is likely to be a good approximation of the pressure Schur complement matrix $\mathcal{A} = \mathcal{B}^\top \tilde{\mathcal{S}}^{-1} \tilde{\mathcal{B}}$. In fact, one Richardson iteration per time step is typically sufficient as long as $\mathcal{C}$ is a good preconditioner. This version of Algorithm 6.1 corresponds to the discrete counterpart of the projection schemes proposed by Chorin [Cho68] ($p^{n+1,0} = 0$) and van Kan [van86] ($p^{n+1,0} = p^n$). Even though the presented approach performs best when the time increment is small, the van Kan-like discrete projection scheme is also able to produce steady state solutions to the discretized Navier-Stokes equations.

Remark 6.2. The saddle point problem (6.11) and, hence, Algorithm 6.1 depend on $\mathcal{B}$ *and* the slightly modified matrix $\tilde{\mathcal{B}}$, in which the last $N_\mathbf{v} - M_\mathbf{v}$ rows are set to zero. In a practical implementation, the storage of $\tilde{\mathcal{B}}$ can be avoided by modifying subroutines for calculating matrix vector-products of the form $\mathcal{B}v$. In a similar vein, the discrete Poisson preconditioner $\mathcal{C} = \mathcal{B}^\top \mathcal{M}_{\mathrm{L}}^{-1} \tilde{\mathcal{B}}$ can be written in the form $\mathcal{C} = \mathcal{B}^\top \tilde{\mathcal{D}}_{\mathrm{L}} \mathcal{B}$, where $\tilde{\mathcal{D}}_{\mathrm{L}} = (\delta_{ij} \tilde{d}_i)_{i,j=1}^N$ is a diagonal matrix defined by

$$\tilde{d}_i = \begin{cases} m_i^{-1} & : i \leqslant M_\mathbf{v}, \\ 0 & : i > M_\mathbf{v} \end{cases} \qquad \forall i \in \{1, \ldots, N_\mathbf{v}\}.$$

Using this representation, the sparse matrix $\mathcal{C}$ can be readily assembled if it is required by an iterative solver or smoother/preconditioner for the discrete Poisson equation.

6.3 Orientation dynamics

The next step toward the simulation of fiber suspensions is to derive a numerical method for equation (6.6b). This high dimensional PDE describes the evolution of a nonnegative and normalized orientation distribution function, which determines the orientation states of the fibers. Therefore, the use of special bound-preserving algorithms is required to preserve the properties of interest and produce reliable approximations.

6.3.1 Orientation discretization and splitting

The orientation distribution function $\psi(\mathbf{x}, \mathbf{p}, t)$ is a scalar function which depends not only on the space location $\mathbf{x}$ and the time variable t but also on the unit vector $\mathbf{p}$ which determines the fiber orientation. Even if there is no dependence on t, the domain of $\psi(\mathbf{x}, \mathbf{p})$ is $2d - 1$ dimensional and, hence, direct finite element discretizations may be very expensive, particularly when $d = 3$. In addition, the rheological behavior of the carrier fluid is governed by (6.6a). Since this constitutive law depends only on the fourth order orientation tensor $\underline{\mathbf{A}}$, the computation of a highly accurate numerical approximation to the orientation distribution ψ might be redundant. Therefore, a common approach to fiber suspension flow modeling is to evolve the second (or fourth) order orientation tensor instead of ψ.

To derive the equation which governs the evolution of the second order orientation tensor, we first notice that [Doi81]

$$\partial_{\mathbf{p},\ell}(p_k) = \delta_{k\ell} - p_k p_\ell \qquad \forall k, \ell \in \{1, \ldots, d\},$$

where $\partial_{\mathbf{p},\ell}(\cdot) := \mathbf{e}_\ell \cdot \mathrm{grad}_{\mathbf{p}}(\cdot)$ and $\mathbf{p} \in \partial B^d$. Then the following identities can be derived

$$\Delta_{\mathbf{p}}(p_k) = \sum_\ell \partial_{\mathbf{p},\ell}(\partial_{\mathbf{p},\ell}(p_k)) = \sum_\ell \partial_{\mathbf{p},\ell}(\delta_{k\ell} - p_k p_\ell)$$

$$= -p_k \sum_\ell \partial_{\mathbf{p},\ell} p_\ell - \sum_\ell p_\ell \partial_{\mathbf{p},\ell} p_k$$

$$= -p_k \sum_\ell (1 - p_\ell^2) - \sum_\ell p_\ell (\delta_{k\ell} - p_k p_\ell) = -(d-1) p_k,$$

$$\Delta_{\mathbf{p}}(p_k p_\ell) = p_\ell \Delta_{\mathbf{p}} p_k + 2 \, \mathrm{grad}_{\mathbf{p}}(p_k) \cdot \mathrm{grad}_{\mathbf{p}}(p_\ell) + p_k \Delta_{\mathbf{p}} p_\ell$$

$$= -2(d-1) p_k p_\ell + 2 \sum_m (\delta_{km} - p_k p_m)(\delta_{\ell m} - p_\ell p_m)$$

$$= -2(d-1) p_k p_\ell + 2\delta_{k\ell}$$

$$\qquad - 2 \sum_m (\delta_{km} p_\ell + \delta_{\ell m} p_k) p_m + 2 p_k p_\ell \sum_m p_m^2$$

$$= -2(d-1) p_k p_\ell + 2\delta_{k\ell} - 4 p_k p_\ell + 2 p_k p_\ell = 2(\delta_{k\ell} - d p_k p_\ell)$$

because $\|\mathbf{p}\|_2^2 = \sum_k p_k^2 = 1$ and, in view of the fact that $\mathbf{p} \cdot \dot{\mathbf{p}} = 0$, integration by parts yields

$$-\oint_{\partial B^d} p_k p_\ell \operatorname{div}_{\mathbf{p}}(\dot{\mathbf{p}}\psi)\, d\mathbf{p}$$

$$= \oint_{\partial B^d} \psi \operatorname{grad}_{\mathbf{p}}(p_k p_\ell) \cdot \dot{\mathbf{p}}\, d\mathbf{p}$$

$$= \oint_{\partial B^d} \psi \sum_m (p_\ell \partial_{\mathbf{p},m} p_k + p_k \partial_{\mathbf{p},m} p_\ell)\dot{p}_m\, d\mathbf{p}$$

$$= \oint_{\partial B^d} \psi \sum_m (p_\ell (\delta_{km} - p_k p_m) + p_k(\delta_{\ell m} - p_\ell p_m))\dot{p}_m\, d\mathbf{p}$$

$$= \oint_{\partial B^d} \psi p_\ell \dot{p}_k\, d\mathbf{p} + \oint_{\partial B^d} \psi p_k \dot{p}_\ell\, d\mathbf{p} - 2\oint_{\partial B^d} \psi p_k p_\ell (\mathbf{p}\cdot\dot{\mathbf{p}})\, d\mathbf{p},$$

$$= \oint_{\partial B^d} \psi p_\ell \dot{p}_k\, d\mathbf{p} + \oint_{\partial B^d} \psi p_k \dot{p}_\ell\, d\mathbf{p}.$$

Recalling definition (6.5a) of the second order orientation tensor $\mathbf{A}$ and differentiating its components with respect to t, we find that

$$\frac{da_{k\ell}}{dt} = \oint_{\partial B^d} p_k p_\ell d_t \psi\, d\mathbf{p}$$

$$= -\oint_{\partial B^d} p_k p_\ell \operatorname{div}_{\mathbf{p}}(\dot{\mathbf{p}}\psi)\, d\mathbf{p} + D_r \oint_{\partial B^d} p_k p_\ell \Delta_p(\psi)\, d\mathbf{p}$$

$$= \oint_{\partial B^d} \psi p_\ell \dot{p}_k\, d\mathbf{p} + \oint_{\partial B^d} \psi p_k \dot{p}_\ell\, d\mathbf{p} + D_r \oint_{\partial B^d} \Delta_{\mathbf{p}}(p_k p_\ell)\psi\, d\mathbf{p}$$

$$= \sum_m (w_{km} + \lambda d_{km})a_{m\ell} + \sum_m (w_{\ell m} + \lambda d_{\ell m})a_{mk}$$

$$\quad - 2\lambda \sum_{m,n} d_{mn} a_{k\ell mn} + 2D_r(\delta_{k\ell} - da_{k\ell})$$

by virtue of (6.2) and (6.3). Therefore, the evolution equation for the second order orientation tensor $\mathbf{A}$ can be written as [Han62; Doi81]

$$\frac{d\mathbf{A}}{dt} = \frac{\partial \mathbf{A}}{\partial t} + \mathbf{v} \cdot \operatorname{grad}(\mathbf{A})$$

$$= \mathbf{W}\mathbf{A} - \mathbf{A}\mathbf{W} + \lambda(\mathbf{D}\mathbf{A} + \mathbf{A}\mathbf{D} - 2\underline{\mathbf{A}}:\mathbf{D}) + 2D_r(\mathbf{I} - d\mathbf{A}). \quad (6.19)$$

This model of fiber orientation dynamics is known as the *Folgar-Tucker equation*. It includes a rotary diffusion term modeling the interaction of fibers and can be interpreted as a low order spectral approximation to the Fokker-Planck equation for ψ [Mon+10].

To simplify the theoretical analysis and the development of property-preserving numerical algorithms, we split (6.19) into the non-conservative form

$$\frac{\partial \mathbf{A}}{\partial t} + \mathbf{v} \cdot \text{grad}(\mathbf{A}) = 0 \qquad (6.20a)$$

of a homogeneous advection equation and the space independent Folgar-Tucker equation

$$\frac{\partial \mathbf{A}}{\partial t} = \mathbf{WA} - \mathbf{AW} + \lambda(\mathbf{DA} + \mathbf{AD} - 2\underline{\underline{\mathbf{A}}} : \mathbf{D}) + 2D_r(\mathbf{I} - d\mathbf{A}). \qquad (6.20b)$$

Unfortunately, the right hand side of the latter equation contains the unknown fourth order orientation tensor which must be approximated using a suitable *closure*.

6.3.2 Closures

Many closures for the fourth order orientation tensor can be found in the literature. Although quite accurate approximations of $\underline{\underline{\mathbf{A}}}$ exist for simple flows, they may fail to preserve certain properties of orientation tensors. As a consequence, the resulting solution to (6.20b) can become unphysical and simulations may blow up. This section is devoted to the derivation of sufficient requirements for obtaining physics-compatible solutions. For better illustration purposes, we verify them for commonly used two dimensional closures and postpone a brief analysis of selected three dimensional closures to Section 6.3.4.

To begin with, let us derive necessary conditions for a tensor $\mathbf{A} \in \mathbb{R}^{d \times d}$ to be an orientation tensor of a nonnegative orientation distribution function ψ defined by (6.5a).

Theorem 6.3. *The second order orientation tensor* $\mathbf{A} \in \mathbb{R}^{d \times d}$ *of an orientation distribution function* $\psi : \partial B^d \to \mathbb{R}_0^+$ *is symmetric and has nonnegative eigenvalues that sum up to unity. The space of tensors possessing these properties is denoted by* $\mathbb{A}_d := \{ \mathbf{A} \in \mathbb{S}_{d,+} \mid \text{tr}(\mathbf{A}) = 1 \}$.

Proof. The symmetry of $\mathbf{A}$ is a direct consequence of its definition (6.5a). Furthermore, the trace of $\mathbf{A}$ is given by

$$\text{tr}(\mathbf{A}) = \sum_k \oint_{\partial B^d} p_k^2 \psi \, d\mathbf{p} = \oint_{\partial B^d} \left(\sum_k p_k^2 \right) \psi \, d\mathbf{p} = \oint_{\partial B^d} \psi \, d\mathbf{p} = 1.$$

For each $\mathbf{v} \in \mathbb{R}^d$, we have

$$\mathbf{v}^\top \mathbf{A} \mathbf{v} = \oint_{\partial B^d} \Big(\sum_{k,\ell} v_k p_k v_\ell p_\ell \Big) \psi \, \mathrm{d}\mathbf{p} = \oint_{\partial B^d} \Big(\sum_k v_k p_k \Big)^2 \psi \, \mathrm{d}\mathbf{p} \geqslant 0$$

by virtue of the fact that $\psi \geqslant 0$. Hence, $\mathbf{A}$ is positive semidefinite. $\qquad\square$

These properties follow from the nonnegativity and normalization of the orientation distribution function. Clearly, a well-designed numerical scheme should comply with these physical requirements when $\mathbf{A}$ is evolved instead of ψ.

Similar properties can be established for the fourth order orientation tensor which is defined by (6.5b).

Theorem 6.4. *The fourth order orientation tensor $\underline{\underline{\mathbf{A}}} \in \mathbb{R}^{d \times d \times d \times d}$ of an orientation distribution function $\psi : \partial B^d \to \mathbb{R}_0^+$ satisfies*

$$\underline{\underline{a}}_{k\ell mn} = \underline{\underline{a}}_{\ell kmn} = \underline{\underline{a}}_{m\ell kn} = \underline{\underline{a}}_{n\ell mk} \qquad \forall k, \ell, m, n \in \{1, \ldots, d\}, \quad (6.21a)$$

$$\mathbf{V} : (\underline{\underline{\mathbf{A}}} : \mathbf{V}) = \sum_{k,\ell,m,n} \underline{\underline{a}}_{k\ell mn} v_{k\ell} v_{mn} \geqslant 0 \qquad \forall \mathbf{V} \in \mathbb{R}^{d \times d}, \quad (6.21b)$$

$$(\underline{\underline{\mathbf{A}}} : \mathbf{I})_{k\ell} = \sum_m \underline{\underline{a}}_{k\ell mm} = a_{k\ell} \qquad \forall k, \ell \in \{1, \ldots, d\}. \quad (6.21c)$$

Proof. Again, property (6.21a) follows by definition of $\underline{\underline{\mathbf{A}}}$. For (6.21b), we have

$$\mathbf{V} : (\underline{\underline{\mathbf{A}}} : \mathbf{V}) = \oint_{\partial B^d} \Big(\sum_{k,\ell,m,n} v_{k\ell} p_k p_\ell v_{mn} p_m p_n \Big) \psi \, \mathrm{d}\mathbf{p}$$

$$= \oint_{\partial B^d} \Big(\sum_{k,\ell} v_{k\ell} p_k p_\ell \Big)^2 \psi \, \mathrm{d}\mathbf{p} \geqslant 0$$

while (6.21c) holds due to

$$\sum_m \underline{\underline{a}}_{k\ell mm} = \oint_{\partial B^d} \psi p_k p_\ell \Big(\sum_m p_m^2 \Big) \, \mathrm{d}\mathbf{p} = \oint_{\partial B^d} \psi p_k p_\ell \, \mathrm{d}\mathbf{p} = a_{k\ell}. \qquad \square$$

Identity (6.21c) establishes a relationship between the second and fourth order orientation tensors. While in two dimensions this property is equivalent to

$$\underline{\underline{a}}_{1122} = a_{11} - \underline{\underline{a}}_{1111} = a_{22} - \underline{\underline{a}}_{2222}, \qquad (6.22)$$

the following equalities hold in the three dimensional case if and only if
(6.21c) is satisfied:

$$2\underline{a}_{1122} = \quad a_{11} + a_{22} - a_{33} - \underline{a}_{1111} - \underline{a}_{2222} + \underline{a}_{3333}, \tag{6.23a}$$

$$2\underline{a}_{1133} = \quad a_{11} - a_{22} + a_{33} - \underline{a}_{1111} + \underline{a}_{2222} - \underline{a}_{3333}, \tag{6.23b}$$

$$2\underline{a}_{2233} = -a_{11} + a_{22} + a_{33} + \underline{a}_{1111} - \underline{a}_{2222} - \underline{a}_{3333}. \tag{6.23c}$$

We adopt the viewpoint that a properly designed closure for reconstruction
of the fourth order orientation tensor, written as $\underline{\mathbf{A}}(\cdot)$, should provide the
properties established in Theorem 6.4. Otherwise, there is no nonnegative
orientation distribution function such that $\underline{\mathbf{A}}(\cdot)$ represents its fourth order
moment, as defined by (6.5).

Definition 6.5. Let $\mathbf{A} \in \mathbb{A}_d$ be a second order orientation tensor. Then we
call a closure $\underline{\mathbf{A}}(\mathbf{A})$ *admissible* if it possesses properties (6.21a)–(6.21c).

If an eigenvalue of a second order orientation tensor vanishes, further
properties of its admissible fourth order counterparts can be established.

Lemma 6.6. *Let $\mathbf{A} \in \mathbb{A}_d$ be a second order orientation tensor with $a_k = 0$
for some $k \in \{1, \ldots, d\}$. Then every admissible closure $\underline{\mathbf{A}}(\mathbf{A})$ satisfies
$\underline{\mathbf{A}} : (\mathbf{q}_k \otimes \mathbf{q}_k) = \mathbf{0}$.*

Proof. We first notice that

$$\mathrm{tr}\big(\underline{\mathbf{A}} : (\mathbf{q}_k \otimes \mathbf{q}_k)\big) = \mathbf{I} : \big(\underline{\mathbf{A}} : (\mathbf{q}_k \otimes \mathbf{q}_k)\big) = \mathbf{A} : (\mathbf{q}_k \otimes \mathbf{q}_k) = \mathbf{q}_k^\top \mathbf{A} \mathbf{q}_k = a_k = 0$$

by (6.21a) and (6.21c). Consequently, the eigenvalues of $\underline{\mathbf{A}} : (\mathbf{q}_k \otimes \mathbf{q}_k)$ sum
up to zero and, hence, vanish because $\underline{\mathbf{A}} : (\mathbf{v}_1 \otimes \mathbf{v}_1)$ is positive semidefinite
for all $\mathbf{v}_1 \in \mathbb{R}^d$. Indeed, we have

$$\mathbf{v}_2^\top \big(\underline{\mathbf{A}} : (\mathbf{v}_1 \otimes \mathbf{v}_1)\big)\mathbf{v}_2 = (\mathbf{v}_2 \otimes \mathbf{v}_1) : \big(\underline{\mathbf{A}} : (\mathbf{v}_2 \otimes \mathbf{v}_1)\big) \geqslant 0 \qquad \forall \mathbf{v}_1, \mathbf{v}_2 \in \mathbb{R}^d \tag{6.24}$$

by virtue of (6.21a) and (6.21b). $\qquad\qquad\square$

After the characterization of second and fourth order orientation tensors,
we analyze the impact of admissible closures on the solution of the space independent Folgar-Tucker equation. The next theorem shows the preservation
of positive semidefinite orientation states for the solution to the continuous
ODE (6.20b) if admissible and Lipschitz continuous closures are employed.
Under these conditions, the existence of a locally unique solution can be
shown, e.g., following [Chi06, Theorem 1.184].

Theorem 6.7. *Let $\underline{\underline{\mathbf{A}}}(\mathbf{A})$ be a closure which is admissible for every second order orientation tensor $\mathbf{A} \in \mathbb{A}_d$ and Lipschitz continuous, that is,*

$$\left\|\underline{\underline{\mathbf{A}}}(\mathbf{A}) - \underline{\underline{\mathbf{A}}}(\bar{\mathbf{A}})\right\|_{\mathrm{F}} \leqslant L_{\mathrm{cl}} \|\mathbf{A} - \bar{\mathbf{A}}\|_{\mathrm{F}} \qquad \forall \mathbf{A}, \bar{\mathbf{A}} \in \mathbb{A}_d. \tag{6.25}$$

Suppose that $\mathbf{A}(t)$ is the solution to the space independent Folgar-Tucker equation (6.20b) with differentiable eigenvalues and eigenvectors. Then $\mathbf{A}(t)$ stays symmetric with unit trace and satisfies $\mathrm{d}_t a_k \geqslant 0$ for every eigenvalue $a_k = 0$.

The Lipschitz continuity condition (6.25) is formulated using the following definition of the Frobenius norm of a fourth order tensor:

$$\|\underline{\underline{\mathbf{V}}}\|_{\mathrm{F}}^2 := \sum_{k,\ell,m,n} \underline{\underline{v}}^2_{k\ell mn} \qquad \forall \underline{\underline{\mathbf{V}}} \in \mathbb{R}^{d \times d \times d \times d}.$$

Proof. We first notice that $\mathbf{A}$ stays symmetric because the right hand side of (6.20b) is symmetric.

Furthermore, the trace of the right hand side is given by

$$\mathrm{tr}\big(\mathbf{WA} - \mathbf{AW} + \lambda(\mathbf{DA} + \mathbf{AD} - 2\underline{\underline{\mathbf{A}}} : \mathbf{D}) + 2D_r(\mathbf{I} - d\mathbf{A})\big)$$

$$= \underbrace{\mathrm{tr}(\mathbf{WA} - \mathbf{AW})}_{=0} + 2D_r \underbrace{\mathrm{tr}(\mathbf{I} - d\mathbf{A})}_{=0} + \lambda \, \mathrm{tr}(\mathbf{DA} + \mathbf{AD} - 2\underline{\underline{\mathbf{A}}} : \mathbf{D})$$

$$= 2\lambda\big(\mathrm{tr}(\mathbf{AD}) - \mathrm{tr}(\underline{\underline{\mathbf{A}}} : \mathbf{D})\big) = 2\lambda\Big(\sum_{k,\ell} a_{k\ell} d_{\ell k} - \sum_{k,\ell,m} \underline{\underline{a}}_{k\ell mm} d_{k\ell}\Big)$$

$$= 2\lambda\Big(\sum_{k,\ell} a_{k\ell} d_{\ell k} - \sum_{k,\ell} a_{k\ell} d_{k\ell}\Big) = 0$$

due to the invariance of the trace (5.8) and property (6.21c). Therefore, the time derivative of $\mathrm{tr}(\mathbf{A})$ vanishes and the trace of $\mathbf{A}$ stays constant.

Finally, eigenvalues cannot become negative. To show that this statement is true, let us focus on the time derivatives of the eigenvalues. Similarly to the proof of Theorem 5.5, we have (cf. [VC03; Loh17b])

$$\frac{\mathrm{d}\tilde{\mathbf{A}}}{\mathrm{d}t} = (\mathrm{d}_t \mathbf{Q}^\top)\mathbf{AQ} + \mathbf{Q}^\top(\mathrm{d}_t \mathbf{A})\mathbf{Q} + \mathbf{Q}^\top \mathbf{A}(\mathrm{d}_t \mathbf{Q})$$

$$= (\mathrm{d}_t \mathbf{Q}^\top)\mathbf{Q}\tilde{\mathbf{A}} + \mathbf{Q}^\top(\mathrm{d}_t \mathbf{A})\mathbf{Q} + \tilde{\mathbf{A}}\mathbf{Q}^\top(\mathrm{d}_t \mathbf{Q})$$

$$= \mathbf{H}\tilde{\mathbf{A}} + \mathbf{Q}^\top(\mathrm{d}_t \mathbf{A})\mathbf{Q} - \tilde{\mathbf{A}}\mathbf{H},$$

where $\mathbf{H} := (\mathrm{d}_t \mathbf{Q}^\top)\mathbf{Q}$ is skew symmetric due to

$$\mathbf{H} = (\mathrm{d}_t \mathbf{Q}^\top)\mathbf{Q} = \mathrm{d}_t(\mathbf{Q}^\top \mathbf{Q}) - \mathbf{Q}^\top(\mathrm{d}_t \mathbf{Q}) = -\mathbf{Q}^\top(\mathrm{d}_t \mathbf{Q}) = -\mathbf{H}^\top.$$

Since $\tilde{\mathbf{A}}$ is the diagonal form of $\mathbf{A}$, the diagonal of $\mathbf{H}\tilde{\mathbf{A}} - \tilde{\mathbf{A}}\mathbf{H}$ vanishes and the time derivative of an arbitrary eigenvalue a_k is given by

$$\frac{\mathrm{d}a_k}{\mathrm{d}t} = \mathbf{q}_k^\top(\mathrm{d}_t\mathbf{A})\mathbf{q}_k = 2\lambda\big(a_k\mathbf{q}_k^\top\mathbf{D}\mathbf{q}_k - \mathbf{q}_k^\top(\underline{\underline{\mathbf{A}}} : \mathbf{D})\mathbf{q}_k\big) + 2D_r(1 - \mathrm{d}a_k).$$

Let us now assume that $a_k = 0$. Then we have

$$\mathbf{q}_k^\top(\underline{\underline{\mathbf{A}}} : \mathbf{D})\mathbf{q}_k = \mathbf{D} : \big(\underline{\underline{\mathbf{A}}} : (\mathbf{q}_k \otimes \mathbf{q}_k)\big) = \mathbf{D} : \mathbf{0} = 0$$

by condition (6.21a) and Lemma 6.6. Therefore, the time derivative of the eigenvalue $a_k = 0$ is given by

$$\frac{\mathrm{d}a_k}{\mathrm{d}t} = 2D_r \geqslant 0$$

and eigenvalues do not become negative.   □

In Theorem 6.7, we have shown the physics-compatibility of the solution to (6.20b) under the assumption that the closure is admissible. If this is not the case, positive semidefinite orientation states may not be preserved.

Example 6.8. The *linear closure*, proposed by Hand [Han62], can be written as [AT87]

$$\begin{aligned}
\underline{\underline{a}}_{klmn} &= \gamma_1(\delta_{k\ell}\delta_{mn} + \delta_{km}\delta_{\ell n} + \delta_{kn}\delta_{\ell m}) \\
&\quad + \gamma_2(a_{k\ell}\delta_{mn} + a_{km}\delta_{\ell n} + a_{kn}\delta_{\ell m} + a_{mn}\delta_{k\ell} + a_{\ell n}\delta_{km} + a_{\ell m}\delta_{kn}),
\end{aligned} \tag{6.26}$$

where $\gamma_1 = -\frac{1}{24}$, $\gamma_2 = \frac{1}{6}$ if $\mathbf{A} \in \mathbb{A}_2$ and $\gamma_1 = -\frac{1}{35}$, $\gamma_2 = \frac{1}{7}$ if $\mathbf{A} \in \mathbb{A}_3$. This closure is exact for random distributions of fibers and the corresponding inner product with the strain rate tensor reads [GR99]

$$\underline{\underline{\mathbf{A}}} : \mathbf{D} = \gamma_1\big(\mathrm{tr}(\mathbf{D})\mathbf{I} + 2\mathbf{D}\big) + \gamma_2\big(\mathrm{tr}(\mathbf{D})\mathbf{A} + 2\mathbf{A}\mathbf{D} + 2\mathbf{D}\mathbf{A} + (\mathbf{A} : \mathbf{D})\mathbf{I}\big). \tag{6.27}$$

This formula simplifies to

$$\underline{\underline{\mathbf{A}}} : \mathbf{D} = 2\gamma_1\mathbf{D} + \gamma_2\big(2\mathbf{A}\mathbf{D} + 2\mathbf{D}\mathbf{A} + (\mathbf{A} : \mathbf{D})\mathbf{I}\big) \tag{6.28}$$

if the velocity field is solenoidal, i.e., $\mathrm{div}(\mathbf{v}) = \mathrm{tr}(\mathbf{D}) = 0$.

While the linear closure possesses properties (6.21a) and (6.21c) due to

$$\begin{aligned}
\underline{\underline{\mathbf{A}}} : \mathbf{I} &= \gamma_1\big(d\mathbf{I} + 2\mathbf{I}\big) + \gamma_2\big(d\mathbf{A} + 2\mathbf{A} + 2\mathbf{A} + \mathbf{I}\big) \\
&= \big(\gamma_1(d + 2) + \gamma_2\big)\mathbf{I} + \gamma_2(d + 4)\mathbf{A} = \mathbf{A},
\end{aligned}$$

condition (6.21b) may be violated and, hence, Theorem 6.7 is not applicable. In fact, unphysical orientation states do occur in the following two dimensional example:

$$\mathbf{A} = \begin{pmatrix} 1 & 0 \\ 0 & 0 \end{pmatrix}, \qquad \nabla \mathbf{v} = \begin{pmatrix} 2 & 0 \\ 0 & -2 \end{pmatrix}$$

$$\implies \mathbf{D} = \begin{pmatrix} 2 & 0 \\ 0 & -2 \end{pmatrix}, \quad \mathbf{W} = \begin{pmatrix} 0 & 0 \\ 0 & 0 \end{pmatrix}.$$

The linear closure $\underline{\underline{\mathbf{A}}}(\mathbf{A})$ yields the entries

$$\underline{\underline{a}}_{1111} = \tfrac{7}{8}, \quad \underline{\underline{a}}_{1112} = 0, \quad \underline{\underline{a}}_{1122} = \tfrac{1}{8}, \quad \underline{\underline{a}}_{1222} = 0, \quad \underline{\underline{a}}_{2222} = -\tfrac{1}{8}$$

and the Folgar-Tucker equation (6.20b) is given by

$$\frac{\partial \mathbf{A}}{\partial t} = \mathbf{W}\mathbf{A} - \mathbf{A}\mathbf{W} + \lambda(\mathbf{D}\mathbf{A} + \mathbf{A}\mathbf{D} - 2\underline{\underline{\mathbf{A}}} : \mathbf{D}) + 2D_r(\mathbf{I} - d\mathbf{A})$$

$$= \lambda \begin{pmatrix} 2 & 0 \\ 0 & 0 \end{pmatrix} + \lambda \begin{pmatrix} 2 & 0 \\ 0 & 0 \end{pmatrix} - \lambda \begin{pmatrix} 3 & 0 \\ 0 & 1 \end{pmatrix} + 2D_r \begin{pmatrix} -1 & 0 \\ 0 & 1 \end{pmatrix}$$

$$= \begin{pmatrix} \lambda - 2D_r & 0 \\ 0 & -\lambda + 2D_r \end{pmatrix}.$$

Therefore, the time derivative of $a_1 = 0$ is $-\lambda + 2D_r$ and the minimal eigenvalue becomes negative if $D_r < \tfrac{\lambda}{2}$.

Although the non-admissible linear closure does not preserve nonnegative eigenvalues, properties (6.21a)–(6.21c) are not mandatory for $\mathbf{A}$ to remain an orientation tensor.

Example 6.9. The quadratic closure [HL76; Doi81] defined by

$$\underline{\underline{\mathbf{A}}} = \mathbf{A} \otimes \mathbf{A} \qquad \Longleftrightarrow \qquad \underline{\underline{a}}_{k\ell mn} = a_{k\ell} a_{mn} \quad \forall 1 \leqslant k, \ell, m, n \qquad (6.29)$$

satisfies (6.21b) and (6.21c) by definition but violates condition (6.21a). However, the results of Theorem 6.7 can still be shown as before using the identities

$$\mathbf{q}_1^\top (\underline{\underline{\mathbf{A}}} : \mathbf{D}) \mathbf{q}_1 = \mathbf{q}_1^\top \mathbf{A} \mathbf{q}_1 (\mathbf{A} : \mathbf{D}) = a_1 (\mathbf{A} : \mathbf{D}),$$

$$\mathrm{tr}(\mathbf{D}\mathbf{A} + \mathbf{A}\mathbf{D} - 2\underline{\underline{\mathbf{A}}} : \mathbf{D}) = 2\,\mathrm{tr}(\mathbf{A}\mathbf{D}) - 2\,\mathrm{tr}(\mathbf{A})(\mathbf{A} : \mathbf{D})$$

$$= 2\big(\mathrm{tr}(\mathbf{A}\mathbf{D}) - \mathrm{tr}(\mathbf{A}\mathbf{D})\big) = 0,$$

where the vanishing trace property is true because $\mathrm{tr}(\mathbf{A}) = 1$ and $\mathrm{tr}(\mathbf{A}\mathbf{D}) = \mathrm{tr}(\mathbf{D}\mathbf{A}) = \mathbf{A} : \mathbf{D}$.

6.3.2.1 Orthotropic closures

Cintra and Tucker [CT95] claimed that the second order orientation tensor $\mathbf{A}$ and its fourth order counterpart $\underline{\underline{\mathbf{A}}}$ defined by a closure should be simultaneously diagonalizable because $\underline{\underline{\mathbf{A}}}$ depends only on the information provided by $\mathbf{A}$.

Definition 6.10 ([CT95]). A closure $\underline{\underline{\mathbf{A}}}(\mathbf{A})$ is called *orthotropic* if the reconstructed tensor $\underline{\underline{\mathbf{A}}}$ has the same principal axes as $\mathbf{A} = \mathbf{Q}\tilde{\mathbf{A}}\mathbf{Q}^{\top}$, i.e., there is a diagonal fourth order orientation tensor $\tilde{\underline{\underline{\mathbf{A}}}} = (\tilde{\underline{\underline{a}}}_{ijkl})^{d}_{i,j,k,\ell=1}$ such that (6.21a) holds for $\tilde{\mathbf{A}}$ and

$$\underline{\underline{a}}_{k_1 k_2 k_3 k_4} = \sum_{\ell_1,\ell_2,\ell_3,\ell_4} \tilde{\underline{\underline{a}}}_{\ell_1 \ell_2 \ell_3 \ell_4} q_{k_1 \ell_1} q_{k_2 \ell_2} q_{k_3 \ell_3} q_{k_4 \ell_4}$$

$$\forall k_1, k_2, k_3, k_4 \in \{1, \ldots, d\}. \quad (6.30)$$

A fourth order tensor $\tilde{\underline{\underline{\mathbf{A}}}} \in \mathbb{R}^{d \times d \times d \times d}$ satisfying (6.21a) is diagonal if and only if

$$\tilde{\underline{\underline{a}}}_{\ell_1 \ell_2 \ell_3 \ell_4} = 0 \qquad \forall \ell_1, \ell_2, \ell_3, \ell_4 \in \{1, \ldots, d\} \text{ s.t. } \ell_1 \neq \ell_2, \ell_3, \ell_4$$

$$\text{or } \ell_2 \neq \ell_1, \ell_3, \ell_4 \text{ or } \ell_3 \neq \ell_1, \ell_2, \ell_4 \text{ or } \ell_4 \neq \ell_1, \ell_2, \ell_3. \quad (6.31)$$

Hence, such a tensor has at most $d + 6\binom{d}{2}$ nonzero entries, that is, 8 for $d = 2$ and 21 for $d = 3$. According to (6.30), the fourth order orientation tensor $\underline{\underline{\mathbf{A}}}$ satisfies (6.21a) and we have

$$
\begin{aligned}
\|\underline{\underline{\mathbf{A}}}\|_{\mathrm{F}}^2 &= \sum_{k_1,\ldots,k_4} \Big(\sum_{\ell_1,\ldots,\ell_4} \tilde{\underline{\underline{a}}}_{\ell_1\ell_2\ell_3\ell_4} q_{k_1\ell_1} q_{k_2\ell_2} q_{k_3\ell_3} q_{k_4\ell_4} \Big)^2 \\
&= \sum_{k_1,\ldots,k_4} \Big(\sum_{\ell_1,\ldots,\ell_8} \tilde{\underline{\underline{a}}}_{\ell_1\ell_2\ell_3\ell_4} q_{k_1\ell_1} q_{k_2\ell_2} q_{k_3\ell_3} q_{k_4\ell_4} \\
&\qquad\qquad\qquad \cdot \tilde{\underline{\underline{a}}}_{\ell_5\ell_6\ell_7\ell_8} q_{k_1\ell_5} q_{k_2\ell_6} q_{k_3\ell_7} q_{k_4\ell_8} \Big) \\
&= \sum_{\ell_1,\ldots,\ell_8} \tilde{\underline{\underline{a}}}_{\ell_1\ell_2\ell_3\ell_4} \tilde{\underline{\underline{a}}}_{\ell_5\ell_6\ell_7\ell_8} \Big(\sum_{k_1} q_{k_1\ell_1} q_{k_1\ell_5} \Big) \Big(\sum_{k_2} q_{k_2\ell_2} q_{k_2\ell_6} \Big) \\
&\qquad\qquad\qquad \cdot \Big(\sum_{k_3} q_{k_3\ell_3} q_{k_3\ell_7} \Big) \Big(\sum_{k_4} q_{k_4\ell_4} q_{k_4\ell_8} \Big) \\
&= \sum_{\ell_1,\ldots,\ell_8} \tilde{\underline{\underline{a}}}_{\ell_1\ell_2\ell_3\ell_4} \tilde{\underline{\underline{a}}}_{\ell_5\ell_6\ell_7\ell_8} \delta_{\ell_1\ell_5} \delta_{\ell_2\ell_6} \delta_{\ell_3\ell_7} \delta_{\ell_4\ell_8} \\
&= \sum_{\ell_1,\ldots,\ell_4} \tilde{\underline{\underline{a}}}_{\ell_1\ell_2\ell_3\ell_4}^2 = \|\tilde{\underline{\underline{\mathbf{A}}}}\|_{\mathrm{F}}^2 .
\end{aligned}
$$

Theorem 6.11. *Let $\mathbf{A}$ be a second order orientation tensor with $a_k = a_\ell$ for some $k \neq \ell$. Then the diagonal tensor $\tilde{\underline{\underline{\mathbf{A}}}}$ occurring in Definition 6.10 satisfies*

$$
\begin{aligned}
\tilde{\underline{\underline{a}}}_{kkkk} = \tilde{\underline{\underline{a}}}_{\ell\ell\ell\ell} = 3\tilde{\underline{\underline{a}}}_{kk\ell\ell} & \qquad \forall m \in \{1,\ldots,d\} \ s.t. \ m \neq k,\ell, & (6.32\mathrm{a}) \\
\tilde{\underline{\underline{a}}}_{kkmm} = \tilde{\underline{\underline{a}}}_{\ell\ell mm} & \qquad \forall m \in \{1,\ldots,d\} \ s.t. \ m \neq k,\ell. & (6.32\mathrm{b})
\end{aligned}
$$

Orthotropic closures violating these conditions are not uniquely determined for some orientation tensors with repeated eigenvalues.

Proof. Without loss of generality, let us assume that $\mathbf{A} = \tilde{\mathbf{A}}$ is diagonal and $a_1 = a_2$ holds. Then we notice that $\mathbf{A} = \mathbf{Q}\tilde{\mathbf{A}}\mathbf{Q}^\top$ is valid for any

$$
\mathbf{Q} = \begin{pmatrix} c_1 & c_2 & & \\ c_2 & -c_1 & & \\ & & 1 & \\ & & & \ddots \end{pmatrix} \qquad s.t. \ c_1, c_2 \in \mathbb{R},\ c_1^2 + c_2^2 = 1. \qquad (6.33)
$$

According to (6.30), the validity of (6.33) implies

$$\underline{\underline{a}}_{k_1 k_2 k_3 k_4} = \underline{\underline{\tilde{a}}}_{k_1 k_2 k_3 k_4} = \sum_{\ell_1, \ell_2, \ell_3, \ell_4} \underline{\underline{\tilde{a}}}_{\ell_1 \ell_2 \ell_3 \ell_4} q_{k_1 \ell_1} q_{k_2 \ell_2} q_{k_3 \ell_3} q_{k_4 \ell_4}$$

$$\forall k_1, k_2, k_3, k_4 \in \{1, \ldots, d\}. \quad (6.34)$$

Choosing $c_1 = 0$ and $c_2 = 1$, we find that

$$\begin{aligned}
\underline{\underline{\tilde{a}}}_{11mm} &= \sum_{\ell_1, \ell_2, \ell_3, \ell_4} \underline{\underline{\tilde{a}}}_{\ell_1 \ell_2 \ell_3 \ell_4} q_{1\ell_1} q_{1\ell_2} q_{m\ell_3} q_{m\ell_4} \\
&= \sum_{\ell_1} \underline{\underline{\tilde{a}}}_{\ell_1 \ell_1 \ell_1 \ell_1} q_{1\ell_1}^2 q_{m\ell_1}^2 \\
&\quad + \sum_{\ell_1, \ell_2, \, \ell_1 < \ell_2} \underline{\underline{\tilde{a}}}_{\ell_1 \ell_1 \ell_2 \ell_2} (q_{1\ell_1}^2 q_{m\ell_2}^2 + q_{m\ell_1}^2 q_{1\ell_2}^2 + 4 q_{1\ell_1} q_{m\ell_1} q_{1\ell_2} q_{m\ell_2}) \\
&= \sum_{\ell_1} \underline{\underline{\tilde{a}}}_{\ell_1 \ell_1 \ell_1 \ell_1} \delta_{2\ell_1} \delta_{m\ell_1} \\
&\quad + \sum_{\ell_1, \ell_2, \, \ell_1 < \ell_2} \underline{\underline{\tilde{a}}}_{\ell_1 \ell_1 \ell_2 \ell_2} (\delta_{2\ell_1} \delta_{m\ell_2} + \delta_{m\ell_1} \delta_{2\ell_2} + 4 \delta_{2\ell_1} \delta_{m\ell_1} \delta_{2\ell_2} \delta_{m\ell_2}) \\
&= \underline{\underline{\tilde{a}}}_{22mm} \qquad \forall m \in \{3, \ldots, d\}
\end{aligned}$$

and

$$\begin{aligned}
\underline{\underline{\tilde{a}}}_{1111} &= \sum_{\ell_1, \ell_2, \ell_3, \ell_4} \underline{\underline{\tilde{a}}}_{\ell_1 \ell_2 \ell_3 \ell_4} q_{1\ell_1} q_{1\ell_2} q_{1\ell_3} q_{1\ell_4} \\
&= \sum_{\ell_1} \underline{\underline{\tilde{a}}}_{\ell_1 \ell_1 \ell_1 \ell_1} q_{1\ell_1}^2 q_{1\ell_1}^2 \\
&\quad + \sum_{\ell_1, \ell_2, \, \ell_1 < \ell_2} \underline{\underline{\tilde{a}}}_{\ell_1 \ell_1 \ell_2 \ell_2} (q_{1\ell_1}^2 q_{1\ell_2}^2 + q_{1\ell_1}^2 q_{1\ell_2}^2 + 4 q_{1\ell_1} q_{1\ell_1} q_{1\ell_2} q_{1\ell_2}) \\
&= \sum_{\ell_1} \underline{\underline{\tilde{a}}}_{\ell_1 \ell_1 \ell_1 \ell_1} \delta_{2\ell_1} \delta_{2\ell_1} \\
&\quad + \sum_{\ell_1, \ell_2, \, \ell_1 < \ell_2} \underline{\underline{\tilde{a}}}_{\ell_1 \ell_1 \ell_2 \ell_2} (\delta_{2\ell_1} \delta_{2\ell_2} + \delta_{2\ell_1} \delta_{2\ell_2} + 4 \delta_{2\ell_1} \delta_{2\ell_1} \delta_{2\ell_2} \delta_{2\ell_2}) \\
&= \underline{\underline{\tilde{a}}}_{2222}.
\end{aligned}$$

On the other hand, identity (6.34) with $c_1 = c_2 = \frac{\sqrt{2}}{2}$ implies the relation

$$
\begin{aligned}
\tilde{\underline{\underline{a}}}_{1111} &= \sum_{\ell_1,\ell_2,\ell_3,\ell_4} \tilde{\underline{\underline{a}}}_{\ell_1\ell_2\ell_3\ell_4} q_{1\ell_1} q_{1\ell_2} q_{1\ell_3} q_{1\ell_4} \\
&= \sum_{\ell_1} \tilde{\underline{\underline{a}}}_{\ell_1\ell_1\ell_1\ell_1} q_{1\ell_1}^2 q_{1\ell_1}^2 \\
&\quad + \sum_{\ell_1,\ell_2,\,\ell_1<\ell_2} \tilde{\underline{\underline{a}}}_{\ell_1\ell_1\ell_2\ell_2} (q_{1\ell_1}^2 q_{1\ell_2}^2 + q_{1\ell_1}^2 q_{1\ell_2}^2 + 4 q_{1\ell_1} q_{1\ell_1} q_{1\ell_2} q_{1\ell_2}) \\
&= \tfrac{1}{4}(\tilde{\underline{\underline{a}}}_{1111} + 6\tilde{\underline{\underline{a}}}_{1122} + \tilde{\underline{\underline{a}}}_{2222}) = \tfrac{1}{2}(\tilde{\underline{\underline{a}}}_{1111} + 3\tilde{\underline{\underline{a}}}_{1122}),
\end{aligned}
$$

which holds if and only if $\tilde{\underline{\underline{a}}}_{1111} = 3\tilde{\underline{\underline{a}}}_{1122}$. $\qquad\square$

Corollary 6.12. *For* $\mathbf{A} = \frac{1}{d}\mathbf{I}$, *every orthotropic closure* $\underline{\underline{A}}(\mathbf{A})$ *providing property (6.21c) produces a fourth order orientation tensor with the following nonzero entries:*

$$
\underline{\underline{a}}_{kkkk} = 3(2d + d^2)^{-1}, \qquad \underline{\underline{a}}_{kk\ell\ell} = \underline{\underline{a}}_{k\ell k\ell} = \underline{\underline{a}}_{k\ell\ell k} = (2d + d^2)^{-1}
$$
$$
\forall k,\ell \in \{1,\ldots,d\},\ k \neq \ell. \quad (6.35)
$$

Orthotropic closures violating these conditions are not well-defined for the isotropic orientation state $\mathbf{A} = \frac{1}{d}\mathbf{I}$.

Proof. Obviously, $\underline{\underline{A}}$ is a diagonal fourth order orientation tensor satisfying (6.21a) because $\mathbf{A}$ is diagonal. Furthermore, we have

$$
\tfrac{1}{d} = a_k = \underline{\underline{a}}_{kkkk} + \sum_{\ell \neq k} \underline{\underline{a}}_{kk\ell\ell} = \tilde{\underline{\underline{a}}}_{kkkk} + (d-1)\tfrac{1}{3}\tilde{\underline{\underline{a}}}_{kkkk} = \tfrac{1}{3}(2 + d)\tilde{\underline{\underline{a}}}_{kkkk}
$$
$$
\forall k \in \{1,\ldots,d\}
$$

by virtue of property (6.21c) and Theorem 6.11. This completes the proof. $\qquad\square$

In the two and three dimensional case, condition (6.35) is equivalent to

$$
\underline{\underline{a}}_{kkkk} = \tfrac{3}{8}, \quad \underline{\underline{a}}_{kk\ell\ell} = \underline{\underline{a}}_{k\ell k\ell} = \underline{\underline{a}}_{k\ell\ell k} = \tfrac{1}{8} \quad \forall k,\ell \in \{1,2\},\ k \neq \ell \quad \text{if } d = 2,
$$
$$
(6.36a)
$$
$$
\underline{\underline{a}}_{kkkk} = \tfrac{1}{5}, \quad \underline{\underline{a}}_{kk\ell\ell} = \underline{\underline{a}}_{k\ell k\ell} = \underline{\underline{a}}_{k\ell\ell k} = \tfrac{1}{15} \quad \forall k,\ell \in \{1,2,3\},\ k \neq \ell \quad \text{if } d = 3.
$$
$$
(6.36b)
$$

In general, it is quite challenging to prove the admissibility of $\underline{\underline{A}}(\mathbf{A})$ for every second order orientation tensor $\mathbf{A}$. However, in the special case of orthotropic closures, it suffices to show this property for diagonal tensors.

Theorem 6.13. *Let* $\underline{\underline{\mathbf{A}}}(\mathbf{A})$ *be an orthotropic closure depending on the second order orientation tensor* $\mathbf{A} = \mathbf{Q}\tilde{\mathbf{A}}\mathbf{Q}^\top \in \mathbb{A}_d$. *Then* $\underline{\underline{\mathbf{A}}}(\mathbf{A})$ *is admissible if and only if* $\underline{\underline{\mathbf{A}}}(\tilde{\mathbf{A}})$ *is admissible.*

Proof. Conditions (6.21a) and (6.21c) carry over trivially from $\underline{\underline{\mathbf{A}}}(\mathbf{A})$ to $\underline{\underline{\mathbf{A}}}(\tilde{\mathbf{A}})$ and vice versa. Additionally, (6.21b) holds for $\underline{\underline{\mathbf{A}}}(\mathbf{A})$ if and only if condition (6.21b) is satisfied for $\underline{\underline{\tilde{\mathbf{A}}}} := \underline{\underline{\mathbf{A}}}(\tilde{\mathbf{A}})$ because

$$
\begin{aligned}
\mathbf{V} : (\underline{\underline{\mathbf{A}}} : \mathbf{V}) &= \sum_{k_1,k_2,k_3,k_4} \underline{\underline{a}}_{k_1 k_2 k_3 k_4} v_{k_1 k_2} v_{k_3 k_4} \\
&= \sum_{k_1,k_2,k_3,k_4} \Big(\sum_{\ell_1,\ell_2,\ell_3,\ell_4} \underline{\underline{\tilde{a}}}_{\ell_1 \ell_2 \ell_3 \ell_4} q_{k_1 \ell_1} q_{k_2 \ell_2} q_{k_3 \ell_3} q_{k_4 \ell_4} \Big) v_{k_1 k_2} v_{k_3 k_4} \\
&= \sum_{\ell_1,\ell_2,\ell_3,\ell_4} \underline{\underline{\tilde{a}}}_{\ell_1 \ell_2 \ell_3 \ell_4} \Big(\big(\sum_{k_1,k_2} v_{k_1 k_2} q_{k_1 \ell_1} q_{k_2 \ell_2} \big) \big(\sum_{k_3,k_4} v_{k_3 k_4} q_{k_3 \ell_3} q_{k_4 \ell_4} \big) \Big) \\
&= (\mathbf{Q}^\top \mathbf{V} \mathbf{Q}) : (\underline{\underline{\tilde{\mathbf{A}}}} : (\mathbf{Q}^\top \mathbf{V} \mathbf{Q})) \qquad \forall \mathbf{V} \in \mathbb{R}^{d \times d}. \qquad \square
\end{aligned}
$$

Furthermore, the following proposition provides a handy tool to verify condition (6.21b).

Proposition 6.14. *A diagonal fourth order orientation tensor* $\underline{\underline{\tilde{\mathbf{A}}}}$ *satisfying* (6.21a) *possesses property* (6.21b) *if and only if*

$$
(\underline{\underline{\tilde{a}}}_{kk\ell\ell})_{k,\ell=1}^d \succcurlyeq \mathbf{0}, \qquad \underline{\underline{\tilde{a}}}_{kk\ell\ell} \geqslant 0 \quad \forall k,\ell \in \{1,\ldots,d\}. \tag{6.37}
$$

Proof. If (6.37) is valid, we have

$$
\begin{aligned}
\mathbf{V} : (\underline{\underline{\tilde{\mathbf{A}}}} : \mathbf{V}) &= \sum_{k,\ell,m,n} \tilde{a}_{k\ell mn} v_{k\ell} v_{mn} \\
&= \sum_k \underline{\underline{\tilde{a}}}_{kkkk} v_{kk}^2 + \sum_{k,\ell,\, k<\ell} \underline{\underline{\tilde{a}}}_{kk\ell\ell} (v_{k\ell}^2 + v_{\ell k}^2 + 2 v_{kk} v_{\ell\ell} + 2 v_{k\ell} v_{\ell k}) \\
&\geqslant \sum_k \underline{\underline{\tilde{a}}}_{kkkk} v_{kk}^2 + \sum_{k,\ell,\, k \neq \ell} \underline{\underline{\tilde{a}}}_{kk\ell\ell} v_{kk} v_{\ell\ell} = \sum_{k,\ell} \underline{\underline{\tilde{a}}}_{kk\ell\ell} v_{kk} v_{\ell\ell} \geqslant 0 \\
&\hspace{6cm} \forall \mathbf{V} \in \mathbb{R}^{d \times d} \quad (6.38)
\end{aligned}
$$

by Young's inequality, i.e., $|v_{k\ell} v_{\ell k}| \leqslant \frac{1}{2}(v_{k\ell}^2 + v_{\ell k}^2)$. Hence, $\underline{\underline{\tilde{\mathbf{A}}}}$ satisfies (6.21b).

On the other hand, if (6.37) is violated then $(\underline{\underline{\tilde{a}}}_{kk\ell\ell})_{k,\ell=1}^d$ possesses either a negative component or a negative eigenvalue. In the former case, there exist $k_0, \ell_0 \in \{1, \ldots, d\}$ such that $\underset{\approx}{\tilde{a}}_{k_0 k_0 \ell_0 \ell_0} < 0$. By definition of $\mathbf{V} = (\delta_{kk_0} \delta_{\ell\ell_0})^d_{k,\ell=1}$, we find that

$$\mathbf{V} : (\underset{\approx}{\tilde{\mathbf{A}}} : \mathbf{V}) = \sum_{k,\ell,m,n} \underset{\approx}{\tilde{a}}_{k\ell mn} v_{k\ell} v_{mn}$$

$$= \sum_{k,\ell,m,n} \underset{\approx}{\tilde{a}}_{k\ell mn} \delta_{kk_0} \delta_{\ell\ell_0} \delta_{mk_0} \delta_{n\ell_0} = \underset{\approx}{\tilde{a}}_{k_0 k_0 \ell_0 \ell_0} < 0$$

and property (6.21b) is violated. Otherwise, there exists $\mathbf{w} \in \mathbb{R}^d$ such that

$$\sum_{k,\ell} \underset{\approx}{\tilde{a}}_{kk\ell\ell} w_k w_\ell < 0.$$

Then the choice of $\mathbf{V} = (\delta_{k\ell} w_k)^d_{k,\ell=1}$ leads to

$$\mathbf{V} : (\underset{\approx}{\tilde{\mathbf{A}}} : \mathbf{V}) = \sum_{k,\ell,m,n} \underset{\approx}{\tilde{a}}_{k\ell mn} v_{k\ell} v_{mn}$$

$$= \sum_k \underset{\approx}{\tilde{a}}_{kkkk} v_{kk}^2 + \sum_{k,\ell,\, k<\ell} \underset{\approx}{\tilde{a}}_{kk\ell\ell} (2 v_{kk} v_{\ell\ell} + 2 v_{k\ell} v_{\ell k} + v_{k\ell}^2 + v_{\ell k}^2)$$

$$= \sum_k \underset{\approx}{\tilde{a}}_{kkkk} w_k^2 + \sum_{k,\ell,\, k\neq\ell} \underset{\approx}{\tilde{a}}_{kk\ell\ell} w_k w_\ell = \sum_{k,\ell} \underset{\approx}{\tilde{a}}_{kk\ell\ell} w_k w_\ell < 0,$$

which again contradicts (6.21b). $\qquad\qquad\square$

In the two dimensional case, condition (6.37) of Proposition 6.14 reads

$$\underset{\approx}{\tilde{a}}_{1111} = a_1 - \underset{\approx}{\tilde{a}}_{1122} \geqslant 0, \qquad \underset{\approx}{\tilde{a}}_{2222} = a_2 - \underset{\approx}{\tilde{a}}_{1122} \geqslant 0, \qquad \underset{\approx}{\tilde{a}}_{1122} \geqslant 0,$$
$$\det\left((\underset{\approx}{\tilde{a}}_{kk\ell\ell})^2_{k,\ell=1}\right) = a_1 a_2 - \underset{\approx}{\tilde{a}}_{1122}(a_1 + a_2) = a_1 a_2 - \underset{\approx}{\tilde{a}}_{1122} \geqslant 0,$$

which boils down to the equivalent requirement

$$0 \leqslant \underset{\approx}{\tilde{a}}_{1122} \leqslant a_1 a_2. \tag{6.39}$$

The following two dimensional example presents an admissible and orthotropic closure.

Example 6.15. Due to Definition 6.10 and (6.22), a two dimensional orthotropic closure is uniquely determined by specifying $\underset{\approx}{\tilde{a}}_{1122}$ in terms of the eigenvalue a_1 (and $a_2 = 1 - a_1$). For instance, we can set

$$\underset{\approx}{\tilde{a}}_{1122} := \gamma a_1 a_2 = \gamma \det(\mathbf{A}) = \gamma(a_{11} a_{22} - a_{12}^2). \tag{6.40}$$

This closure is admissible for all $\gamma \in [0,1]$ due to condition (6.39) but not well-defined for $\mathbf{A} = \frac{1}{2}\mathbf{I}$ and $\gamma \neq \frac{1}{2}$ by virtue of Corollary 6.12 or, more specifically, (6.36a).

For $\gamma = \frac{1}{2}$, the closure at hand is called the *natural closure* [VD93; DV99]. The corresponding fourth order orientation tensor can be equivalently expressed by

$$
\begin{aligned}
&\underline{\underline{a}}_{1112} = a_{11}a_{12}, \qquad \underline{\underline{a}}_{1122} = \tfrac{1}{2}(a_{11}a_{22} + a_{12}^2), \qquad \underline{\underline{a}}_{1222} = a_{12}a_{22}, \\
&\underline{\underline{a}}_{1111} = \tfrac{1}{2}(a_{11} + a_{11}^2 - a_{12}^2), \qquad \underline{\underline{a}}_{2222} = \tfrac{1}{2}(a_{22} + a_{22}^2 - a_{12}^2).
\end{aligned}
\tag{6.41}
$$

Therefore, the calculation of eigenvectors (and eigenvalues) becomes redundant although orthotropic closures are defined in terms of spectral decompositions.

To show (6.41), we first notice that

$$
\begin{aligned}
a_1 + a_2 = 1, \qquad \det(\mathbf{A}) &= a_1 a_2 = a_{11}a_{22} - a_{12}^2, \\
\mathbf{q}_k \cdot \mathbf{q}_\ell = q_{1k}q_{1\ell} + q_{2k}q_{2\ell} = \delta_{k\ell}, \qquad &|q_{11}| = |q_{22}|, \qquad |q_{12}| = |q_{21}|.
\end{aligned}
$$

Then, according to Definition 6.10, (6.22), and (6.40), we have

$$
\begin{aligned}
\underline{\underline{a}}_{1122} &= \underline{\tilde{\underline{a}}}_{1111} q_{11}^2 q_{21}^2 + \underline{\tilde{\underline{a}}}_{2222} q_{12}^2 q_{22}^2 \\
&\quad + \underline{\tilde{\underline{a}}}_{1122}(q_{11}^2 q_{22}^2 + q_{12}^2 q_{21}^2 + 4q_{11}q_{12}q_{21}q_{22}) \\
&= \underline{\tilde{\underline{a}}}_{1111} q_{11}^2 q_{21}^2 + \underline{\tilde{\underline{a}}}_{2222} q_{11}^2 q_{21}^2 + \underline{\tilde{\underline{a}}}_{1122}(q_{11}^4 + q_{21}^4 - 4q_{11}^2 q_{21}^2) \\
&= (a_1 + a_2) q_{11}^2 q_{21}^2 + \underline{\tilde{\underline{a}}}_{1122}(q_{11}^4 + q_{21}^4 - 6q_{11}^2 q_{21}^2) \\
&= q_{11}^2 q_{21}^2 + \underline{\tilde{\underline{a}}}_{1122}(q_{11}^2 + q_{21}^2)^2 - 8\underline{\tilde{\underline{a}}}_{1122} q_{11}^2 q_{21}^2 \\
&= \underline{\tilde{\underline{a}}}_{1122} + (1 - 8\underline{\tilde{\underline{a}}}_{1122}) q_{11}^2 q_{21}^2 \\
&= \underline{\tilde{\underline{a}}}_{1122} + ((a_1 + a_2)^2 - 4a_1 a_2) q_{11}^2 q_{21}^2 \\
&= \underline{\tilde{\underline{a}}}_{1122} + (a_1 - a_2)^2 q_{11}^2 q_{21}^2 \\
&= \underline{\tilde{\underline{a}}}_{1122} + (a_1 q_{11}q_{21} + a_2 q_{12}q_{22})^2 \\
&= \underline{\tilde{\underline{a}}}_{1122} + a_{12}^2 = \tfrac{1}{2}(a_{11}a_{22} + a_{12}^2).
\end{aligned}
\tag{6.42}
$$

Similar transformations can be used to show that

$$
\begin{aligned}
\underline{\underline{a}}_{1112} &= (a_1 - \underline{\underline{\tilde{a}}}_{1122})q_{11}^3 q_{21} - (a_2 - \underline{\underline{\tilde{a}}}_{1122})q_{11}q_{21}^3 + \underline{\underline{\tilde{a}}}_{1122}(3q_{11}q_{21}^3 - 3q_{11}^3 q_{21}) \\
&= a_1 q_{11}^3 q_{21} - a_2 q_{11}q_{21}^3 + 4\underline{\underline{\tilde{a}}}_{1122}(q_{11}q_{21}^3 - q_{11}^3 q_{21}) \\
&= (a_1 + a_2)a_1 q_{11}^3 q_{21} - (a_1 + a_2)a_2 q_{11}q_{21}^3 + 2a_1 a_2(q_{11}q_{21}^3 - q_{11}^3 q_{21}) \\
&= a_1^2 q_{11}^3 q_{21} - a_2^2 q_{11}q_{21}^3 + a_1 a_2(q_{11}q_{21}^3 - q_{11}^3 q_{21}) \\
&= (a_1 q_{11}^2 + a_2 q_{21}^2)(a_1 q_{11}q_{21} - a_2 q_{11}q_{21}) = a_{11}a_{12}.
\end{aligned}
$$

This result directly implies the Lipschitz continuity of the natural closure. For instance,

$$
\begin{aligned}
& a_1 a_2 \leqslant \tfrac{1}{4}, \qquad 0 \leqslant a_{11}, a_{22} \leqslant 1, \\
& 0 \leqslant \det(\mathbf{A}) = a_{11}a_{22} - a_{12}^2 \quad \Longrightarrow \quad |a_{12}| \leqslant \sqrt{a_{11}a_{22}} \leqslant \tfrac{1}{2}
\end{aligned}
\tag{6.43}
$$

holds for any second order orientation tensor $\mathbf{A}$ and, hence,

$$
\begin{aligned}
|\underline{\underline{a}}_{1112} - \underline{\underline{\bar{a}}}_{1112}| &= |a_{11}a_{12} - \bar{a}_{11}\bar{a}_{12}| \\
&\leqslant |a_{11}||a_{12} - \bar{a}_{12}| + |a_{11} - \bar{a}_{11}||\bar{a}_{12}| \\
&\leqslant |a_{12} - \bar{a}_{12}| + \tfrac{1}{2}|a_{11} - \bar{a}_{11}|,
\end{aligned}
$$

where $\mathbf{A}, \bar{\mathbf{A}}$ are arbitrary second order orientation tensors while $\underline{\underline{\mathbf{A}}} := \underline{\underline{\mathbf{A}}}(\mathbf{A})$ and $\underline{\underline{\bar{\mathbf{A}}}} := \underline{\underline{\mathbf{A}}}(\bar{\mathbf{A}})$ denote their fourth order counterparts defined by (6.41).

However, this straightforward proof yields just a rough estimate of the Lipschitz constant. A more advanced analysis leads to the improved result $L_{\mathrm{cl}} \leqslant \sqrt{7}$. This bound can be derived as follows: Without loss of generality, let us assume that $\bar{\mathbf{A}}$ is diagonal. Then, according to (6.42) and (6.43), we have

$$
\begin{aligned}
16(\underline{\underline{a}}_{1122} - \underline{\underline{\bar{a}}}_{1122})^2 &= 4(a_{11}a_{22} + a_{12}^2 - \bar{a}_{11}\bar{a}_{22})^2 \\
&\leqslant \left(|a_{11} - \bar{a}_{11}|(a_{22} + \bar{a}_{22}) + |a_{22} - \bar{a}_{22}|(a_{11} + \bar{a}_{11}) + 2a_{12}^2\right)^2 \\
&\leqslant \left(2|a_{11} - \bar{a}_{11}| + 2|a_{22} - \bar{a}_{22}| + |a_{12}|\right)^2 \\
&\leqslant 3\left(4(a_{11} - \bar{a}_{11})^2 + 4(a_{22} - \bar{a}_{22})^2 + a_{12}^2\right) \\
&= 12\|\mathbf{A} - \bar{\mathbf{A}}\|_{\mathrm{F}}^2 - 21a_{12}^2.
\end{aligned}
\tag{6.44}
$$

Furthermore, the Frobenius norm of the orthotropic fourth order orientation tensor $\underline{\underline{\mathbf{A}}}$ can be written as

$$
\begin{aligned}
\|\underline{\underline{\mathbf{A}}}\|_F^2 = \|\tilde{\underline{\underline{\mathbf{A}}}}\|_F^2 &= \tilde{\underline{\underline{a}}}_{1111}^2 + \tilde{\underline{\underline{a}}}_{2222}^2 + 6\tilde{\underline{\underline{a}}}_{1122}^2 \\
&= (a_1 - \tilde{\underline{\underline{a}}}_{1122})^2 + (a_2 - \tilde{\underline{\underline{a}}}_{1122})^2 + 6\tilde{\underline{\underline{a}}}_{1122}^2 \\
&= a_1^2 + a_2^2 - 2\tilde{\underline{\underline{a}}}_{1122}(a_1 + a_2) + 8\tilde{\underline{\underline{a}}}_{1122}^2 \\
&= \|\mathbf{A}\|_F^2 - 2\tilde{\underline{\underline{a}}}_{1122} + 8\tilde{\underline{\underline{a}}}_{1122}^2
\end{aligned}
$$

due to the frame invariance of the Frobenius norm. Additionally, we have

$$
\begin{aligned}
(\underline{\underline{\mathbf{A}}}, \bar{\underline{\underline{\mathbf{A}}}})_F &= \underline{\underline{a}}_{1111}\bar{\underline{\underline{a}}}_{1111} + \underline{\underline{a}}_{2222}\bar{\underline{\underline{a}}}_{2222} + 6\underline{\underline{a}}_{1122}\bar{\underline{\underline{a}}}_{1122} \\
&= \underline{\underline{a}}_{1111}(\bar{a}_{11} - \bar{\underline{\underline{a}}}_{1122}) + \underline{\underline{a}}_{2222}(\bar{a}_{22} - \bar{\underline{\underline{a}}}_{1122}) + 6\underline{\underline{a}}_{1122}\bar{\underline{\underline{a}}}_{1122} \\
&= \underline{\underline{a}}_{1111}\bar{a}_{11} + \underline{\underline{a}}_{2222}\bar{a}_{22} + (6\underline{\underline{a}}_{1122} - \underline{\underline{a}}_{1111} - \underline{\underline{a}}_{2222})\bar{\underline{\underline{a}}}_{1122} \\
&= a_{11}\bar{a}_{11} + a_{22}\bar{a}_{22} - \underline{\underline{a}}_{1122}(\bar{a}_{11} + \bar{a}_{22}) \\
&\quad + (8\underline{\underline{a}}_{1122} - a_{11} - a_{22})\bar{\underline{\underline{a}}}_{1122} \\
&= (\mathbf{A}, \bar{\mathbf{A}})_F - \underline{\underline{a}}_{1122} - \bar{\underline{\underline{a}}}_{1122} + 8\underline{\underline{a}}_{1122}\bar{\underline{\underline{a}}}_{1122}
\end{aligned}
$$

because $\bar{a}_1 + \bar{a}_2 = a_{11} + a_{22} = 1$. Using this result, we find that

$$
\begin{aligned}
\|\underline{\underline{\mathbf{A}}} - \bar{\underline{\underline{\mathbf{A}}}}\|_F^2 &= \|\underline{\underline{\mathbf{A}}}\|_F^2 + \|\bar{\underline{\underline{\mathbf{A}}}}\|_F^2 - 2(\underline{\underline{\mathbf{A}}}, \bar{\underline{\underline{\mathbf{A}}}})_F \\
&= \|\mathbf{A}\|_F^2 + \|\bar{\mathbf{A}}\|_F^2 - 2(\mathbf{A}, \bar{\mathbf{A}})_F - 2\tilde{\underline{\underline{a}}}_{1122} + 8\tilde{\underline{\underline{a}}}_{1122}^2 \\
&\quad - 2\bar{\underline{\underline{a}}}_{1122} + 8\bar{\underline{\underline{a}}}_{1122}^2 + 2\underline{\underline{a}}_{1122} + 2\bar{\underline{\underline{a}}}_{1122} - 16\underline{\underline{a}}_{1122}\bar{\underline{\underline{a}}}_{1122} \\
&= \|\mathbf{A} - \bar{\mathbf{A}}\|_F^2 + 2(\underline{\underline{a}}_{1122} - \tilde{\underline{\underline{a}}}_{1122}) + 8\tilde{\underline{\underline{a}}}_{1122}^2 + 8\bar{\underline{\underline{a}}}_{1122}^2 \\
&\quad - 16\underline{\underline{a}}_{1122}\bar{\underline{\underline{a}}}_{1122} \\
&= \|\mathbf{A} - \bar{\mathbf{A}}\|_F^2 + 2(\underline{\underline{a}}_{1122} - \tilde{\underline{\underline{a}}}_{1122}) + 8(\tilde{\underline{\underline{a}}}_{1122}^2 - \underline{\underline{a}}_{1122}^2) \\
&\quad + 8(\underline{\underline{a}}_{1122} - \bar{\underline{\underline{a}}}_{1122})^2 \\
&= \|\mathbf{A} - \bar{\mathbf{A}}\|_F^2 + 2(\underline{\underline{a}}_{1122} - \tilde{\underline{\underline{a}}}_{1122}) \\
&\quad + 8(\tilde{\underline{\underline{a}}}_{1122} - \underline{\underline{a}}_{1122})(\tilde{\underline{\underline{a}}}_{1122} + \underline{\underline{a}}_{1122}) + 8(\underline{\underline{a}}_{1122} - \bar{\underline{\underline{a}}}_{1122})^2 \\
&= \|\mathbf{A} - \bar{\mathbf{A}}\|_F^2 + 2a_{12}^2\big(1 - 4(\tilde{\underline{\underline{a}}}_{1122} + \underline{\underline{a}}_{1122})\big) + 8(\underline{\underline{a}}_{1122} - \bar{\underline{\underline{a}}}_{1122})^2 \\
&\leqslant \|\mathbf{A} - \bar{\mathbf{A}}\|_F^2 + 2a_{12}^2 + 8(\underline{\underline{a}}_{1122} - \bar{\underline{\underline{a}}}_{1122})^2 \leqslant 7\|\mathbf{A} - \bar{\mathbf{A}}\|_F^2
\end{aligned}
$$

by virtue of (6.42), (6.44), and the fact that $\tilde{\underline{\underline{a}}}_{1122} + \underline{\underline{a}}_{1122} \geqslant 0$.

6.3.3 Temporal discretization

Theorem 6.7 shows that the solution to the continuous Folgar-Tucker equation (6.20b) is physics-compatible for any nonnegative value of the rotary diffusion coefficient if an admissible closure is employed. It is essential to ensure that this property carries over to the temporally discretized evolution problem, possibly under a time step restriction.

To study the preservation of positive semidefiniteness in the fully discrete case, we first apply the *forward Euler method* to (6.20b) and obtain

$$
\begin{aligned}
\mathbf{A}^{n+1} = \mathbf{A}^n &+ \Delta t(\mathbf{W}^n \mathbf{A}^n - \mathbf{A}^n \mathbf{W}^n) \\
&+ \Delta t \lambda (\mathbf{D}^n \mathbf{A}^n + \mathbf{A}^n \mathbf{D}^n - 2\underline{\underline{\mathbf{A}}}(\mathbf{A}^n) : \mathbf{D}^n) + 2\Delta t D_r (\mathbf{I} - d\mathbf{A}^n), \quad (6.45)
\end{aligned}
$$

where Δt is the time increment while $\mathbf{D}^n$ and $\mathbf{W}^n$ are the strain rate and spin tensors at the time level t^n.

This discretization is easy to analyze and implement due to the explicit treatment of inherent nonlinearities. In particular, there is no need to use a Lipschitz continuous closure and computable bounds for admissible time steps can be inferred from the coefficients of the discrete problem by exploiting the techniques of Section 5.4.1. However, the following example illustrates that unphysical orientation states may arise for any $\Delta t > 0$ if the rotary diffusion coefficient vanishes.

Example 6.16. Let us consider the following setup:

$$
\mathbf{A}^n = \begin{pmatrix} 1 & 0 \\ 0 & 0 \end{pmatrix}, \qquad \nabla \mathbf{v} = \begin{pmatrix} 0 & 1 \\ -1 & 0 \end{pmatrix}
$$

$$
\implies \quad \mathbf{D} = \begin{pmatrix} 0 & 0 \\ 0 & 0 \end{pmatrix}, \quad \mathbf{W} = \begin{pmatrix} 0 & 1 \\ -1 & 0 \end{pmatrix}.
$$

Then $\mathbf{A}^{n+1}$ corresponding to (6.45) with $\lambda = 1$, $D_r = 0$ is given by

$$
\begin{aligned}
\mathbf{A}^{n+1} &= \mathbf{A}^n + \Delta t(\mathbf{W}\mathbf{A}^n - \mathbf{A}^n\mathbf{W}) + \Delta t(\mathbf{D}\mathbf{A}^n + \mathbf{A}^n\mathbf{D} - 2\underline{\underline{\mathbf{A}}}(\mathbf{A}^n) : \mathbf{D}) \\
&= \begin{pmatrix} 1 & 0 \\ 0 & 0 \end{pmatrix} + \Delta t \begin{pmatrix} 0 & -1 \\ -1 & 0 \end{pmatrix} = \begin{pmatrix} 1 & -\Delta t \\ -\Delta t & 0 \end{pmatrix},
\end{aligned}
$$

and its minimal eigenvalue $a_1 = \frac{1}{2}(1 - \sqrt{1 + 4\Delta t^2})$ is strictly negative for all $\Delta t > 0$. Therefore, the forward Euler method cannot preserve physics-compatible solutions for any $\Delta t > 0$ if the rotary diffusion coefficient vanishes.

However, in the presence of rotary diffusivity, the time increment can be chosen so that $\mathbf{A}^{n+1}$ is positive semidefinite for any positive semidefinite input $\mathbf{A}^n$.

Theorem 6.17. *Let $\mathbf{A}^n \in \mathbb{A}_d$ be a second order orientation tensor and $\underline{\underline{\mathbf{A}}}(\mathbf{A}^n)$ an admissible closure. Then the tensor $\mathbf{A}^{n+1}$ defined by (6.45) with $D_r > 0$ possesses the properties established in Theorem 6.3, i.e., $\mathbf{A}^{n+1} \in \mathbb{A}_d$, provided that*

$$\left(\tfrac{1}{2} D_r^{-1} \|\mathbf{W}^n + \lambda \mathbf{D}^n\|_F^2 + 2|\lambda|\, \|\mathbf{D}^n\|_F + 2dD_r\right)\Delta t \leqslant 1. \tag{6.46}$$

Proof. For brevity, we drop the superscript n in this proof.

Obviously, the solution $\mathbf{A}^{n+1}$ of (6.45) is symmetric with unit trace, which can be shown as in the proof of Theorem 6.7. We prove the validity of $\mathbf{A}^{n+1} \succcurlyeq \mathbf{0}$ by showing that

$$\mathbf{q}^\top \mathbf{A}^{n+1}\mathbf{q} \geqslant 0 \qquad \forall \mathbf{q} \in \mathbb{R}^d \text{ s.t. } \|\mathbf{q}\|_2 = 1. \tag{6.47}$$

Due to the admissibility of the closure $\underline{\underline{\mathbf{A}}} := \underline{\underline{\mathbf{A}}}(\mathbf{A})$, the tensor $\underline{\underline{\mathbf{A}}} : (\mathbf{q} \otimes \mathbf{q})$ is positive semidefinite by (6.24) and satisfies

$$\operatorname{tr}\big(\underline{\underline{\mathbf{A}}} : (\mathbf{q} \otimes \mathbf{q})\big) = (\mathbf{I} : \underline{\underline{\mathbf{A}}}) : (\mathbf{q} \otimes \mathbf{q}) = \mathbf{q}^\top \mathbf{A}\mathbf{q} \leqslant 1.$$

Thus, all eigenvalues of $\underline{\underline{\mathbf{A}}} : (\mathbf{q} \otimes \mathbf{q})$ are nonnegative and bounded above by 1. Hence, we have

$$\begin{aligned}
\mathbf{q}^\top (\underline{\underline{\mathbf{A}}} : \mathbf{D})\mathbf{q} &= \mathbf{D} : \big(\underline{\underline{\mathbf{A}}} : (\mathbf{q} \otimes \mathbf{q})\big) \\
&\leqslant \|\mathbf{D}\|_F \big\|\underline{\underline{\mathbf{A}}} : (\mathbf{q} \otimes \mathbf{q})\big\|_F = \|\mathbf{D}\|_F \sum_k \lambda_k \big(\underline{\underline{\mathbf{A}}} : (\mathbf{q} \otimes \mathbf{q})\big)^2 \\
&\leqslant \|\mathbf{D}\|_F \sum_k \lambda_k \big(\underline{\underline{\mathbf{A}}} : (\mathbf{q} \otimes \mathbf{q})\big) \\
&= \|\mathbf{D}\|_F \operatorname{tr}\big(\underline{\underline{\mathbf{A}}} : (\mathbf{q} \otimes \mathbf{q})\big) = \|\mathbf{D}\|_F \mathbf{q}^\top \mathbf{A}\mathbf{q}.
\end{aligned}$$

To show (6.47), let us first treat the special case $\mathbf{W} + \lambda \mathbf{D} = \mathbf{0}$. Then $\mathbf{A}^{n+1}$ satisfies

$$\begin{aligned}
\mathbf{q}^\top \mathbf{A}^{n+1}\mathbf{q} &= \mathbf{q}^\top \mathbf{A}\mathbf{q} - 2\Delta t \lambda \mathbf{q}^\top (\underline{\underline{\mathbf{A}}} : \mathbf{D})\mathbf{q} + 2\Delta t D_r (1 - d\mathbf{q}^\top \mathbf{A}\mathbf{q}) \\
&\geqslant \mathbf{q}^\top \mathbf{A}\mathbf{q} - 2\Delta t |\lambda|\, \|\mathbf{D}\|_F \mathbf{q}^\top \mathbf{A}\mathbf{q} + 2\Delta t D_r (1 - d\mathbf{q}^\top \mathbf{A}\mathbf{q}) \\
&= \big(1 - 2\Delta t |\lambda|\, \|\mathbf{D}\|_F - 2d\Delta t D_r\big)\mathbf{q}^\top \mathbf{A}\mathbf{q} + 2\Delta t D_r.
\end{aligned}$$

Both terms on the right hand side are nonnegative if $2\Delta t|\lambda|\,\|\mathbf{D}\|_{\mathrm{F}}+2d\Delta t D_r \leqslant 1$ and, hence, condition (6.46) guarantees the preservation of positive semidefiniteness.

In the case $\mathbf{W}+\lambda\mathbf{D}\neq\mathbf{0}$, we write the normalized vector $\mathbf{q}\in\mathbb{R}^d$ as

$$\mathbf{q}=\sum_k \alpha_k\mathbf{q}_k,\quad\text{where }\alpha_1,\ldots,\alpha_d\in\mathbb{R}\text{ s.t. }\sum_k \alpha_k^2=1$$

and $\mathbf{q}_1,\ldots,\mathbf{q}_d$ are the normalized eigenvectors of $\mathbf{A}$. By Young's inequality, we find that

$$\mathbf{q}^{\top}\mathbf{A}\mathbf{V}^{\top}\mathbf{q}=\mathbf{q}^{\top}\mathbf{V}\mathbf{A}\mathbf{q}=\sum_{k,\ell}\alpha_k\alpha_{\ell}\mathbf{q}_k^{\top}\mathbf{V}\mathbf{A}\mathbf{q}_{\ell}=\sum_{k,\ell}\alpha_k\alpha_{\ell}a_{\ell}\mathbf{q}_k^{\top}\mathbf{V}\mathbf{q}_{\ell}$$

$$\leqslant\frac{1}{2\gamma}\sum_{k,\ell}\alpha_k^2\alpha_{\ell}^2 a_{\ell}+\frac{\gamma}{2}\sum_{k,\ell}a_{\ell}(\mathbf{q}_k^{\top}\mathbf{V}\mathbf{q}_{\ell})^2$$

$$\leqslant\frac{1}{2\gamma}\sum_{\ell}\alpha_{\ell}^2 a_{\ell}+\frac{\gamma}{2}\sum_{k,\ell}(\mathbf{q}_k^{\top}\mathbf{V}\mathbf{q}_{\ell})^2=\frac{1}{2\gamma}\mathbf{q}^{\top}\mathbf{A}\mathbf{q}+\frac{\gamma}{2}\|\mathbf{V}\|_{\mathrm{F}}^2$$

$$\forall\mathbf{V}\in\mathbb{R}^{d\times d}$$

for every $\gamma>0$. Therefore, multiplication of (6.45) from left and right by $\mathbf{q}^{\top}$ and $\mathbf{q}$ results in

$$\mathbf{q}^{\top}\mathbf{A}^{n+1}\mathbf{q}=\mathbf{q}^{\top}\mathbf{A}\mathbf{q}+\Delta t(\mathbf{q}^{\top}(\mathbf{W}+\lambda\mathbf{D})\mathbf{A}\mathbf{q}+\mathbf{q}^{\top}\mathbf{A}(\mathbf{W}+\lambda\mathbf{D})^{\top}\mathbf{q})$$

$$-2\Delta t\lambda\mathbf{q}^{\top}(\underline{\underline{\mathbf{A}}}:\mathbf{D})\mathbf{q}+2\Delta t D_r(1-d\mathbf{q}^{\top}\mathbf{A}\mathbf{q})$$

$$\geqslant\mathbf{q}^{\top}\mathbf{A}\mathbf{q}-\Delta t\gamma^{-1}\mathbf{q}^{\top}\mathbf{A}\mathbf{q}-\Delta t\gamma\|\mathbf{W}+\lambda\mathbf{D}\|_{\mathrm{F}}^2$$

$$-2\Delta t|\lambda|\,\|\mathbf{D}\|_{\mathrm{F}}\mathbf{q}^{\top}\mathbf{A}\mathbf{q}+2\Delta t D_r(1-d\mathbf{q}^{\top}\mathbf{A}\mathbf{q})$$

$$=\big(1-\tfrac{1}{2}\Delta t D_r^{-1}\|\mathbf{W}+\lambda\mathbf{D}\|_{\mathrm{F}}^2-2\Delta t|\lambda|\,\|\mathbf{D}\|_{\mathrm{F}}-2d\Delta t D_r\big)\mathbf{q}^{\top}\mathbf{A}\mathbf{q},$$

where $\gamma=2D_r\|\mathbf{W}+\lambda\mathbf{D}\|_{\mathrm{F}}^{-2}$ is used. Thus, the solution $\mathbf{A}^{n+1}$ proves positive semidefinite by virtue of (6.46). $\qquad\square$

According to condition (6.46), the time increment must behave like D_r for (6.45) to produce physics-compatible solutions. Especially in the case of small rotary diffusion coefficients, this a priori criterion for the choice of Δt may turn out very pessimistic. To avoid the use of impractically small Δt without generating unphysical orientation states, an a posteriori time step control may be employed. In the numerical studies below, we use a simple adaptive substepping approach to guarantee the positive semidefiniteness of $\mathbf{A}^{n+1}$.

Algorithm 6.18. *Given* $\mathbf{A}^n \in \mathbb{A}_d$ *and* $D_r > 0$, *set* $\delta t = \Delta t$ *and* $\tilde{t} = t^n$. *Using* $\mathbf{A}^{n+1} = \mathbf{A}^n$ *as an initial guess, perform the following steps until reaching* $\tilde{t} = t^{n+1}$:

1. *Calculate* $\bar{\mathbf{A}}^{n+1}$ *using* (6.45) *with the time increment* δt *and the old solution* $\mathbf{A}^{n+1}$.

2. *If* $\bar{\mathbf{A}}^{n+1} \succcurlyeq 0$, *set* $\mathbf{A}^{n+1} = \bar{\mathbf{A}}^{n+1}$ *and* $\tilde{t} = \tilde{t} + \delta t$. *Otherwise, set* $\delta t = \frac{1}{2}\delta t$.

This algorithm terminates after a finite number of cycles due to Theorem 6.17. However, it requires that the parameter D_r be positive. If this condition does not hold, the Folgar-Tucker equation (6.20b) may be discretized using the *backward Euler method*

$$
\begin{aligned}
\mathbf{A}^{n+1} &- \Delta t (\mathbf{W}^{n+1}\mathbf{A}^{n+1} - \mathbf{A}^{n+1}\mathbf{W}^{n+1}) \\
&- \Delta t \lambda (\mathbf{D}^{n+1}\mathbf{A}^{n+1} + \mathbf{A}^{n+1}\mathbf{D}^{n+1} - 2\underline{\underline{\mathbf{A}}}(\mathbf{A}^{n+1}) : \mathbf{D}^{n+1}) \\
&- 2\Delta t D_r (\mathbf{I} - d\mathbf{A}^{n+1}) = \mathbf{A}^n, \quad (6.48)
\end{aligned}
$$

where $\mathbf{D}^{n+1}$ and $\mathbf{W}^{n+1}$ are the strain rate and spin tensors evaluated at time level t^{n+1}.

In the following theorem, we formulate sufficient requirements on the closure and the time increment for problem (6.48) to be well-posed.

Theorem 6.19. *Let* $\underline{\underline{\mathbf{A}}}(\cdot)$ *be an admissible closure satisfying* (6.25). *Suppose that* $\mathbf{A}^n \in \mathbb{A}_d$ *and*

$$
2\big(\|\mathbf{W}^{n+1} + \lambda\mathbf{D}^{n+1}\|_2 + |\lambda|\,\|\mathbf{D}^{n+1}\|_{\mathrm{F}} L - dD_r\big)\Delta t < 1 \qquad (6.49)
$$

holds, where $L := 1 + (1 + \sqrt{d})(2 + \sqrt{d} + L_{\mathrm{cl}})$. *Then there exists a unique second order orientation tensor* $\mathbf{A}^{n+1} \in \mathbb{A}_d$ *such that* (6.48) *is satisfied.*

Proof. This proof is composed of 4 parts.

1. To show that the statement of the theorem holds for a given closure, we first extend $\underline{\underline{\mathbf{A}}}(\cdot)$ to symmetric tensors that possess the unit trace property but are possibly not positive semidefinite. For this purpose, we define the orientation tensor

$$
\mathbf{V}' := \frac{\max(\mathbf{0}, \mathbf{V})}{\mathrm{tr}\big(\max(\mathbf{0}, \mathbf{V})\big)} \in \mathbb{A}_d \qquad \forall \mathbf{V} \in \mathbb{S}_{d,1} := \big\{\mathbf{V} \in \mathbb{S}_d \,\big|\, \mathrm{tr}(\mathbf{V}) = 1\big\},
$$
$$
(6.50)
$$

which is well-defined because $\mathrm{tr}(\mathbf{V}) = 1$ and, hence,

$$\mathrm{tr}\big(\max(\mathbf{0}, \mathbf{V})\big) \geqslant 1.$$

Then an auxiliary closure $\underline{\underline{\mathbf{A}}}'(\cdot)$ for tensors that are possibly not positive semidefinite can be written as

$$\underline{\underline{\mathbf{A}}}'(\mathbf{V}) := \underline{\underline{\mathbf{A}}}(\mathbf{V}') + \mathbf{V}' \otimes (\mathbf{V} - \mathbf{V}') \qquad \forall \mathbf{V} \in \mathbb{S}_{d,1}. \tag{6.51}$$

2. This closure is Lipschitz continuous due to the Lipschitz continuity of $\underline{\underline{\mathbf{A}}}(\cdot)$. To prove this statement, we note that

$$\frac{\big\|\max(\mathbf{0}, \mathbf{V})\big\|_\mathrm{F}^2}{\mathrm{tr}(\max(\mathbf{0}, \mathbf{V}))^2} = \|\mathbf{V}'\|_\mathrm{F}^2 = \sum_k (v'_k)^2 \leqslant \sum_k v'_k = 1 \quad \forall \mathbf{V} \in \mathbb{S}_{d,1} \tag{6.52}$$

because $v'_1, \ldots, v'_d \in [0,1]$. Furthermore, the Cauchy-Schwarz inequality yields

$$\big(\mathrm{tr}(\mathbf{V}) - \mathrm{tr}(\bar{\mathbf{V}})\big)^2 \leqslant \Big(\sum_k 1 \cdot \big|\lambda_k(\mathbf{V} - \bar{\mathbf{V}})\big|\Big)^2$$
$$\leqslant d \sum_k \lambda_k(\mathbf{V} - \bar{\mathbf{V}})^2 = d\|\mathbf{V} - \bar{\mathbf{V}}\|_\mathrm{F}^2 \qquad \forall \mathbf{V}, \bar{\mathbf{V}} \in \mathbb{S}_d, \tag{6.53}$$

where $\lambda_k(\cdot)$ denotes the k-th eigenvalue of the tensor quantity in the brackets. For any $\mathbf{V}, \bar{\mathbf{V}} \in \mathbb{S}_{d,1}$, we find that

$$\mathrm{tr}\big(\max(\mathbf{0}, \mathbf{V})\big)\|\mathbf{V}' - \bar{\mathbf{V}}'\|_\mathrm{F}$$
$$= \Big\|\max(\mathbf{0}, \mathbf{V}) - \frac{\mathrm{tr}\big(\max(\mathbf{0}, \mathbf{V})\big)}{\mathrm{tr}\big(\max(\mathbf{0}, \bar{\mathbf{V}})\big)} \max(\mathbf{0}, \bar{\mathbf{V}})\Big\|_\mathrm{F}$$
$$\leqslant \big\|\max(\mathbf{0}, \mathbf{V}) - \max(\mathbf{0}, \bar{\mathbf{V}})\big\|_\mathrm{F}$$
$$+ \big\|\max(\mathbf{0}, \bar{\mathbf{V}})\big\|_\mathrm{F} \frac{\big|\mathrm{tr}(\max(\mathbf{0}, \mathbf{V})) - \mathrm{tr}(\max(\mathbf{0}, \bar{\mathbf{V}}))\big|}{\mathrm{tr}\big(\max(\mathbf{0}, \bar{\mathbf{V}})\big)}$$
$$\leqslant \big\|\max(\mathbf{0}, \mathbf{V}) - \max(\mathbf{0}, \bar{\mathbf{V}})\big\|_\mathrm{F} + \big|\mathrm{tr}(\max(\mathbf{0}, \mathbf{V})) - \mathrm{tr}(\max(\mathbf{0}, \bar{\mathbf{V}}))\big|$$
$$\leqslant \big(1 + \sqrt{d}\big)\big\|\max(\mathbf{0}, \mathbf{V}) - \max(\mathbf{0}, \bar{\mathbf{V}})\big\|_\mathrm{F}$$
$$\leqslant \big(1 + \sqrt{d}\big)\|\mathbf{V} - \bar{\mathbf{V}}\|_\mathrm{F}$$

by virtue of (5.13c), (6.52), and (6.53). It follows that

$$
\begin{aligned}
\|\underline{\underline{\mathbf{A}}}'&(\mathbf{V}) - \underline{\underline{\mathbf{A}}}'(\bar{\mathbf{V}})\|_\mathrm{F} \\
&\leqslant \|\underline{\underline{\mathbf{A}}}(\mathbf{V}') - \underline{\underline{\mathbf{A}}}(\bar{\mathbf{V}}')\|_\mathrm{F} + \|\mathbf{V}' \otimes (\mathbf{V} - \mathbf{V}') - \bar{\mathbf{V}}' \otimes (\bar{\mathbf{V}} - \bar{\mathbf{V}}')\|_\mathrm{F} \\
&\leqslant L_\mathrm{cl}\|\mathbf{V}' - \bar{\mathbf{V}}'\|_\mathrm{F} + \|(\mathbf{V}' - \bar{\mathbf{V}}') \otimes \mathbf{V}\|_\mathrm{F} + \|\bar{\mathbf{V}}' \otimes (\mathbf{V} - \bar{\mathbf{V}})\|_\mathrm{F} \\
&\quad + \|(\mathbf{V}' - \bar{\mathbf{V}}') \otimes \mathbf{V}'\|_\mathrm{F} + \|\bar{\mathbf{V}}' \otimes (\mathbf{V}' - \bar{\mathbf{V}}')\|_\mathrm{F} \\
&= L_\mathrm{cl}\|\mathbf{V}' - \bar{\mathbf{V}}'\|_\mathrm{F} + \|\mathbf{V}\|_\mathrm{F}\|\mathbf{V}' - \bar{\mathbf{V}}'\|_\mathrm{F} + \|\bar{\mathbf{V}}'\|_\mathrm{F}\|\mathbf{V} - \bar{\mathbf{V}}\|_\mathrm{F} \\
&\quad + \|\mathbf{V}'\|_\mathrm{F}\|\mathbf{V}' - \bar{\mathbf{V}}'\|_\mathrm{F} + \|\bar{\mathbf{V}}'\|_\mathrm{F}\|\mathbf{V}' - \bar{\mathbf{V}}'\|_\mathrm{F} \\
&\leqslant L_\mathrm{cl}\|\mathbf{V}' - \bar{\mathbf{V}}'\|_\mathrm{F} + \sqrt{d}\,\mathrm{tr}\big(\max(\mathbf{0}, \mathbf{V})\big)\|\mathbf{V}' - \bar{\mathbf{V}}'\|_\mathrm{F} \\
&\quad + \|\mathbf{V} - \bar{\mathbf{V}}\|_\mathrm{F} + \|\mathbf{V}' - \bar{\mathbf{V}}'\|_\mathrm{F} + \|\mathbf{V}' - \bar{\mathbf{V}}'\|_\mathrm{F} \\
&\leqslant \big(2 + \sqrt{d} + L_\mathrm{cl}\big)\,\mathrm{tr}\big(\max(\mathbf{0}, \mathbf{V})\big)\|\mathbf{V}' - \bar{\mathbf{V}}'\|_\mathrm{F} + \|\mathbf{V} - \bar{\mathbf{V}}\|_\mathrm{F} \\
&\leqslant \big(1 + (1 + \sqrt{d})(2 + \sqrt{d} + L_\mathrm{cl})\big)\|\mathbf{V} - \bar{\mathbf{V}}\|_\mathrm{F}
\end{aligned}
$$

in view of (6.25), (6.52), $\mathrm{tr}\big(\max(0, \mathbf{V})\big) \geqslant 1$, and the fact that

$$
\begin{aligned}
\|\mathbf{V}\|_\mathrm{F} \leqslant \sqrt{d}\|\mathbf{V}\|_2 &\leqslant \sqrt{d}\max\Big(\sum_k \max(0, v_k), \sum_k \max(0, -v_k)\Big) \\
&= \sqrt{d}\sum_k \max(0, v_k) = \sqrt{d}\,\mathrm{tr}\big(\max(\mathbf{0}, \mathbf{V})\big)
\end{aligned}
$$

because

$$
\begin{aligned}
1 = \mathrm{tr}(\mathbf{V}) &= \sum_k \max(0, v_k) + \sum_k \min(0, v_k) \\
&= \sum_k \max(0, v_k) - \sum_k \max(0, -v_k).
\end{aligned}
$$

3. In this step, we focus on the existence and uniqueness of a solution to (6.48) employing the extended closure $\underline{\underline{\mathbf{A}}}'(\cdot)$ defined by (6.51). To show the existence of a solution under the time step restriction (6.49), we consider the mapping $\mathscr{F} : \mathbb{S}_{d,1} \to \mathbb{S}_{d,1}$ defined by

$$
\begin{aligned}
\mathscr{F}(\mathbf{V}) = \omega\big(\mathbf{A}^n &+ \Delta t(\mathbf{W}^{n+1}\mathbf{V} - \mathbf{V}\mathbf{W}^{n+1}) \\
&+ \Delta t\lambda\big(\mathbf{D}^{n+1}\mathbf{V} + \mathbf{V}\mathbf{D}^{n+1} - 2\underline{\underline{\mathbf{A}}}'(\mathbf{V}) : \mathbf{D}^{n+1}\big) \qquad (6.54) \\
&+ 2\Delta t D_r(\mathbf{I} - d\mathbf{V})\big) + (1 - \omega)\mathbf{V} \qquad \forall \mathbf{V} \in \mathbb{S}_{d,1},
\end{aligned}
$$

where $\omega \in (0,1]$ is a relaxation parameter to be defined later. The mapping $\mathscr{F}$ is well-defined by definition of $\mathbb{S}_{d,1}$. For example, the crucial auxiliary result

$$\mathrm{tr}\big(\underline{\mathbf{A}}'(\mathbf{V}) : \mathbf{D}^{n+1}\big) = \mathrm{tr}\big(\underline{\mathbf{A}}(\mathbf{V}') : \mathbf{D}^{n+1}\big) + \mathrm{tr}(\mathbf{V}')\big((\mathbf{V} - \mathbf{V}') : \mathbf{D}^{n+1}\big)$$
$$= \mathbf{V}' : \mathbf{D}^{n+1} + (\mathbf{V} - \mathbf{V}') : \mathbf{D}^{n+1} = \mathrm{tr}(\mathbf{V}\mathbf{D}^{n+1})$$

can be obtained by using the unit trace property of $\mathbf{V}'$, the definition of the extended closure $\underline{\mathbf{A}}'(\cdot)$, and property (6.21c) of $\underline{\mathbf{A}}(\cdot)$.

For $\omega \in \big(0, (1 + 2\Delta t d D_r)^{-1}\big]$ and $\mathbf{V}, \bar{\mathbf{V}} \in \mathbb{S}_{d,1}$, we have

$$\big\|\mathscr{F}(\mathbf{V}) - \mathscr{F}(\bar{\mathbf{V}})\big\|_{\mathrm{F}}$$
$$\leqslant |1 - \omega - 2\Delta t d D_r \omega| \|\mathbf{V} - \bar{\mathbf{V}}\|_{\mathrm{F}}$$
$$+ 2\Delta t \omega \|\mathbf{W}^{n+1} + \lambda \mathbf{D}^{n+1}\|_2 \|\mathbf{V} - \bar{\mathbf{V}}\|_{\mathrm{F}}$$
$$+ 2\Delta t |\lambda| \omega| \|\mathbf{D}^{n+1}\|_{\mathrm{F}} \big\|\underline{\mathbf{A}}'(\mathbf{V}) - \underline{\mathbf{A}}'(\bar{\mathbf{V}})\big\|_{\mathrm{F}}$$
$$\leqslant (1 - \omega - 2\Delta t d D_r \omega) \|\mathbf{V} - \bar{\mathbf{V}}\|_{\mathrm{F}}$$
$$+ 2\Delta t \omega \big(\|\mathbf{W}^{n+1} + \lambda \mathbf{D}^{n+1}\|_2 + |\lambda| \|\mathbf{D}^{n+1}\|_{\mathrm{F}} L\big) \|\mathbf{V} - \bar{\mathbf{V}}\|_{\mathrm{F}}$$
$$= \Big(1 - \omega - 2\Delta t \omega \big(d D_r - \|\mathbf{W}^{n+1} + \lambda \mathbf{D}^{n+1}\|_2$$
$$- |\lambda| \|\mathbf{D}^{n+1}\|_{\mathrm{F}} L\big)\Big) \|\mathbf{V} - \bar{\mathbf{V}}\|_{\mathrm{F}}.$$

If (6.49) holds, the mapping $\mathscr{F}$ is a contraction and the *damped fixed point iteration* $\mathbf{V}^{s+1} = \mathscr{F}(\mathbf{V}^s)$, $s = 0, 1, \ldots$, converges to the unique solution of (6.48) for any initial condition $\mathbf{V}^0$ by the Banach Fixed Point Theorem.

4. To show that the obtained solution $\mathbf{V} = \lim_{s\to\infty} \mathbf{V}^s$ is positive semidefinite, let us suppose that $v_1 < 0$. Then we have

$$v_1' = 0, \qquad \underline{\mathbf{A}}(\mathbf{V}') \otimes (\mathbf{q}_1 \otimes \mathbf{q}_1) = \mathbf{0}$$

by definition of $\mathbf{V}'$ and Lemma 6.6, where $\mathbf{V} = \mathbf{Q}\tilde{\mathbf{V}}\mathbf{Q}^\top$. Therefore,

$$\mathbf{q}_1^\top \big(\underline{\mathbf{A}}'(\mathbf{V}) : \mathbf{D}^{n+1}\big) \mathbf{q}_1$$
$$= \mathbf{q}_1^\top \big(\underline{\mathbf{A}}(\mathbf{V}') : \mathbf{D}^{n+1}\big) \mathbf{q}_1 + \mathbf{q}_1^\top \mathbf{V}' \mathbf{q}_1 \big((\mathbf{V} - \mathbf{V}') : \mathbf{D}^{n+1}\big)$$
$$= \big(\underline{\mathbf{A}}(\mathbf{V}') \otimes (\mathbf{q}_1 \otimes \mathbf{q}_1)\big) : \mathbf{D}^{n+1} + v_1'\big((\mathbf{V} - \mathbf{V}') : \mathbf{D}^{n+1}\big) = 0$$

holds because condition (6.21a) is satisfied for $\underline{\underline{\mathbf{A}}}(\cdot)$. Multiplying (6.48) from left and right by $\mathbf{q}_1^\top$ and $\mathbf{q}_1$, we find that

$$\mathbf{q}_1^\top \mathbf{A}^n \mathbf{q}_1 = v_1 - 2\Delta t \mathbf{q}_1^\top (\mathbf{W}^{n+1} + \lambda \mathbf{D}^{n+1}) \mathbf{q}_1 v_1$$
$$+ 2\Delta t \lambda \mathbf{q}_1^\top (\underline{\underline{\mathbf{A}}}'(\mathbf{V}) : \mathbf{D}^{n+1}) \mathbf{q}_1 - 2\Delta t D_r (1 - dv_1)$$
$$\leqslant v_1 + 2\Delta t \|\mathbf{W}^{n+1} + \lambda \mathbf{D}^{n+1}\|_2 |v_1| - 2\Delta t D_r (1 - dv_1)$$
$$= (1 - 2\Delta t \|\mathbf{W}^{n+1} + \lambda \mathbf{D}^{n+1}\|_2 + 2\Delta t D_r d) v_1 - 2\Delta t D_r$$

and $v_1 \geqslant 0$ if $\mathbf{A}^n \succcurlyeq \mathbf{0}$ because

$$2(\|\mathbf{W}^{n+1} + \lambda \mathbf{D}^{n+1}\|_2 - dD_r)\Delta t < 1$$

holds by condition (6.49). This contradicts the assumption of a negative minimal eigenvalue $v_1 < 0$. Hence, $\mathbf{V} \in \mathbb{A}_d$ and satisfies (6.48). $\qquad\square$

According to Theorem 6.19, the existence of a unique orientation tensor $\mathbf{A}^{n+1}$ solving (6.48) is guaranteed for the whole range of nonnegative rotary diffusivity coefficients if the time step condition (6.49) is met. In particular, the special case $D_r = 0$ is included and does not imply $\Delta t = 0$ as in Theorem 6.17. However, the practical utility of this result is restricted by the need to use a closure satisfying (6.25) where L_{cl} is the Lipschitz constant that occurs in (6.49). Even if the closure under consideration is Lipschitz continuous, the calculation of (a sharp bound for) L_{cl} can be very challenging.

The following example illustrates that a time step restriction like (6.49) is, indeed, a necessary condition for the well-posedness of the Folgar-Tucker equation (6.20b) discretized in time by the backward Euler method.

Example 6.20. In this two dimensional example, we show that problem (6.48) equipped with the natural closure does not provide uniqueness for every $\Delta t > 0$. For this purpose, let us consider the setup

$$\mathbf{A}^n = \begin{pmatrix} 0 & 0 \\ 0 & 1 \end{pmatrix}, \qquad \nabla \mathbf{v} = \begin{pmatrix} 1 & 0 \\ 0 & -1 \end{pmatrix}$$

$$\implies \quad \mathbf{D} = \begin{pmatrix} 1 & 0 \\ 0 & -1 \end{pmatrix}, \quad \mathbf{W} = \begin{pmatrix} 0 & 0 \\ 0 & 0 \end{pmatrix}$$

and assume that the solution $\mathbf{A}^{n+1} = \mathrm{diag}(a, 1 - a)$ is diagonal with $a \in [0, 1]$. Then the fourth order orientation tensor $\underline{\underline{\mathbf{A}}}^{n+1}$ defined by the natural closure (6.41) can be written as

$$\underline{\underline{a}}_{1122}^{n+1} = \tfrac{1}{2} a(1 - a), \quad \underline{\underline{a}}_{1111}^{n+1} = \tfrac{1}{2} a(1 + a), \quad \underline{\underline{a}}_{2222}^{n+1} = \tfrac{1}{2}(1 - a)(2 - a).$$

All other entries vanish while satisfying condition (6.21a). For $D_r = 0$, we have

$$\mathbf{A}^{n+1} - \Delta t(\mathbf{W}^{n+1}\mathbf{A}^{n+1} - \mathbf{A}^{n+1}\mathbf{W}^{n+1})$$
$$- \Delta t\lambda(\mathbf{D}^{n+1}\mathbf{A}^{n+1} + \mathbf{A}^{n+1}\mathbf{D}^{n+1} - 2\underline{\underline{\mathbf{A}}}^{n+1} : \mathbf{D}^{n+1})$$
$$- 2\Delta t D_r(\mathbf{I} - d\mathbf{A}^{n+1})$$
$$= \begin{pmatrix} a & 0 \\ 0 & 1-a \end{pmatrix} - 2\Delta t\lambda \begin{pmatrix} a & 0 \\ 0 & a-1 \end{pmatrix}$$
$$+ \Delta t\lambda \begin{pmatrix} a(1+a) - a(1-a) & 0 \\ 0 & a(1-a) - (1-a)(2-a) \end{pmatrix}$$
$$= \begin{pmatrix} a - 2\Delta t\lambda a(1-a) & 0 \\ 0 & 1-a+2\Delta t\lambda a(1-a) \end{pmatrix} = \begin{pmatrix} 0 & 0 \\ 0 & 1 \end{pmatrix},$$

which holds if and only if the unknown value a satisfies

$$a - 2\Delta t\lambda a(1-a) = 0$$
$$\Longleftrightarrow \quad a\Big(a - \big(1 - (2\Delta t\lambda)^{-1}\big)\Big) = 0 \quad \text{if } \Delta t\lambda \neq 0.$$

Therefore, we have the two distinct solutions $a_1 = 0$ and $a_2 = 1 - (2\Delta t\lambda)^{-1}$ if $\Delta t\lambda > \frac{1}{2}$.

6.3.4 Selected three dimensional closures

In the previous section, we analyzed time-discrete counterparts of the space independent Folgar-Tucker equation (6.20b) and established sufficient conditions for the methods under investigation to produce physics-compatible solutions. We verified these conditions for some commonly employed two dimensional closures and illustrated the benefits of using closures that satisfy our assumptions. Unfortunately, an in-depth analysis of three dimensional closures seems to be far more challenging (especially when it comes to proofs of the Lipschitz continuity) and is, therefore, beyond the scope of this work. However, the following remark summarizes the basic requirements for the successful design of admissible and orthotropic closures.

Remark 6.21.

- Let the eigenvalues of the second order orientation tensor $\mathbf{A} \in \mathbb{A}_3$ be arranged in a non-decreasing order, i.e., $0 \leqslant a_1 \leqslant a_2 \leqslant a_3 \leqslant 1$. These eigenvalues can be interpreted as barycentric coordinates of a point (a_1, a_2, a_3) on the shaded area as shown in Fig. 6.1. Therefore, it suffices to define the diagonal form $\underline{\underline{\tilde{\mathbf{A}}}}(\tilde{\mathbf{A}})$ of the closure only for orientation states corresponding to this region [CT95]. Additionally, the unit trace condition $a_1 + a_2 + a_3 = 1$ implies that one eigenvalue (e.g., a_1) and the entries of $\underline{\underline{\tilde{\mathbf{A}}}}(\tilde{\mathbf{A}})$ are functions of the other two eigenvalues.

- Condition (6.21c) of Theorem 6.4 must hold for the unit trace property of the solution to be preserved in the process of solving the Folgar-Tucker equation. If formula (6.23) is exploited to express $\tilde{\underline{\underline{a}}}_{1122}$, $\tilde{\underline{\underline{a}}}_{1133}$, and $\tilde{\underline{\underline{a}}}_{2233}$ in terms of $\tilde{\underline{\underline{a}}}_{1111}$, $\tilde{\underline{\underline{a}}}_{2222}$, and $\tilde{\underline{\underline{a}}}_{3333}$, the validity of (6.21c) is guaranteed. Moreover, only three independent components of $\underline{\underline{\tilde{\mathbf{A}}}}$ remain undefined.

- A necessary condition for the admissibility of fourth order orientation tensors is presented in Lemma 6.6. It leads to $\tilde{\underline{\underline{a}}}_{1111} = \tilde{\underline{\underline{a}}}_{1122} = \tilde{\underline{\underline{a}}}_{1133} = 0$ if $a_1 = 0$. Then $\underline{\underline{\tilde{\mathbf{A}}}}$ is uniquely defined by specifying $\tilde{\underline{\underline{a}}}_{2222}$ *or* $\tilde{\underline{\underline{a}}}_{3333}$ (edge $\overline{BU}$ in Fig. 6.1).

- If $a_2 = 0$ holds as well, we have $\tilde{\underline{\underline{a}}}_{2233} = 0$ and the only nonzero entry of $\underline{\underline{\tilde{\mathbf{A}}}}$ is given by $\tilde{\underline{\underline{a}}}_{3333} = 1$ (point U in Fig. 6.1).

- To make sure that the closure is uniquely determined for all orientation states, Theorem 6.11 can be invoked to impose the constraints

$$\tilde{\underline{\underline{a}}}_{1111} = \tilde{\underline{\underline{a}}}_{2222} = 3\tilde{\underline{\underline{a}}}_{1122}, \quad \tilde{\underline{\underline{a}}}_{1133} = \tilde{\underline{\underline{a}}}_{2233} \qquad \text{if } a_1 = a_2,$$

$$\tilde{\underline{\underline{a}}}_{2222} = \tilde{\underline{\underline{a}}}_{3333} = 3\tilde{\underline{\underline{a}}}_{2233}, \quad \tilde{\underline{\underline{a}}}_{1122} = \tilde{\underline{\underline{a}}}_{1133} \qquad \text{if } a_2 = a_3,$$

(edge $\overline{UT}$ and $\overline{TB}$ in Fig. 6.1).

- In the special case of an isotropic second order orientation tensor $\mathbf{A} = \frac{1}{3}\mathbf{I}$ (point T in Fig. 6.1), the above set of conditions for $a_1 = a_2 = a_3$ implies $\tilde{\underline{\underline{a}}}_{1111} = \tilde{\underline{\underline{a}}}_{2222} = \tilde{\underline{\underline{a}}}_{3333} = \frac{1}{5}$ (cf. (6.36b)).

- The biaxial orientation state corresponding to a tensor $\mathbf{A}$ with eigenvalues $a_1 = 0$ and $a_2 = a_3 = \frac{1}{2}$ (point B in Fig. 6.1) is uniquely characterized by $\underline{\underline{\tilde{\mathbf{A}}}}$ with entries $\tilde{\underline{\underline{a}}}_{1111} = 0$ and $\tilde{\underline{\underline{a}}}_{2222} = \tilde{\underline{\underline{a}}}_{3333} = \frac{3}{8}$.

In summary, the fourth order orientation tensor of an orthotropic and admissible closure has at most 3 independent components (shaded area in Fig. 6.1). If two eigenvalues coincide or the smallest one vanishes, there is only one unknown left (boundary of the shaded area in Fig. 6.1). The value of this degree of freedom is predefined for the three basic orientation states (corners of the shaded area in Fig. 6.1).

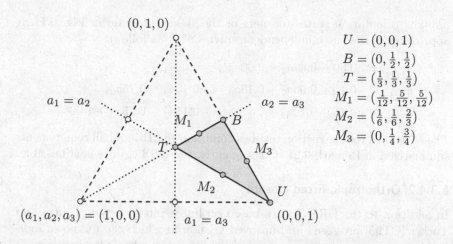

$$U = (0,0,1)$$
$$B = (0,\tfrac{1}{2},\tfrac{1}{2})$$
$$T = (\tfrac{1}{3},\tfrac{1}{3},\tfrac{1}{3})$$
$$M_1 = (\tfrac{1}{12},\tfrac{5}{12},\tfrac{5}{12})$$
$$M_2 = (\tfrac{1}{6},\tfrac{1}{6},\tfrac{2}{3})$$
$$M_3 = (0,\tfrac{1}{4},\tfrac{3}{4})$$

Figure 6.1: Domain of eigenvalues for possible orientation states in barycentric coordinate system (triangle with dashed sides if eigenvalues are not sorted; shaded area if eigenvalues are arranged in non-decreasing order, i.e., $0 \leqslant a_1 \leqslant a_2 \leqslant a_3$).

In the literature on orthotropic closures, the unknown components $\underline{\tilde{a}}_{1111}$, $\underline{\tilde{a}}_{2222}$, and $\underline{\tilde{a}}_{3333}$ are frequently approximated by polynomials in the eigenvalues a_2 and a_3 (see below). A (piecewise) polynomial closure approximation $\underline{\tilde{\mathbf{A}}}$ can be written in terms of scalar Bernstein basis functions with coefficients that are fourth order tensors. Due to the nonnegativity of the Bernstein basis functions, condition (6.21b) holds for every orientation state if all Bernstein coefficients satisfy (6.21b).

In what follows, we briefly present selected three dimensional orthotropic closures, two of which are used in the numerical study below.

6.3.4.1 Orthotropic smooth closure

One of the simplest representatives of orthotropic closures is the *orthotropic smooth (ORS) closure presented in [CT95, Section II C]*. Using linear interpolation of the uniquely defined values for the isotropic, biaxial, and uniaxial orientation states (corners of the shaded domain in Fig. 6.1), it approximates the three independent entries of $\underline{\tilde{\mathbf{A}}}$ as follows:

$$\tilde{\underline{a}}_{1111} = 0.60 - 0.60a_2 - 0.60a_3 = 0.60a_1,$$
$$\tilde{\underline{a}}_{2222} = -0.15 + 0.90a_2 + 0.15a_3 = -0.15a_1 + 0.75a_2,$$
$$\tilde{\underline{a}}_{3333} = -0.15 - 0.10a_2 + 1.15a_3 = -0.15a_1 - 0.25a_2 + 1.00a_3.$$

The so defined reconstruction satisfies condition (6.21b) and all requirements summarized in Remark 6.21. Consequently, the ORS closure is admissible.

6.3.4.2 Orthotropic fitted closure

In addition to the ORS closure based on linear interpolation, Cintra and Tucker [CT95] proposed an improved version in which the unknown independent components of $\underline{\tilde{\mathbf{A}}}$ are determined using a quadratic least-squares fit to numerical solutions of the Fokker-Planck equation for selected flow fields. This procedure results in the *orthotropic fitted (ORF) closure*

$$\tilde{\underline{a}}_{1111} = 1.228982 - 2.260574a_2 + 1.053907a_2^2$$
$$- 2.054116a_3 + 0.821548a_3^2 + 1.819756a_2a_3,$$
$$\tilde{\underline{a}}_{2222} = 0.124711 + 0.086169a_2 + 0.796080a_2^2$$
$$- 0.389402a_3 + 0.258844a_3^2 + 0.544992a_2a_3,$$
$$\tilde{\underline{a}}_{3333} = 0.060964 - 0.369160a_2 + 0.318266a_2^2$$
$$+ 0.371243a_3 + 0.555301a_3^2 + 0.371218a_2a_3.$$

Although the accuracy of this closure is likely to be higher than that of the ORS closure, requirement (6.36b) is not met because

$$\tilde{\underline{a}}_{2222} = 0.124711 + (0.086169 - 0.389402)\tfrac{1}{3}$$
$$+ (0.796080 + 0.258844 + 0.544992)\tfrac{1}{9} \approx 0.201402,$$
$$\tilde{\underline{a}}_{3333} = 0.060964 + (0.371243 - 0.369160)\tfrac{1}{3}$$
$$+ (0.318266 + 0.555301 + 0.371218)\tfrac{1}{9} \approx 0.199968.$$

Moreover, condition (6.37) is violated in the case $a_1 = 0$, $a_2 = 0$, and $a_3 = 1$, in which the fitted closure yields

$$\tilde{\underline{a}}_{1111} = -0.003586, \qquad \tilde{\underline{a}}_{2222} = -0.005847, \qquad \tilde{\underline{a}}_{3333} = 0.987508,$$
$$\tilde{\underline{a}}_{1122} = -0.0015295, \qquad \tilde{\underline{a}}_{1133} = 0.0051155, \qquad \tilde{\underline{a}}_{2233} = 0.0073765.$$

We conclude that the ORF closure is neither well-defined for isotropic second order tensors nor admissible for uniaxial orientations. As a consequence, Theorems 6.17 and 6.19 do not guarantee that models based on the ORF closure produce physically realistic results for arbitrary orientation states even if the corresponding time step condition is met. For that reason, we will not consider this closure in the numerical studies below.

6.3.4.3 Natural B closure

Another way to construct orthotropic closures based on quadratic and higher order polynomial approximations was proposed in [Kuz18b, Section 4.5.2]. Similarly to the ORS closure, it is based on the idea of interpolating values of the independent components of fourth order orientation tensors at sample points on the triangle representing the set of possible orientation states (cf. Fig. 6.1). The author proposed an orthotropic closure based on the following values:

$$\underset{=}{\tilde{a}}_{1111} = 1.1880 - 2.1264a_2 + 0.9384a_2^2 - 2.0136a_3 + 0.8256a_3^2 + 1.7640a_2a_3,$$
$$\underset{=}{\tilde{a}}_{2222} = 0.0708 + 0.2252a_2 + 0.7040a_2^2 - 0.2792a_3 + 0.2084a_3^2 + 0.4124a_2a_3,$$
$$\underset{=}{\tilde{a}}_{3333} = 0.0708 - 0.3776a_2 + 0.3068a_2^2 + 0.3236a_3 + 0.6056a_3^2 + 0.4124a_2a_3.$$

By using the representation in terms of Bernstein basis functions, one can show the admissibility of the so defined fourth order orientation tensors. However, as in the case of the ORF closure, this closure is not well-defined for every second order orientation tensor. For example, the eigenvalues $a_1 = \frac{1}{6}$, $a_2 = \frac{1}{6}$, and $a_3 = \frac{2}{3}$ lead to

$$\underset{=}{\tilde{a}}_{1111} = 0.0802, \qquad \underset{=}{\tilde{a}}_{2222} = 0.0802, \qquad \underset{=}{\tilde{a}}_{3333} = 0.5471,$$
$$\underset{=}{\tilde{a}}_{1122} \approx 0.02668, \qquad \underset{=}{\tilde{a}}_{1133} \approx 0.05978, \qquad \underset{=}{\tilde{a}}_{2233} \approx 0.05978$$

and condition (6.32) is violated since $0.0802 = \underset{=}{\tilde{a}}_{1111} \neq 3\underset{=}{\tilde{a}}_{1122} = 0.08005$.

6.4 Numerical examples

In this section, we perform numerical studies of the flow behavior of fiber reinforced composites in an axisymmetric contraction geometry. For this purpose, we solve the system of PDEs (6.6) with suitable initial and boundary conditions using the techniques presented in Chapters 5 and 6. The results are compared with those of VerWeyst and Tucker [VT02] and experimental data provided in [Lip+88].

Several authors recognized that physical admissibility of numerical approximations to evolving orientation tensors is a vitally important criterion for the design of closures and discretization techniques for the Folgar-Tucker equation. For example, Reddy and Mitchell [RM01] found that an upwind-type discretization of the convective term is essential to prevent the occurrence of unphysical orientation states in the case of small rotary diffusivity. For that reason, the authors used the upwinding methodology developed by Lasaint and Raviart [LR74] in the context of discontinuous finite elements. The need for upwinding was also emphasized by VerWeyst and Tucker [VT02] who employed the SUPG scheme proposed in [BH82] and additionally corrected the eigenvalues of unphysical orientation tensors in a post-processing step.

The method presented in this section exploits the theoretical results obtained and summarized in the previous chapters. In particular, we use closures and time steps which guarantee the preservation of positive semidefiniteness. No further correction techniques are needed to ensure that the results are physically realistic. Optional parameters can be adapted to improve the accuracy of the solution without making the method unstable or losing the benefit of provable physical admissibility. However, the main focus of the numerical study presented in this section is not on the detailed assessment of modeling and discretization errors. Instead, it is supposed to illustrate how numerical tools developed for property-preserving treatment of different subproblems can be combined to solve the nonlinear coupled problem.

To simulate the flow of a fiber suspension through an axisymmetric $4.5 : 1$ contraction, we use a computational domain Ω consisting of a long wide pipe (radius $R_1 = 0.0225$; length $L_1 = 0.09$) joined to a shorter narrow one (radius $R_2 = 0.005$; length $L_2 = 0.02$) as shown in Fig. 6.2. That is,

$$\Omega = \{\mathbf{x} \in \mathbb{R}^3 \mid -L_1 < x_1 < 0, \, x_2^2 + x_3^2 < R_1^2\}$$
$$\cup \{\mathbf{x} \in \mathbb{R}^3 \mid 0 \leqslant x_1 < L_2, \, x_2^2 + x_3^2 < R_2^2\}.$$

We strongly enforce a parabolic velocity profile $\mathbf{v}_D$ and an isotropic orientation state on the inflow boundary $\Gamma_{\text{in}} = \{\mathbf{x} \in \mathbb{R}^3 \mid x_1 = -L_1, \, x_2^2 + x_3^2 \leqslant R_1^2\}$. The fluid-fiber mixture flows out of the domain at $\Gamma_{\text{out}} = \{\mathbf{x} \in \mathbb{R}^3 \mid x_1 = L_2, \, x_2^2 + x_3^2 \leqslant R_2^2\}$, where we impose the vanishing traction condition $\mathbf{h} = \mathbf{0}$ in a weak sense. On the remaining part of the boundary $\Gamma_{\text{Wall}} = \partial\Omega \backslash (\Gamma_{\text{in}} \cup \Gamma_{\text{out}})$, we impose no-slip boundary conditions by setting $\mathbf{v} = \mathbf{0}$. Due to the hyperbolicity of the evolution equation (6.19) for the second order orientation tensor, no boundary conditions for $\mathbf{A}$ need to be prescribed on $\Gamma_{\text{out}} \cup \Gamma_{\text{Wall}}$.

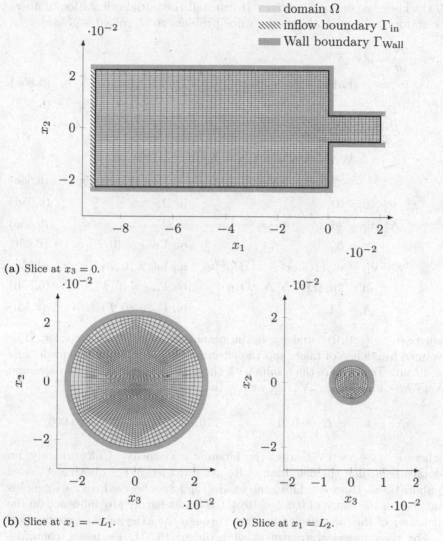

Figure 6.2: Axisymmetric contraction: Geometry of domain and computational mesh (courtesy of Omid Ahmadi).

If the fiber suspension is initially at rest and the initial orientation of fibers is isotropic, the initial-boundary value problem to be solved is given by

$$\rho\frac{\partial \mathbf{v}}{\partial t} + \rho(\mathbf{v}\cdot\nabla)\mathbf{v}$$
$$= -\operatorname{grad}(p) + 2\mu_I\nabla\cdot(\mathbf{D} + N_p\underline{\underline{\mathbf{A}}} : \mathbf{D}) \quad \text{in } \Omega\times(0,T), \quad (6.55a)$$
$$\operatorname{div}(\mathbf{v}) = 0 \quad\quad\quad\quad\quad\quad\quad \text{in } \Omega\times(0,T), \quad (6.55b)$$

$$\frac{\partial \mathbf{A}}{\partial t} + \mathbf{v}\cdot\operatorname{grad}(\mathbf{A})$$
$$= \mathbf{WA} - \mathbf{AW} + \lambda(\mathbf{DA} + \mathbf{AD} - 2\underline{\underline{\mathbf{A}}} : \mathbf{D})$$
$$+ 2D_r(\mathbf{I} - d\mathbf{A}) \quad\quad \text{in } \Omega\times(0,T), \quad (6.55c)$$

$$\mathbf{v}(\cdot,0) = \mathbf{0} \quad\quad\quad\quad\quad \text{in } \Omega, \quad\quad\quad\quad\quad (6.55d)$$
$$\mathbf{A}(\cdot,0) = \tfrac{1}{d}\mathbf{I} \quad\quad\quad\quad \text{in } \Omega, \quad\quad\quad\quad\quad (6.55e)$$
$$\mathbf{v} = \mathbf{0} \quad\quad\quad\quad\quad\quad \text{on } \Gamma_{\text{Wall}}\times(0,T), \quad (6.55f)$$
$$\mathbf{v} = \mathbf{v}_{\text{D}} = v_\infty\big(1 - (x_1^2 + x_2^2)R_1^{-2}\big)\mathbf{e}_1 \quad \text{on } \Gamma_{\text{in}}\times(0,T), \quad (6.55g)$$
$$p\mathbf{n} = 2\mu_I(\mathbf{D} + N_p\underline{\underline{\mathbf{A}}} : \mathbf{D})\mathbf{n} \quad \text{on } \Gamma_{\text{out}}\times(0,T), \quad (6.55h)$$
$$\mathbf{A} = \tfrac{1}{d}\mathbf{I} \quad\quad\quad\quad\quad \text{on } \Gamma_{\text{in}}\times(0,T), \quad (6.55i)$$

where $\mathbf{e}_1 = (1,0,0)^\top$ and v_∞ is the maximum velocity at the inlet. The volume fraction ϕ of fibers and the effective viscosity μ_I are assumed to be constant. To compare the results with those of Lipscomb et al. [Lip+88] and VerWeyst and Tucker [VT02], we use the parameters

$$N_p = 6, \quad C_I = 0.01 \quad r_e = 276, \quad \text{Re} = \frac{v_\infty R_1}{\nu} = 0.005,$$

where $\nu = \frac{\mu_I}{\rho} = 0.1$ denotes the kinematic viscosity. Unfortunately, no details regarding the definition of Re are provided in the above mentioned publications. However, Lipscomb et al. [Lip+88] observed that a Reynolds number in the range of $0.005 \leqslant \text{Re} \leqslant 0.036$ has hardly any influence on the behavior of the fiber suspension flow through the axisymmetric contraction.

For the numerical treatment of problem (6.55), we use a computational mesh composed of 132 096 hexahedra with a total of 155 950 nodes (cf. Fig. 6.2). The union of all mesh elements defines a polyhedral approximation Ω_h of the domain Ω. Each component of the velocity field $\mathbf{v}$ is approximated by a continuous and piecewise triquadratic function (denoted by $\mathbf{v}_h$), while the pressure p_h is discontinuous and piecewise linear in each element. Every entry of the orientation tensor $\mathbf{A}$ is discretized using a

Q_1-iso-Q_2 finite element approximation (i.e., Q_1 discretization on a submesh defined by the nodal points of the Q_2 discretization). This discretized counterpart of **A** is denoted by $\mathbf{A}_h$.

The finite element approximations of primary variables $(\mathbf{v}_h, p_h, \mathbf{A}_h)$ are updated step-by-step using an operator splitting approach to decompose the coupled problem into

1. the space independent Folgar-Tucker equation (6.20b) for each nodal orientation tensor $\mathbf{A}_i(t) = \mathbf{A}_h(\mathbf{x}_i, t)$, i.e., the subproblem which was analyzed in Section 6.3;

2. the homogeneous advection equation (6.20a) for $\mathbf{A}_h$, whose numerical treatment was discussed in Chapter 5;

3. the Navier-Stokes equations (6.7a) which can be solved using Algorithm 6.1.

In the case of the former two subproblems, a tensor quantity has to be evolved while preserving the physical properties of orientation tensors. If the trace does not stay constant or $\mathbf{A}_h$ ceases to be positive semidefinite, the assumptions behind the derivation of closures for reconstructing the fourth order orientation tensor become invalid and simulations may break down or produce physically meaningless results.

The space independent Folgar-Tucker equation calls for the use of an admissible and orthotropic closure. As shown in Section 6.3, such closures are guaranteed to produce physics-compatible solutions for sufficiently small time steps.

For the time dependent hyperbolic problem (6.20a), each bound-preserving method of Chapter 5 guarantees that the solution is an orientation tensor. Indeed, eigenvalues cannot become negative by definition of the global discrete maximum principles and the trace stays constant if scalar correction factors are used to limit diffusive fluxes of the form $(1 - \alpha_{ij})(\mathbf{U}_j - \mathbf{U}_i)$ with a vanishing trace. Algebraic corrections of this kind preserve the trace of the Galerkin solution which is in turn preserved for the non-conservative form of PDE (6.20a) even if the velocity field is not exactly solenoidal.

In light of the above, admissible approximations to problem (6.55) can be computed as follows:

Algorithm 6.22. *Set* $\mathbf{v}_h^0 = \mathbf{0}$ *and* $\mathbf{A}_h^0 = \frac{1}{d}\mathbf{I}$. *For* $n = 0, 1, \ldots$, *perform the following steps:*

1. *For each nodal point $\mathbf{x}_i$ of $\mathbf{A}_h$, calculate $\bar{\mathbf{A}}_i^{n+1}$ by solving the space independent Folgar-Tucker equation (6.20b). Use Algorithm 6.18, the averaged nodal gradients of $\mathbf{v}_h^n$, and the input data $\mathbf{A}_i^n$.*

2. *Advect the second order orientation tensor $\mathbf{A}_h$ by solving (5.78). Use Algorithm 5.21, the velocity field $\mathbf{v}_h^n$, the input data $\bar{\mathbf{A}}_h^{n+1}$, and the parameters $\theta = 1$, $q_2 = 4$, $p_2 = 1$.*

3. *Reconstruct the fourth order orientation tensor $\underline{\underline{\mathbf{A}}}_i^{n+1} = \underline{\underline{\mathbf{A}}}(\mathbf{A}_i^{n+1})$ in each node i.*

4. *Calculate the velocity field $\mathbf{v}_h^{n+1}$ (and the pressure p_h^{n+1}) by performing (one cycle of) Algorithm 6.1. Use $\theta = 1$, $p^{n+1,0} = p^n$, $\omega = 1$, and τ defined in terms of $(\underline{\underline{\mathbf{A}}}_i^{n+1})_{i=1}^N$.*

Due to the a posteriori time step control involved in Algorithm 6.18 and the fully implicit treatment of the advection equation (6.20a), steps 1 and 2 preserve the properties established in Theorem 6.3 no matter how the time increment is chosen. If the parameter $\dot{\beta}_{ij}$ of Algorithm 5.21 is set to zero and condition (6.46) holds, the orientation tensor $\mathbf{A}_h^{n+1}$ evaluated at node i is given by

$$
\begin{aligned}
m_i \mathbf{A}_i^{n+1} &= \underline{\mathbf{G}}_i + m_i \bar{\mathbf{A}}_i^{n+1} + \Delta t \mathbf{Z}_i^{n+1}(A^{n+1}) \\
&= \Delta t \mathbf{G}_i^{n+1} + m_i \mathbf{A}_i^n + \Delta t \sum_{j=1}^N k_{ij}^{n+1} \mathbf{A}_j^{n+1} \\
&\quad + \Delta t \sum_{j \neq i} d_{ij}\big(\mathscr{I} - \mathscr{A}_{ij}(A^{n+1})\big)[\mathbf{A}_j^{n+1} - \mathbf{A}_i^{n+1}] \\
&\quad + \Delta t m_i(\mathbf{W}_i^n \mathbf{A}_i^n - \mathbf{A}_i^n \mathbf{W}_i^n) \\
&\quad + \Delta t m_i \lambda\big(\mathbf{D}_i^n \mathbf{A}_i^n + \mathbf{A}_i^n \mathbf{D}_i^n - 2\underline{\underline{\mathbf{A}}}(\mathbf{A}_i^n) : \mathbf{D}_i^n\big) \\
&\quad + 2\Delta t m_i D_r(\mathbf{I} - d\mathbf{A}_i^n).
\end{aligned}
$$

Marching the primary variables to the steady state, we find that the stationary solutions $(\mathbf{v}_h, p_h, \mathbf{A}_h)$ must satisfy

$$
\begin{aligned}
\sum_{j=1}^N (-k_{ij})\mathbf{A}_j &- \sum_{j \neq i} d_{ij}\big(\mathscr{I} - \mathscr{A}_{ij}(A)\big)[\mathbf{A}_j - \mathbf{A}_i] \\
&= \mathbf{G}_i + m_i(\mathbf{W}_i \mathbf{A}_i - \mathbf{A}_i \mathbf{W}_i) \\
&\quad + m_i \lambda\big(\mathbf{D}_i \mathbf{A}_i + \mathbf{A}_i \mathbf{D}_i - 2\underline{\underline{\mathbf{A}}}(\mathbf{A}_i) : \mathbf{D}_i\big) + 2m_i D_r(\mathbf{I} - d\mathbf{A}_i) \\
&\hspace{6cm} \forall i \in \{1, \ldots, N\},
\end{aligned}
$$

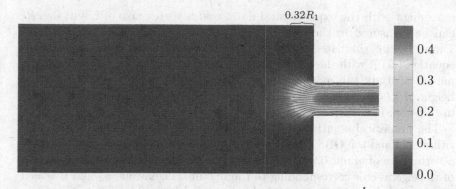

Figure 6.3: Streamlines and magnitude of velocity field for Newtonian fluid flow
($N_p = 0$) through axisymmetric contraction. The curly brace denotes
the reference vortex size of [Ver98].

where $\mathcal{K}$ is the convection matrix based on the steady state velocity field $\mathbf{v}_h$.
Therefore, the steady variables approximate the solution to the stationary
counterpart of (6.55) and are independent of Δt because a van Kan-like
discrete projection scheme is employed. In the numerical computations below,
we ignore the CFL-like condition (6.46) and calculate the solution using
the pseudo time increment $\Delta t = 10^{-2}$ until the changes of the approximate
solution become sufficiently small.

We first study the flow behavior of a Newtonian fluid in the 4.5 : 1
axisymmetric contraction geometry. In this numerical experiment, the fiber
induced stresses are deactivated by setting $N_p = 0$. The velocity profile of
the resulting steady state approximation to (6.55) exhibits a small vortex in
the corner of the wide pipe (cf. Fig. 6.3). The vortex detachment point in
the slice at $x_3 = 0$ is in an excellent agreement with the one presented, e.g.,
in [Ver98].

Next, we simulate the axisymmetric contraction flow of a dilute fiber
suspension using $N_p = 6$ and the above parameter settings and compare the
results to those reported by [Lip+88] and [Ver98]. When the carrier fluid is
loaded with small rigid fibers, the vortex detachment point moves upstream
and a significant increase in the size of the recirculating corner vortex can
be observed. To simulate this behavior without solving the Folgar-Tucker
equation, Lipscomb et al. [Lip+88] simplified problem (6.55) by using the
quadratic closure and the assumption of fully aligned orientation states,
i.e., $\mathbf{A} = (\mathbf{v} \otimes \mathbf{v})\|\mathbf{v}\|_2^{-2}$. The corresponding numerical results are in good

agreement with the experimental data, and a vortex size of about $0.92R_1$ can be measured in the case $N_p = 6$ [Lip+88, Fig. 17]. VerWeyst and Tucker [VT02] simulated the orientation dynamics using the Folgar-Tucker equation (6.19) with the so called ORT closure presented in [Ver98]. In their numerical study, this model produced a flow pattern with a secondary vortex height of $1.06R_1$, which is very similar to the one measured experimentally in [Lip+88, Fig. 10a].

The numerical solution presented in Fig. 6.4 was obtained with Algorithm 6.22 and the ORS closure. It produces a stationary velocity field with a vortex size of about $0.96R_1$. The line plot in Fig. 6.4c shows the direction of the eigenvector corresponding to the maximal eigenvalue a_d and is scaled with $a_d - \frac{1}{3}$ (note that $a_d \geqslant \frac{1}{3}$ holds by definition). These lines illustrate sample fibers entering the domain at Γ_{in} with a random orientation. As the fibers move with the carrier fluid and rotate under influence of velocity gradients, they tend to align themselves with the flow direction. However, no full alignment of fibers with the streamlines occurs within the secondary vortex. This qualitative observation matches the findings of VerWeyst [Ver98], who reported that the orientation of the fibers follows the flow direction with a lag caused by abrupt changes of the velocity gradients along closed streamlines with locally high curvature.

Even though the results shown in Fig. 6.4 exhibit the correct qualitative behavior, the maximal eigenvalue of $\mathbf{A}$ is smaller than the one obtained in the simulation of VerWeyst [Ver98]. This discrepancy can be attributed to the fact that the reconstruction of $\underline{\mathbf{A}}$ using the ORS closure is not very accurate. When the natural B closure is employed instead, the mismatch between the simulation results for the maximal eigenvalue becomes less pronounced and the size of the recirculating corner vortex increases to $1.04R_1$. The numerical solutions presented in Fig. 6.5 are in excellent agreement with the experimental results [Lip+88, Fig. 10a]. In this simulation, the minimal eigenvalue reaches the value $3.38 \cdot 10^{-3}$ without becoming negative and no ad hoc corrections of $\mathbf{A}_h$ are needed to preserve the properties of orientation tensors.

The presented case study for the axisymmetric contraction flow problem indicates that a well-tuned combination of physics-compatible numerical methods for pure advection of tensor fields and the space independent Folgar-Tucker equation makes it possible to simulate the evolution of fiber orientation states without producing unphysical artifacts or inordinately large amounts of artificial diffusion. Moreover, the proposed methodology is backed by theoretical results that guarantee well-posedness and physical admissibility. Although the numerical study in this section was restricted to

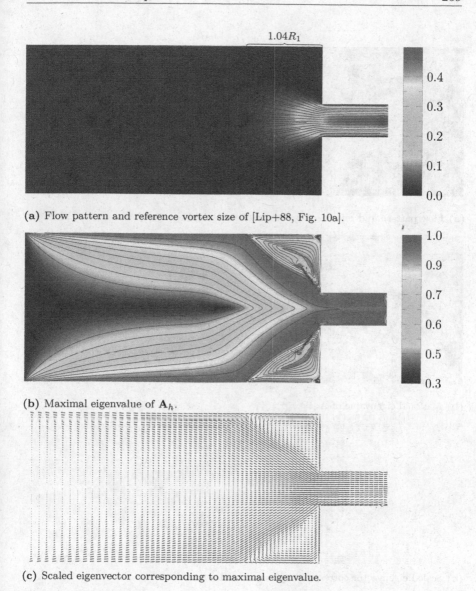

(a) Flow pattern and reference vortex size of [Lip+88, Fig. 10a].

(b) Maximal eigenvalue of $\mathbf{A}_h$.

(c) Scaled eigenvector corresponding to maximal eigenvalue.

Figure 6.4: Numerical simulation of fiber suspension flow through axisymmetric contraction using ORS closure.

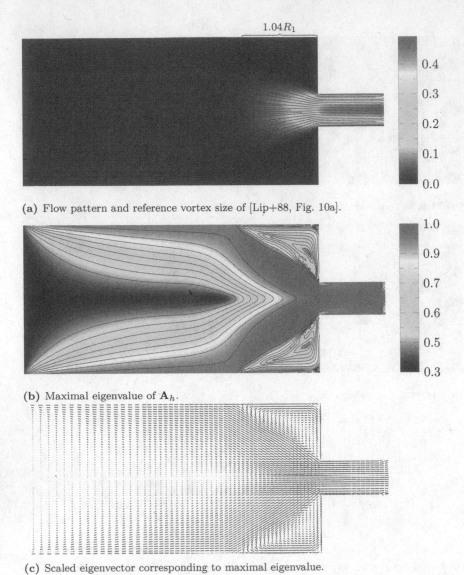

(a) Flow pattern and reference vortex size of [Lip+88, Fig. 10a].

(b) Maximal eigenvalue of $\mathbf{A}_h$.

(c) Scaled eigenvector corresponding to maximal eigenvalue.

Figure 6.5: Numerical simulation of fiber suspension flow through axisymmetric contraction using natural B closure.

steady fiber suspension flows, the employed numerical techniques are ideally suited for solving transient problems subject to corresponding time step restrictions. Unfortunately, no well-documented test benchmark seems to exist for validation and verification of numerical methods for unsteady flows of fiber suspensions. The development of such a benchmark would provide a very useful tool for the design of improved computational techniques and evaluation of closures.

7 Conclusions

Motivated by applications to fiber suspension flows, the research conducted in this thesis was aimed at the development of bound-preserving finite element discretizations of hyperbolic problems. In the context of algebraic flux correction schemes for scalar linear advection models, we formulated new generalized discrete maximum principles and derived sufficient conditions for their validity. Numerical analysis of nonlinear schemes based on these criteria was performed for FCT-type post-processing procedures and monolithic AFC discretizations. These theoretical investigations shed additional light on the numerical behavior of existing limiting techniques and provide useful design criteria for the development of new ones. Equipped with a better understanding of AFC approaches for scalar quantities, we extended some of the presented schemes to symmetric tensor fields and embedded them into physics-compatible numerical algorithms for simulation of fiber suspensions. The proposed approach to numerical treatment of orientation tensors is based on the principle of eigenvalue range preservation. In addition to designing property-preserving AFC schemes for general symmetric tensors, we performed theoretical investigations of commonly used fiber orientation dynamics models and formulated sufficient conditions for the involved closures to be physically admissible. The main highlights of the developed methodology, important theoretical results, key findings of our numerical investigations, and some promising directions for further development of the presented approaches are summarized in this chapter.

7.1 Summary

Many components of the proposed bound-preserving numerical schemes for the tensorial advection equation represent relatively straightforward generalizations of the AFC methodology for constraining Galerkin finite element discretizations of scalar conservation laws. The basic concepts and the design philosophy of AFC schemes were reviewed in Chapter 4. To guarantee the absence of unphysical overshoots and undershoots in a provable manner, the Galerkin system matrices of spatially discretized stationary and

© Springer Fachmedien Wiesbaden GmbH, part of Springer Nature 2019
C. Lohmann, *Physics-Compatible Finite Element Methods for Scalar and Tensorial Advection Problems*, https://doi.org/10.1007/978-3-658-27737-6_7

transient problems are modified so as to satisfy the inequality constraints of Lemmas 4.16 and 4.24. The underlying design criteria include local and global discrete maximum principles (DMPs) as well as the boundedness of nodal function values in terms of (weakly or strongly) imposed boundary data. For the transient problem discretized in time using the θ-scheme, we derived a CFL-like time step restriction which guarantees the validity of DMPs under the assumption of a bound-preserving space discretization. In the AFC framework, a low order method with desired properties is commonly constructed by adding a great amount of artificial diffusion to the Galerkin discretization. The employed discrete diffusion operator rules out violations of local bounds but the resulting scheme is too inaccurate for practical purposes. Following existing approaches to algebraic flux correction, two ways to improve the accuracy of the solution were explored in this work:

The first approach to reducing the levels of numerical diffusion is based on the FCT methodology. The nodal values of a non-oscillatory low order solution are corrected by adding limited antidiffusive fluxes. The correction factors for these fluxes are defined in a manner which guarantees boundedness by the local maxima and minima of the low order predictor. The FCT limiting strategy has been in use since the early 1970s and proved its worth in many real-life applications. However, no theoretical investigations of its convergence behavior can be found in the literature. In Section 4.4.1.3, we performed a basic analysis of the FCT correction step and proved that the addition of limited antidiffusive terms cannot prevent convergence to the exact solution regardless of the way in which the local bounds are enforced. Moreover, we found that the two-step nature of FCT implies the validity of local DMPs with respect to extended stencils. The presented theoretical and numerical results indicate that such schemes are ideally suited for the numerical treatment of reaction-dominated or transient problems provided that the low order predictor is already a reasonable approximation of the exact solution.

The second approach to performing bound-preserving high order corrections in AFC schemes incorporates nonlinear antidiffusive terms into the residual of the low order method. This monolithic limiting strategy was presented in Section 4.5. No general proof for the existence of a unique solution to the nonlinear discrete problem is currently available for such schemes. In Section 4.5.1.1, we proved the corresponding result for the steady model problem assuming that a sufficiently large amount of reactivity is treated in a lumped manner and the limited antidiffusive part is Lipschitz continuous. A proof of convergence with order $\frac{1}{2}$ with respect to a mesh-dependent norm $\| \cdot \|_h$ was obtained in Section 4.5.1.3 under the assumption that the coercivity condition (2.14) is satisfied. The presented analysis is based on the proof of a similar result for elliptic problems in the seminal work of Barrenechea, John, and Knobloch [Bar+16]. The worst case a priori error estimate holds for any choice of correction factors and for the low order approximation in particular. This result is of great importance not only for the theoretical justification of monolithic AFC discretizations but also for the analysis of FCT approaches because the convergence of the low order method in $\| \cdot \|_h$ implies that of its flux-corrected counterpart in $\| \cdot \|_{L^2(\Omega)}$ (as shown in Section 4.4.1.3). In contrast to FCT, the limiting procedures of Section 4.5 are designed to satisfy local DMPs for compact stencils. For transient advection problems discretized in time using the θ-scheme, our DMP analysis of the fully discrete problem yields a computable upper bound for admissible time steps. In the absence of antidiffusive corrections, the general a priori time step restriction for bound-preserving AFC schemes reduces to the CFL-like condition of the low order method.

Chapter 5 extends the above limiting approaches to symmetric tensor fields. The idea of limiting the eigenvalue range was originally introduced in [Loh17b] in which it was motivated by a simple example and used to design tensorial extensions of FCT. The imposition of (global) bounds on the maximal and minimal eigenvalues implies the preservation of positive semidefiniteness and, hence, is particularly well-suited for the treatment of orientation tensors in fiber suspension flow models. In Section 5.3, we discretized tensorial versions of the steady and transient advection equations using the low order method originally developed for bound-preserving advection of scalar quantities. The eigenvalue range preservation property was shown building on the proofs of discrete maximum principles for the scalar case. The two approaches to improving the accuracy of constrained solutions were revisited and eigenvalue-based limiters were developed for tensor fields. The tensorial version of the FCT corrector step constrains the eigenvalue range using correction factors that guarantee semidefiniteness of auxiliary tensors. Approximate or exact solutions to corresponding local semidefinite programming problems can be calculated efficiently using the algorithms presented in Section 5.4.1. As an alternative to FCT, we introduced a monolithic limiting approach and supplemented the theory developed in [Loh19] with new results. The presented analysis is focused on the proofs of generalized discrete maximum principles for tensor limiters that guarantee Lipschitz continuity of the residuals. Lemma 5.16 provides a new set of practical criteria for the design of such limiters. Another novelty compared to [Loh19] is the extension of the method to unsteady problems. Following the analysis of AFC schemes for transient advection of scalar quantities, we proved the validity of eigenvalue-based DMPs for symmetric tensors under similar conditions.

As a practical application involving advection of positive semidefinite tensor fields, we considered a fiber suspension flow model in Chapter 6. After a brief introduction to discrete projection schemes for the incompressible Navier-Stokes equations, we focused on the numerical solution of the Folgar-Tucker equation for the second order orientation tensor. The frequently observed occurrence of unphysical orientation states in numerical simulations with this model may be caused not only by a poor numerical treatment of the advective term but also by modeling errors associated with the need to reconstruct a fourth order orientation tensor using closures. Making use of operator splitting, we decomposed the problem at hand into a pure advection equation and an ODE with a closure-dependent source term. The former subproblem was discretized using a bound-preserving numerical method developed in Chapter 5. An in-depth analysis of the space independent Folgar-Tucker equation was performed in Section 6.3. To ensure physical admissibility of closure-based approximations, we derived necessary conditions for a symmetric tensor to be a second or fourth order moment of a nonnegative probability distribution function. The modeling framework proposed by Cintra and Tucker [CT95] provides a set of additional conditions for construction of closures. Using the forward or backward Euler method for time integration purposes and assuming that the chosen closure is physically admissible, we proved preservation of positive semidefiniteness under certain time step restrictions. The importance of this theoretical result lies in the fact that numerical solutions are guaranteed to possess essential properties of orientation tensors at least for properly designed closures and sufficiently small positive time steps. If the a priori upper bound for property-preserving time steps turns out too restrictive, significant speedups can be achieved using a posteriori time step control.

The combination of presented methods for evolving orientation tensors and solving the Navier-Stokes equations leads to a physics-compatible numerical algorithm for the simulation of time dependent fluid-fiber mixtures. Despite the use of operator splitting, converged steady state solutions of the full Folgar-Tucker equation are independent of the (pseudo-)time increment if the fully implicit treatment of the advection equation is combined with the forward Euler time discretization of the space independent Folgar-Tucker equation or vice versa. In particular, such property-preserving stationary approximations to the second order orientation tensor could be obtained in our numerical studies for the axisymmetric contraction benchmark.

7.2 Outlook

The advent of nonlinear bound-preserving algebraic stabilizations backed
by solid mathematical theory is one of the most recent trends in the field
of finite element methods for computational fluid dynamics. Solvability
and uniqueness results, rigorous proofs of discrete maximum principles,
proofs of convergence, and a priori error estimates significantly enhance
the fidelity of numerical simulation tools and open new avenues for further
development of robust discretization techniques. We have made every effort
to obtain such theoretical results at least for simplified models that may
represent subproblems of a fractional-step algorithm for a complex practical
application like the fiber suspension flow model considered in this thesis.
However, even the theory of algebraic flux correction schemes for scalar
hyperbolic conservation laws is still in its infancy. Many aspects remain
untouched and require further analysis for a better understanding and
construction of more efficient or accurate methods. Some open problems
which might be worth investigating in further studies are as follows:

In Section 4.5.1.1, the existence of a unique solution was shown for a monolithic AFC discretization of a (nominally) stationary reaction-dominated
model problem. The assumption of a dominant reaction term is quite restrictive. It is rarely satisfied in practical applications other than L^2-projections
and discrete problems in which such terms result from discretizations of
time derivatives. For example, the test case considered in Section 4.6.1 did
not contain any reactivity at all and the coercivity condition (2.14) was
violated. However, the obtained AFC finite element approximations were still
reasonable. This indicates that the requirements made in Theorem 4.68 can
possibly be weakened and the uniqueness of the solution might be provable
for more general problems.

Error analysis of the monolithic limiting approach was performed in
Section 4.5.1.3 for the steady advection-reaction equation by adapting the
proof of Barrenechea, John, and Knobloch [Bar+16] for elliptic problems. We
showed that, under suitable assumptions, the numerical solution converges
to the exact one with order at least $\frac{1}{2}$ no matter how the correction factors
are chosen. The numerical experiments in Section 4.6.1 indicate that the
validity of the coercivity condition (2.14) might not be mandatory and the
AFC scheme remains well-posed even if the associated high order target does
not converge to the exact solution. Furthermore, there is a strong numerical
evidence that the linearity preservation property of AFC methods implies
an improved order of convergence. Barrenechea et al. [Bar+18] proved the
corresponding theoretical result for elliptic problems. The extension of their

analysis to the hyperbolic limit would provide strong additional motivation for the use and further development of linearity-preserving limiters.

In this work, nonlinear AFC problems were solved iteratively using a pseudo time stepping approach as described in Section 4.6.1. This way to compute steady state solutions is typically very expensive, especially for large values of the parameters p and q. As mentioned in Remark 4.80, examples of potentially more efficient solvers include quasi-Newton methods [Möl08], Anderson acceleration, and suitable regularization techniques [BB17]. The use of differentiable limiters like the one proposed in Remark 4.80 might also improve the convergence behavior of fixed-point iterations and facilitate the development of Newton-like solvers for the AFC system.

In Chapter 4, the existence of a (unique) solution was shown and error analysis was performed for steady problems. An extension of this analysis to the time dependent case would be an important step toward further theoretical justification of monolithic AFC schemes. It might also lead to better algorithms for limiting the antidiffusive fluxes that depend on the (discretized) time derivatives. Furthermore, general proofs of well-posedness are currently restricted to the scalar case. For symmetric tensors, only the existence of a steady state solution was shown in Chapter 5 assuming the validity of the coercivity condition (2.14). Extensions of the scalar AFC theory to monolithic schemes that limit tensors using scalar correction factors are feasible and sometimes remarkably straightforward. However, the analysis becomes more involved when it comes to theoretical investigations of tensorial eigenvalue range limiters like Algorithm 5.18. The analysis of FCT schemes for symmetric tensor quantities is also far more challenging than in the scalar case. The sandwich proof techniques of Section 4.4.1.3 are not readily applicable because the desired convergence result does not seem to follow from the eigenvalue range preservation property.

In Section 6.4, we proposed a physics-compatible numerical algorithm for simulating unsteady fiber suspension flows. A key ingredient of this method is the segregated treatment of subproblems that describe changes of the second order orientation tensor due to advection and rotation of fibers. It is worth investigating if proofs of eigenvalue-based discrete maximum principles for steady state problems can be obtained without using operator splitting. In the framework of fractional-step algorithms, further efforts may be invested in circumventing the time step restrictions under which we proved physical admissibility of numerical solutions. The use of Patankar-type positivity-preserving time integrators [Bur+03] for the space independent Folgar-Tucker equation appears to be a promising approach to accomplishing this task. In the three dimensional case, we did not investigate the existence of a provably admissible, orthotropic, and Lipschitz continuous closure. The development of such a closure and estimation of the Lipschitz constant which determines the time step restriction of the backward Euler method would be another important milestone in the design of physics-compatible simulation tools based on our theoretical results.

It is hoped that our work will motivate the interested reader to join the quest and contribute to the development of property-preserving numerical schemes based on the above ideas.

Bibliography

[Abg17]   R. Abgrall. "High order schemes for hyperbolic problems using globally continuous approximation and avoiding mass matrices". In: *Journal of Scientific Computing* 73.2-3 (2017), pp. 461–494.

[AT87]   S. G. Advani and C. L. Tucker III. "The use of tensors to describe and predict fiber orientation in short fiber composites". In: *Journal of Rheology* 31.8 (1987), pp. 751–784.

[AT93]   M. C. Altan and L. Tang. "Orientation tensors in simple flows of dilute suspensions of non-Brownian rigid ellipsoids, comparison of analytical and approximate solutions". In: *Rheologica Acta* 32.3 (1993), pp. 227–244.

[AD89]   P. Arminjon and A. Dervieux. *Construction of TVD-like artificial viscosities on 2-dimensional arbitrary FEM grids*. Research Report RR-1111. INRIA, 1989.

[Axe94]   O. Axelsson. "Reducible and Irreducible Matrices and the Perron-Frobenius Theory for Nonnegative Matrices". In: *Iterative Solution Methods*. Cambridge University Press, 1994, pp. 122–157.

[BB17]   S. Badia and J. Bonilla. "Monotonicity-preserving finite element schemes based on differentiable nonlinear stabilization". In: *Computer Methods in Applied Mechanics and Engineering* 313 (2017), pp. 133–158.

[Bar+17a]   G. R. Barrenechea, E. Burman, and F. Karakatsani. "Edge-based nonlinear diffusion for finite element approximations of convection–diffusion equations and its relation to algebraic flux-correction schemes". In: *Numerische Mathematik* 135.2 (02/2017), pp. 521–545.

[Bar+15]   G. R. Barrenechea, V. John, and P. Knobloch. "Some analytical results for an algebraic flux correction scheme for a steady convectiondiffusion equation in one dimension". In: *IMA Journal of Numerical Analysis* 35.4 (2015), pp. 1729–1756.

© Springer Fachmedien Wiesbaden GmbH, part of Springer Nature 2019
C. Lohmann, *Physics-Compatible Finite Element Methods for Scalar and Tensorial Advection Problems*, https://doi.org/10.1007/978-3-658-27737-6

[Bar+16]   G. R. Barrenechea, V. John, and P. Knobloch. "Analysis of Algebraic Flux Correction Schemes". In: *SIAM Journal on Numerical Analysis* 54.4 (2016), pp. 2427–2451.

[Bar+17b] G. R. Barrenechea, V. John, and P. Knobloch. "An algebraic flux correction scheme satisfying the discrete maximum principle and linearity preservation on general meshes". In: *Mathematical Models and Methods in Applied Sciences* 27.03 (2017), pp. 525–548.

[Bar+18]   G. R. Barrenechea, V. John, P. Knobloch, and R. Rankin. "A unified analysis of algebraic flux correction schemes for convection–diffusion equations". In: *SeMA Journal* 75 (05/2018), pp. 655–685.

[Boc+12]   P. Bochev, D. Ridzal, G. Scovazzi, and M. Shashkov. "Constrained-Optimization Based Data Transfer". In: *Flux-Corrected Transport: Principles, Algorithms, and Applications*. Ed. by D. Kuzmin, R. Löhner, and S. Turek. Springer Netherlands, 2012, pp. 345–398.

[Boo+75]   D. L. Book, J. P. Boris, and K. Hain. "Flux-corrected transport II: Generalizations of the method". In: *Journal of Computational Physics* 18.3 (1975), pp. 248–283.

[BB73]   J. P. Boris and D. L. Book. "Flux-corrected transport. I. SHASTA, a fluid transport algorithm that works". In: *Journal of Computational Physics* 11.1 (1973), pp. 38–69.

[BB76]   J. P. Boris and D. L. Book. "Flux-corrected transport. III. Minimal-error FCT algorithms". In: *Journal of Computational Physics* 20.4 (1976), pp. 397–431.

[BS07]   S. Brenner and R. Scott. *The mathematical theory of finite element methods*. Vol. 15. Springer Science & Business Media, 2007.

[BF91]   F. Brezzi and M. Fortin. "Incompressible Materials and Flow Problems". In: *Mixed and Hybrid Finite Element Methods*. Ed. by F. Brezzi and M. Fortin. Springer New York, 1991, pp. 200–273.

[BH82] A. N. Brooks and T. J. Hughes. "Streamline upwind/Petrov-Galerkin formulations for convection dominated flows with particular emphasis on the incompressible Navier-Stokes equations". In: *Computer Methods in Applied Mechanics and Engineering* 32.1 (1982), pp. 199–259.

[Bur+03] H. Burchard, E. Deleersnijder, and A. Meister. "A high-order conservative Patankar-type discretisation for stiff systems of productiondestruction equations". In: *Applied Numerical Mathematics* 47.1 (2003), pp. 1–30.

[Bur+09] B. Burgeth, M. BreuSS, S. Didas, and J. Weickert. "PDE-based Morphology for Matrix Fields: Numerical Solution Schemes". In: *Tensors in Image Processing and Computer Vision.* Ed. by S. Aja-Fernández, R. de Luis García, D. Tao, and X. Li. Springer London, 2009, pp. 125–150.

[Bur+07] B. Burgeth, A. Bruhn, S. Didas, J. Weickert, and M. Welk. "Morphology for matrix data: Ordering versus PDE-based approach". In: *Image and Vision Computing* 25.4 (2007), pp. 496–511.

[Car+05] R. J. Caron, H. Song, and T. Traynor. *Positive semidefinite intervals for matrix pencils.* Tech. rep. Internal Report, Department of Mathematics and Statistics, University of Windsor, Ontario, Canada, 2005.

[Chi06] C. Chicone. *Ordinary differential equations with applications.* Vol. 34. Springer Science & Business Media, 2006.

[Cho68] A. J. Chorin. "Numerical Solution of the Navier-Stokes Equations". In: *Mathematics of Computation* 22.104 (10/1968), pp. 745–762.

[CR73a] P. Ciarlet and P.-A. Raviart. "Maximum principle and uniform convergence for the finite element method". In: *Computer Methods in Applied Mechanics and Engineering* 2.1 (1973), pp. 17–31.

[Cia02] P. G. Ciarlet. *Finite Element Method for Elliptic Problems.* Society for Industrial and Applied Mathematics, 2002.

[CT95] J. S. Cintra Jr and C. L. Tucker III. "Orthotropic closure approximations for flow-induced fiber orientation". In: *Journal of Rheology* 39.6 (1995), pp. 1095–1122.

[CK16]   C. Cotter and D. Kuzmin. "Embedded discontinuous Galerkin transport schemes with localised limiters". In: *Journal of Computational Physics* 311 (2016), pp. 363–373.

[Cou+98]   P.-H. Cournède, C. Debiez, and A. Dervieux. *A Positive MUSCL Scheme for Triangulations*. Tech. rep. RR-3465. INRIA, 07/1998.

[CR73b]   M. Crouzeix and P.-A. Raviart. "Conforming and nonconforming finite element methods for solving the stationary Stokes equations I". In: *R.A.I.R.O.* 7 (1973), pp. 33–75.

[DE11]   D. A. Di Pietro and A. Ern. *Mathematical aspects of discontinuous Galerkin methods*. Vol. 69. Springer Science & Business Media, 2011.

[Doi81]   M. Doi. "Molecular dynamics and rheological properties of concentrated solutions of rodlike polymers in isotropic and liquid crystalline phases". In: *Journal of Polymer Science: Polymer Physics Edition* 19.2 (1981), pp. 229–243.

[DH05]   J. Donea and A. Huerta. *Finite Element Methods for Flow Problems*. Wiley-Blackwell, 2005.

[DV99]   F. Dupret and V. Verleye. "Modelling the flow of fiber suspensions in narrow gaps". In: *Advances in the Flow and Rheology of Non-Newtonian Fluids*. Ed. by D. Siginer, D. D. Kee, and R. Chhabra. Vol. 8. Rheology Series. Elsevier, 1999, pp. 1347–1398.

[EG06]   A. Ern and J.-L. Guermond. "Discontinuous Galerkin methods for Friedrichs' systems. I. General theory". In: *SIAM Journal on Numerical Analysis* 44.2 (2006), pp. 753–778.

[EG13]   A. Ern and J.-L. Guermond. *Theory and practice of finite elements*. Vol. 159. Springer Science & Business Media, 2013.

[FT84]   F. Folgar and C. L. Tucker III. "Orientation behavior of fibers in concentrated suspensions". In: *Journal of Reinforced Plastics and Composites* 3.2 (1984), pp. 98–119.

[Gal11]   G. P. Galdi. *An introduction to the mathematical theory of the Navier-Stokes equations: Steady-state problems*. Springer Science & Business Media, 2011.

[GR99]   G. P. Galdi and B. D. Reddy. "Well-posedness of the problem of fiber suspension flows". In: *Journal of Non-Newtonian Fluid Mechanics* 83.3 (1999), pp. 205–230.

[GT15]   D. Gilbarg and N. S. Trudinger. *Elliptic partial differential equations of second order.* Springer, 2015.

[GR86]   V. Girault and P.-A. Raviart. *Finite Element Methods for Navier-Stokes Equations: Theory and Algorithms.* Springer Berlin Heidelberg, 1986.

[God59]   S. K. Godunov. "A difference method for numerical calculation of discontinuous solutions of the equations of hydrodynamics". In: *Matematicheskii Sbornik* 89.3 (1959), pp. 271–306.

[Got+11]   S. Gottlieb, D. Ketcheson, and C.-W. Shu. *Strong Stability Preserving Runge-Kutta and Multistep Time Discretizations.* World Scientific, 2011.

[Got+01]   S. Gottlieb, C.-W. Shu, and E. Tadmor. "Strong stability-preserving high-order time discretization methods". In: *SIAM Review* 43 (2001), pp. 89–112.

[Gue+18]   J. Guermond, M. Nazarov, B. Popov, and I. Tomas. "Second-Order Invariant Domain Preserving Approximation of the Euler Equations Using Convex Limiting". In: *SIAM Journal on Scientific Computing* 40.5 (2018), A3211–A3239.

[Gue+14]   J. Guermond, M. Nazarov, B. Popov, and Y. Yang. "A Second-Order Maximum Principle Preserving Lagrange Finite Element Technique for Nonlinear Scalar Conservation Equations". In: *SIAM Journal on Numerical Analysis* 52.4 (2014), pp. 2163–2182.

[Han62]   G. L. Hand. "A theory of anisotropic fluids". In: *Journal of Fluid Mechanics* 13.1 (1962), pp. 33–46.

[Han94]   P. Hansbo. "Aspects of conservation in finite-element flow computations". In: *Computer Methods in Applied Mechanics and Engineering* 117.3-4 (1994), pp. 423–437.

[HS88]   C. Harris and M. Stephens. "A combined corner and edge detector." In: *Alvey vision conference.* Vol. 15. 50. Citeseer. 1988, pp. 10–5244.

[Har84]   A. Harten. "On a Class of High Resolution Total-Variation-Stable Finite-Difference Schemes". In: *SIAM Journal on Numerical Analysis* 21.1 (1984), pp. 1–23.

[Har83]   A. Harten. "High resolution schemes for hyperbolic conservation laws". In: *Journal of Computational Physics* 49.3 (1983), pp. 357–393.

[Has+01] K. M. Hasan, P. J. Basser, D. L. Parker, and A. L. Alexander. "Analytical computation of the eigenvalues and eigenvectors in DT-MRI". In: *Journal of Magnetic Resonance* 152.1 (2001), pp. 41–47.

[HL73] E. J. Hinch and L. G. Leal. "Time-dependent shear flows of a suspension of particles with weak Brownian rotations". In: *Journal of Fluid Mechanics* 57.4 (1973), pp. 753–767.

[HL76] E. Hinch and L. Leal. "Constitutive equations in suspension mechanics. Part 2. Approximate forms for a suspension of rigid particles affected by Brownian rotations". In: *Journal of Fluid Mechanics* 76.1 (1976), pp. 187–208.

[Hir07] C. Hirsch. "The Basic Equations of Fluid Dynamics". In: *Numerical Computation of Internal and External Flows (Second Edition)*. Ed. by C. Hirsch. Second Edition. Butterworth-Heinemann, 2007. Chap. 1, pp. 27–64.

[HW53] A. J. Hoffman and H. W. Wielandt. "The variation of the spectrum of a normal matrix". In: *Duke Math. J.* 20.1 (03/1953), pp. 37–39.

[Hub07] M. Hubbard. "Non-oscillatory third order fluctuation splitting schemes for steady scalar conservation laws". In: *Journal of Computational Physics* 222.2 (2007), pp. 740–768.

[HV03] W. Hundsdorfer and J. Verwer. "Basic Concepts and Discretizations". In: *Numerical Solution of Time-Dependent Advection-Diffusion-Reaction Equations*. Springer Berlin Heidelberg, 2003. Chap. 1, pp. 1–138.

[Jam93] A. Jameson. "Computational algorithms for aerodynamic analysis and design". In: *Applied Numerical Mathematics* 13.5 (1993), pp. 383–422.

[Jam95] A. Jameson. "Analysis and design of numerical schemes for gas dynamics, 1: artificial diffusion, upwind biasing, limiters and their effect on accuracy and multigrid convergence". In: *International Journal of Computational Fluid Dynamics* 4.3-4 (1995), pp. 171–218.

[Jef22] G. B. Jeffery. "The motion of ellipsoidal particles immersed in a viscous fluid". In: *Proc. R. Soc. Lond. A* 102.715 (1922), pp. 161–179.

[Kat95]   T. Kato. "Perturbation theory in a finite-dimensional space". In: *Perturbation Theory for Linear Operators*. Springer Berlin Heidelberg, 1995, pp. 62–126.

[Klí+17]   M. Klíma, M. Kuchaík, M. J. Shashkov, and J. Velechovský. *Bound-Preserving Reconstruction of Tensor Quantities for Remap in ALE Fluid Dynamics*. Tech. rep. LA-UR-17-20068, Proceedings of XVI International Conference on Hyperbolic Problems Theory, Numerics and Applications, Aachen (Germany), Aug. 1-5, 2016. Los Alamos National Laboratory (LANL), 2017.

[Kno10]   P. Knobloch. "Numerical Solution of Convection–Diffusion Equations Using a Nonlinear Method of Upwind Type". In: *Journal of Scientific Computing* 43.3 (06/2010), pp. 454–470.

[Kno17]   P. Knobloch. "On the Discrete Maximum Principle for Algebraic Flux Correction Schemes with Limiters of Upwind Type". In: *Boundary and Interior Layers, Computational and Asymptotic Methods BAIL 2016*. Ed. by Z. Huang, M. Stynes, and Z. Zhang. Springer International Publishing, 2017, pp. 129–139.

[Kop08]   J. Kopp. "Efficient numerical diagonalization of hermitian 3×3 matrices". In: *International Journal of Modern Physics C* 19.03 (2008), pp. 523–548.

[Kuz12a]   D. Kuzmin. "Algebraic Flux Correction I. Scalar Conservation Laws". In: *Flux-Corrected Transport*. Ed. by D. Kuzmin, R. Löhner, and S. Turek. Scientific Computation. Springer Netherlands, 2012, pp. 145–192.

[Kuz18a]   D. Kuzmin. *Gradient–based limiting and stabilization of continuous Galerkin methods*. Tech. rep. Ergebnisberichte des Instituts für Angewandte Mathematik, Nummer 589. Fakultät für Mathematik, TU Dortmund, 07/2018.

[Kuz18b]   D. Kuzmin. "Planar and Orthotropic Closures for Orientation Tensors in Fiber Suspension Flow Models". In: *SIAM Journal on Applied Mathematics* 78.6 (2018), pp. 3040–3059.

[KT02]   D. Kuzmin and S. Turek. "Flux Correction Tools for Finite Elements". In: *Journal of Computational Physics* 175.2 (2002), pp. 525–558.

[KT04]   D. Kuzmin and S. Turek. "High-resolution FEM-TVD schemes based on a fully multidimensional flux limiter". In: *Journal of Computational Physics* 198.1 (2004), pp. 131–158.

[Kuz07]   D. Kuzmin. "Algebraic flux correction for finite element discretizations of coupled systems". In: *Computational Methods for Coupled Problems in Science and Engineering II, CIMNE, Barcelona* (2007), pp. 653–656.

[Kuz09]   D. Kuzmin. "Explicit and implicit FEM-FCT algorithms with flux linearization". In: *Journal of Computational Physics* 228.7 (2009), pp. 2517–2534.

[Kuz10]   D. Kuzmin. "A Guide to Numerical Methods for Transport Equations". URL: http://www.mathematik.uni-dortmund. de/~kuzmin/Transport.pdf. 2010.

[Kuz12b]   D. Kuzmin. "Linearity-preserving flux correction and convergence acceleration for constrained Galerkin schemes". In: *Journal of Computational and Applied Mathematics* 236.9 (2012), pp. 2317–2337.

[Kuz+17]   D. Kuzmin, S. Basting, and J. N. Shadid. "Linearity-preserving monotone local projection stabilization schemes for continuous finite elements". In: *Computer Methods in Applied Mechanics and Engineering* 322 (2017), pp. 23–41.

[KH14]   D. Kuzmin and J. Hämäläinen. *Finite Element Methods for Computational Fluid Dynamics: A Practical Guide*. Society for Industrial and Applied Mathematics, 2014.

[KM05]   D. Kuzmin and M. Möller. "Algebraic Flux Correction I. Scalar Conservation Laws". In: *Flux-Corrected Transport*. Ed. by D. Kuzmin, R. Löhner, and S. Turek. Scientific Computation. Springer Berlin Heidelberg, 2005, pp. 155–206.

[Kuz+12]   D. Kuzmin, M. Möller, and M. Gurris. "Algebraic Flux Correction II". In: *Flux-Corrected Transport: Principles, Algorithms, and Applications*. Ed. by D. Kuzmin, R. Löhner, and S. Turek. Springer Netherlands, 2012, pp. 193–238.

[LT08]   S. Larsson and V. Thomée. *Partial differential equations with numerical methods*. Vol. 45. Springer Science & Business Media, 2008.

[LR74]   P. Lasaint and P. Raviart. "On a Finite Element Method for Solving the Neutron Transport Equation". In: *Mathematical Aspects of Finite Elements in Partial Differential Equations.* Ed. by C. de Boor. Academic Press, 1974, pp. 89–123.

[Lax54]   P. D. Lax. "Weak solutions of nonlinear hyperbolic equations and their numerical computation". In: *Communications on Pure and Applied Mathematics* 7.1 (1954), pp. 159–193.

[LeV96]   R. J. LeVeque. "High-resolution conservative algorithms for advection in incompressible flow". In: *SIAM Journal on Numerical Analysis* 33.2 (1996), pp. 627–665.

[LeV92]   R. J. LeVeque. *Numerical Methods for Conservation Laws.* Birkhäuser Basel, 1992.

[Lip+88]   G. Lipscomb, M. Denn, D. Hur, and D. Boger. "The flow of fiber suspensions in complex geometries". In: *Journal of Non-Newtonian Fluid Mechanics* 26.3 (1988), pp. 297–325.

[Loh19]   Lohmann, Christoph. "Algebraic flux correction schemes preserving the eigenvalue range of symmetric tensor fields". In: *ESAIM: M2AN* 53.3 (2019), pp. 833–867.

[Loh17a]   C. Lohmann. "Eigenvalue range limiters for tensors in flux-corrected transport algorithms". MultiMat 2017, Santa Fe, USA. URL: https : / / custom . cvent . com / F6288ADDEF3C4A6CBA 5358DAE922C966/files/e1c3aodf74394cb1a33e141f57f33b 2e.pdf. 09/2017.

[Loh17b]   C. Lohmann. "Flux-corrected transport algorithms preserving the eigenvalue range of symmetric tensor quantities". In: *Journal of Computational Physics* 350 (2017), pp. 907–926.

[Loh+17]   C. Lohmann, D. Kuzmin, J. N. Shadid, and S. Mabuza. "Flux-corrected transport algorithms for continuous Galerkin methods based on high order Bernstein finite elements". In: *Journal of Computational Physics* 344 (2017), pp. 151–186.

[Löh+88]   R. Löhner, K. Morgan, M. Vahdati, J. P. Boris, and D. L. Book. "FEM-FCT: Combining unstructured grids with high resolution". In: *Communications in Applied Numerical Methods* 4.6 (1988), pp. 717–729.

[Löh08]   R. Löhner. *Applied Computational Fluid Dynamics Techniques.* Wiley-Blackwell, 2008.

[Löh+87]   R. Löhner, K. Morgan, J. Peraire, and M. Vahdati. "Finite element flux-corrected transport (FEM–FCT) for the Euler and Navier–Stokes equations". In: *International Journal for Numerical Methods in Fluids* 7.10 (1987), pp. 1093–1109.

[Löw34]   K. Löwner. "Über monotone Matrixfunktionen". In: *Mathematische Zeitschrift* 38.1 (1934), pp. 177–216.

[Lut16]   G. Luttwak. "On the Extension of Monotonicity to Multi-Dimensional Flows". In: (2016).

[LF10]   G. Luttwak and J. Falcovitz. "Vector Image Polygon (VIP) limiters in ALE Hydrodynamics". In: *EPJ Web of Conferences*. Vol. 10. EDP Sciences. 2010, p. 00020.

[LF11]   G. Luttwak and J. Falcovitz. "Slope limiting for vectors: A novel vector limiting algorithm". In: *International Journal for Numerical Methods in Fluids* 65.11-12 (2011), pp. 1365–1375.

[Lyr+94]   P. R. M. Lyra, K. Morgan, J. Peraire, and J. Peiró. "TVD algorithms for the solution of the compressible Euler equations on unstructured meshes". In: *International Journal for Numerical Methods in Fluids* 19.9 (1994), pp. 827–847.

[Lyr95]   P. R. M. Lyra. "Unstructured grid adaptive algorithms for fluid dynamics and heat conduction". PhD thesis. University College of Swansea, 1995.

[Mai+13]   P.-H. Maire, R. Abgrall, J. Breil, R. Loubère, and B. Rebourcet. "A nominally second-order cell-centered Lagrangian scheme for simulating elastic–plastic flows on two-dimensional unstructured grids". In: *Journal of Computational Physics* 235 (2013), pp. 626–665.

[MR73]   C. Micchelli and T. Rivlin. "Numerical integration rules near Gaussian quadrature". In: *Israel Journal of Mathematics* 16.3 (1973), pp. 287–299.

[Möl08]   M. Möller. "On an efficient solution strategy of Newton type for implicit finite element schemes based on algebraic flux correction". In: *International Journal for Numerical Methods in Fluids* 56.8 (2008), pp. 1085–1091.

[Mon+10]   S. Montgomery-Smith, D. A. Jack, and D. E. Smith. "A systematic approach to obtaining numerical solutions of Jefferys type equations using Spherical Harmonics". In: *Composites Part A: Applied Science and Manufacturing* 41.7 (2010), pp. 827–835.

[Nit71]   J. Nitsche. "Über ein Variationsprinzip zur Lösung von Dirichlet-Problemen bei Verwendung von Teilräumen, die keinen Randbedingungen unterworfen sind". In: *Abhandlungen aus dem mathematischen Seminar der Universität Hamburg*. Vol. 36. 1. Springer. 1971, pp. 9–15.

[PA87]   T. C. Papanastasiou and A. Alexandrou. "Isothermal extrusion of non-dilute fiber suspensions". In: *Journal of Non-Newtonian Fluid Mechanics* 25.3 (1987), pp. 313–328.

[PC86]   A. Parrott and M. Christie. "FCT applied to the 2-D finite element solution of tracer transport by single phase flow in a porous medium". In: *Proc. ICFD Conf. on Numerical Methods in Fluid Dynamics, Oxford University Press*. Vol. 609. 1986.

[PW12]   M. H. Protter and H. F. Weinberger. *Maximum principles in differential equations*. Springer Science & Business Media, 2012.

[QV94]   A. Quarteroni and A. Valli. *Numerical Approximation of Partial Differential Equations*. Vol. 23. Springer Berlin Heidelberg, 01/1994.

[RT92]   R. Rannacher and S. Turek. "Simple nonconforming quadrilateral Stokes element". In: *Numerical Methods for Partial Differential Equations* 8.2 (1992), pp. 97–111.

[RM01]   B. Reddy and G. Mitchell. "Finite element analysis of fibre suspension flows". In: *Computer Methods in Applied Mechanics and Engineering* 190.18-19 (2001), pp. 2349–2367.

[RW17]   A. Rösch and G. Wachsmuth. "Mass Lumping for the Optimal Control of Elliptic Partial Differential Equations". In: *SIAM Journal on Numerical Analysis* 55.3 (2017), pp. 1412–1436.

[Ros+90]   J. Rosenberg, M. Denn, and R. Keunings. "Simulation of non-recirculating flows of dilute fiber suspensions". In: *Journal of Non-Newtonian Fluid Mechanics* 37.2 (1990), pp. 317–345.

[Sam+13]   S. K. Sambasivan, M. J. Shashkov, and D. E. Burton. "Exploration of new limiter schemes for stress tensors in Lagrangian and ALE hydrocodes". In: *Computers & Fluids* 83 (2013), pp. 98–114.

[Sel93]   V. Selmin. "The node-centred finite volume approach: Bridge between finite differences and finite elements". In: *Computer Methods in Applied Mechanics and Engineering* 102.1 (1993), pp. 107–138.

[SF96]   V. Selmin and L. Formaggia. "Unified construction of finite element and finite volume discretizations for compressible flows". In: *International Journal for Numerical Methods in Engineering* 39.1 (1996), pp. 1–32.

[Sel87a]   V. Selmin. *Finite element solution of hyperbolic equations I.One-dimensional case*. Research Report RR-0655. INRIA, 1987.

[Sel87b]   V. Selmin. *Finite element solution of hyperbolic equations II. Two dimensional case*. Research Report RR-0708. INRIA, 1987.

[Smi61]   O. K. Smith. "Eigenvalues of a Symmetric 3 × 3 Matrix". In: *Commun. ACM* 4.4 (04/1961), p. 168.

[Spe81]   R. P. Sperb. *Maximum Principles and Their Applications*. Vol. 157. Mathematics in Science and Engineering. Elsevier, 1981.

[Tem77]   R. Temam. *NavierStokes Equations*. Vol. 2. Studies in Mathematics and Its Applications. Elsevier, 1977.

[Tuc91]   C. L. Tucker III. "Flow regimes for fiber suspensions in narrow gaps". In: *Journal of Non-Newtonian Fluid Mechanics* 39.3 (1991), pp. 239–268.

[Tur97]   S. Turek. "On discrete projection methods for the incompressible Navier-Stokes equations: an algorithmical approach". In: *Computer Methods in Applied Mechanics and Engineering* 143.3 (1997), pp. 271–288.

[Tur99]   S. Turek. *Efficient Solvers for Incompressible Flow Problems: An Algorithmic and Computational Approach*. Springer Berlin Heidelberg, 1999.

[VC03]   T. Vaithianathan and L. R. Collins. "Numerical Approach to Simulating Turbulent Flow of a Viscoelastic Polymer Solution". In: *Journal of Computational Physics* 187.1 (05/2003), pp. 1–21.

[van86]   J. van Kan. "A Second-Order Accurate Pressure-Correction Scheme for Viscous Incompressible Flow". In: *SIAM Journal on Scientific and Statistical Computing* 7.3 (1986), pp. 870–891.

[Var09]   R. S. Varga. *Matrix iterative analysis*. Vol. 27. Springer Science & Business Media, 2009.

[VD93]   D. Verleye and F. Dupret. "Prediction of fiber orientation in complex injection molded parts". In: *Developments in Non-Newtonian Flows*. Ed. by D. A. Siginer, W. E. VanArsdale, M. C. Altan, and A. N. Alexandrou. Vol. 175. AMD. presented at the 1993 ASME Winter Annual Meeting, New Orleans, Louisiana, November 28-December 3, 1993. New York: American Society of Mechanical Engineers, 1993, pp. 139–163.

[VT02]   B. E. VerWeyst and C. L. Tucker III. "Fiber Suspensions in Complex Geometries: Flow/Orientation Coupling". In: *The Canadian Journal of Chemical Engineering* 80.6 (2002), pp. 1093–1106.

[Ver98]   B. E. VerWeyst. "Numerical predictions of flow-induced fiber orientation in three-dimensional geometries". PhD thesis. University of Illinois at Urbana-Champaign, 1998.

[Wes01]   P. Wesseling. "Scalar conservation laws". In: *Principles of Computational Fluid Dynamics*. Springer Berlin Heidelberg, 2001, pp. 339–396.

[Wil88]   J. H. Wilkinson. *The Algebraic Eigenvalue Problem*. Oxford University Press, Inc., 1988.

[Yee87]   H. Yee. "Construction of explicit and implicit symmetric TVD schemes and their applications". In: *Journal of Computational Physics* 68.1 (1987), pp. 151–179.

[Zal79]   S. T. Zalesak. "Fully Multidimensional Flux-Corrected Transport Algorithms for Fluids". In: *Journal of Computational Physics* 31.3 (1979), pp. 335–362.

Printed in the United States
By Bookmasters